Versatility
of Proteins

ACADEMIC PRESS RAPID MANUSCRIPT REPRODUCTION

Proceedings of the International Symposium on Proteins,
March 6-8, 1978, held in Taipei

VERSATILITY OF PROTEINS

Edited by CHOH HAO LI

The Hormone Research Laboratory
University of California
San Francisco, California

ACADEMIC PRESS   New York   San Francisco   London   1978
A Subsidiary of Harcourt Brace Jovanovich, Publishers

ACADEMIC PRESS, INC.
111 Fifth Avenue, New York, New York 10003

United Kingdom Edition published by
ACADEMIC PRESS, INC. (LONDON) LTD.
24/28 Oval Road, London NW1   7DX

Library of Congress Cataloging in Publication Data

International Symposium on Proteins, T'ai-pei, 1978.
    Versatility of proteins.

    1.   Proteins—Congresses.   2.   Protein binding—
Congresses.   I.   Li, Choh Hao, Date   II.   Title.
[DNLM: 1.   Proteins—Congresses.   2.   Enzymes—Congresses
QU55.3 I61v 1978]
QP551.I554 1978   574.1'9245   78-13214
ISBN 0-12-44750-X

PRINTED IN THE UNITED STATES OF AMERICA

78 79 80 81 82   9 8 7 6 5 4 3 2 1

Contents

4. Regulatory Proteins

List of Contributors

Numbers in parentheses indicate the pages on which the authors' contributions begin.

J. O. ALABA (133), Department of Biochemistry, University of Washington, Seattle, Washington

JAMES BLAKE (81), Hormone Research Laboratory, University of California, San Francisco, California

WILLIAM F. BOSRON (253), Departments of Medicine and Biochemistry, Indiana University School of Medicine, Indianapolis, Indiana

D. L. BRAUTIGAN (133), Department of Biochemistry, University of Washington, Seattle, Washington

P. M. COLMAN (293), Max-Planck-Institut für Biochemie and Physikalisch-Chemisches Institut der Technischen Universität, D-8033, Martinsried, Munich, Germany

B. A. COTTRELL (393), Department of Chemistry, University of California, San Diego, La Jolla, California

WERNER P. DAFELDECKER (253), Biophysics Research Laboratory, Department of Biological Chemistry, Harvard Medical School, Boston, Massachusetts

J. DEISENHOFER (293), Max-Planck-Institut für Biochemie and Physikalisch-Chemisches Institut der Technischen Universität, D-8033 Martinsried, Munich, Germany

R. F. DOOLITTLE (393), Department of Chemistry, University of California, San Diego, La Jolla, California

DAVID EAKER (413), Institute of Biochemistry, University of Uppsala, Uppsala, Sweden

LOWELL H. ERICSSON (39), Department of Biochemistry, University of Washington, Seattle, Washington

BEVERLY J. ERREDE (229), University of Southern California, Los Angeles, California and University of California, San Diego, La Jolla, California

E. H. FISCHER (133), Department of Biochemistry, University of Washington, Seattle, Washington

DAVID GIVOL (295), Department of Chemical Immunology, The Weizmann Institute of Science, Rehovot, Israel

O. HAYAISHI (151), Department of Medical Chemistry, Kyoto University Faculty of Medicine, Kyoto, Japan

C. L. Ho (433), Pharmacological Institute, College of Medicine, National Taiwan University and Institute of Biological Chemistry, Academia Sinica, Taipei, Taiwan, Republic of China

R. HUBER (293), Max-Planck-Institut für Biochemie and Physikalisch-Chemisches Institut der Technischen Universität, D-8033 Martinsried, Munich, Germany

K. IKAI (151), Department of Medical Chemistry, Kyoto University Faculty of Medicine, Kyoto, Japan

KAZUTOMO IMAHORI (167), Department of Biochemistry, Faculty of Medicine, The University of Tokyo, Tokyo, Japan

S. ITO (151), Department of Medical Chemistry, Kyoto University Faculty of Medicine, Kyoto, Japan

MARTIN D. KAMEN (229), University of Southern California, Los Angeles, California and University of California, San Diego, La Jolla, California

IKUNOSHIN KATO (307), Department of Chemistry, Purdue University, West Lafayette, Indiana

M. KAWAICHI (151), Department of Medical Chemistry, Kyoto University Faculty of Medicine, Kyoto, Japan

W. G. L. KERRICK (133), Department of Biochemistry, University of Washington, Seattle, Washington

TE PIAO KING (335), The Rockefeller University, New York, New York

WILLIAM J. KOHR (307), Department of Chemistry, Purdue University, West Lafayette, Indiana

MICHAEL LASKOWSKI, Jr. (307), Department of Chemistry, Purdue University, West Lafayette, Indiana

C. Y. LEE (433), Pharmacological Institute, College of Medicine, National Taiwan University and Institute of Biological Chemistry, Academia Sinica, Taipei, Taiwan, Republic of China

CHOH HAO LI (353), Hormone Research Laboratory, University of California, San Francisco, California

TING-KAI LI (253), Departments of Medicine and Biochemistry, Indiana University School of Medicine, Indianapolis, Indiana

D. A. MALENCIK (133), Department of Biochemistry, University of Washington, Seattle, Washington

M. MARQUART (293), Max-Planck-Institut für Biochemie and Physikalisch-Chemisches Institut der Technischen Universität, D-8033 Martinsried, Munich, Germany

M. MATSUSHIMA (293), Max-Planck-Institut für Biochemie and Physikalisch-Chemisches Institut der Technischen Universität, D-8033 Martinsried, Munich, Germany

H. J. MOESCHLER (133), Department of Biochemistry, University of Washington, Seattle, Washington

EDNA MOZES (269), Department of Chemical Immunology, The Weizmann Institute of Science, Rehovot, Israel

HANS J. MÜLLER-EBERHARD (373), Department of Molecular Immunology, Research Institute of Scripps Clinic, La Jolla, California

TAKASHI MURACHI (183), Department of Clinical Science, Kyoto University Faculty of Medicine, Kyoto, Japan

HANS NEURATH (1), Department of Biochemistry, University of Washington, Seattle, Washington

N. OGATA (151), Department of Medical Chemistry, Kyoto University Faculty of Medicine, Kyoto, Japan

SHIGEO OHNO (167), Department of Biochemistry, Faculty of Medicine, The University of Tokyo, Tokyo, Japan

YOSHIKO OHNO-IWASHITA (167), Department of Biochemistry, Faculty of Medicine, The University of Tokyo, Tokyo, Japan

J. OKA (151), Department of Medical Chemistry, Kyoto University Faculty of Medicine, Kyoto, Japan

H. OKAYAMA (151), Department of Medical Chemistry, Kyoto University Faculty of Medicine, Kyoto, Japan

W. PALM (293), Institut für Medicinifische Biochemie der Universitat, A-8010, Graz, Austria

C. PICTON (133), Department of Biochemistry, University of Washington, Seattle, Washington

SITIVAD POCINWONG (133), Department of Biochemistry, University of Washington, Seattle, Washington

JERKER PORATH (15), Institute of Biochemistry, University of Uppsala, Uppsala, Sweden

JAMES F. RIORDAN (203), Biophysics Research Laboratory, Department of Biological Chemistry, Harvard Medical School and the Division of Medical Biology, Peter Bent Brigham Hospital, Boston, Massachusetts

HAROLD A. SCHERAGA (119), Department of Chemistry, Cornell University, Ithaca, New York and Biophysics Department, Weizmann Institute of Science, Rehovot, Israel

MICHAL SCHWARTZ (269), Department of Chemical Immunology, The Weizmann Institute of Science, Rehovot, Israel

MICHAEL SELA (269), Department of Chemical Immunology, The Weizmann Institute of Science, Rehovot, Israel

Y. SHIZUTA (151), Department of Medical Chemistry, Kyoto University Faculty of Medicine, Kyoto, Japan

KOICHI SUZUKI (167), Department of Biochemistry, Faculty of Medicine, The University of Tokyo, Tokyo, Japan

YUKIO SUZUKI (183), Department of Clinical Science, Kyoto University Faculty of Medicine, Kyoto, Japan

T. TAKAGI (393), Department of Chemistry, University of California, San Diego, La Jolla, California

KOITI TITANI (39), Department of Biochemistry, University of Washington, Seattle, Washington

SIDNEY UDENFRIEND (23), Roche Institute of Molecular Biology, Nutley, New Jersey

K. UEDA (151), Department of Medical Chemistry, Kyoto University Faculty of Medicine, Kyoto, Japan

BERT L. VALLEE (203, 253), Biophysics Research Laboratory, Department of Biological Chemistry, Harvard Medical School and the Division of Medical Biology, Peter Bent Bringham Hospital, Boston, Massachusetts

KENNETH A. WALSH (39), Department of Biochemistry, University of Washington, Seattle, Washington

KUNG-TSUNG WANG (447), Institute of Biological Chemistry, Academia Sinica, Taipei Taiwan, Republic of China

K. W. K. WATT (393), Department of Chemistry, University of California, San Diego, La Jolla, California

THEODOR WIELAND (59), Max-Planck-Institut für Medizinische, Forschung, Abteilung Naturstoff-Chemie, Heidelberg, West Germany

H. G. WITTMANN (319), Max-Planck-Institut für Molekulare Genetik, Berlin-Dahlem, Germany

CHUEN-SHANG C. WU (99), Department of Biochemistry and Biophysics and Cardiovascular Research Institute, University of California, San Francisco, California

DONALD YAMASHIRO (81), Hormone Research Laboratory, University of California, San Francisco, California

JEN TSI YANG (99), Department of Biochemistry and Biophysics and Cardiovascular Research Institute, University of California, San Francisco, California

Preface

This volume is a collection of invited papers presented at the International Symposium of Proteins held March 6–8, 1978, in Taipei. The symposium was organized by the four cochairmen: H. Neurath, J. Porath, M. Sela, and C. H. Li, and was sponsored by the National Science Council of the Republic of China in connection with the celebration of the 50th anniversary of the founding of Academia Sinica.

The symposium was divided into four sessions: techniques in protein chemistry, enzymes, protein–protein interactions, and regulatory proteins. It was attended by 140 participants from laboratories and institutions in Europe, Japan, Canada, Hong Kong, Taiwan, and the United States.

Proteins occupy a central position in the architecture and functioning of living matter. The structures of proteins enable them to act as enzymes that control the rates of all biological reactions. Proteins also serve as structural, extracellular elements, such as hair, wool, and the collagen of connective tissue. Proteins perform a variety of the body's functions, including hormones, toxins, coagulation factors, complement components, association with the genes, participation in muscular contraction, and immunological defense as antibodies. Therefore, proteins play an important and essential role in life processes.

The symposium was opened by H. Neurath, one of the pioneers and builders of modern protein chemistry. The theme of his address, the evolution of proteins, demonstrated his immense knowledge of proteases. This was followed by the first session on techniques in protein chemistry with introductory remarks by J. Porath, whose work has been so important in the development of separation methods. S. Udenfriend lucidly described fluorescent methods for isolation, characterization, and assay of peptides and proteins at the nanomole level. K. A. Walsh analyzed the strategies of amino acid sequencing. Peptide synthetic methodology was separately reviewed by T. Wieland and D. Yamashiro. J. T. Yang discussed in detail his recent work on surfactant-induced conformations of various biologically active peptides. This session was concluded by H. A. Scheraga, who eloquently reported his studies on protein folding.

The second session centered on enzymes, with contributions by E. H. Fischer, O. Hayaishi, K. Imahori, T. Murachi, B. L. Vallee, M. D. Kamen, and T. K. Li. Phosphorylase kinase, colicin, cytochromes, and alcohol dehydrogenase were discussed. Vallee delivered a superb lecture on his fundamental work on local conformation and functional properties of enzymes using ingenious methods of organic and inorganic probes.

The third session on protein–protein interaction was opened by M. Sela, one of the founders of modern immunology. In this lecture, Sela presented data showing that a minimal change in synthetic antigens causes major differences in the biological properties of the immunogens. This was followed by R. Huber on three-dimensional structure of an antibody molecule, D. Givol on structure and function of the antibody combining site, M. Laskowski, Jr., on protein inhibitors of serine proteinases, and T. P. King on ragweed pollen allergen. The structure and function of ribosomes were presented by H. G. Wittman, who has done brilliant work on the isolation and primary structure of ribosomal proteins from *E. coli*.

The final session was on regulatory proteins, including β-lipotropin by C. H. Li, complement systems by H. J. Müller-Eberhard, fibrinogen by R. F. Doolittle, chemistry of snake toxins by D. Eaker, pharmacology of snake toxins by C. Y. Lee, and synthesis of cobrotoxin and cardiotoxin by K. T. Wang. Müller-Eberhard has made major and decisive contributions to the difficult problems of the biology and chemistry of the complement. He reviewed here the molecular dynamics and regulation of the complement system.

The symposium could not have taken place without the support of Dr. S. S. Shu, Chairman of the National Science Council and Dr. S. L. Chien, President of the Academic Sinica. Doctors Neurath, Porath, and Sela join me in thanking the invited speakers for their contributions. We also thank Dr. Tung-Bin Lo and Mr. Chi-wu Wang for their untiring skill in making the symposium a success on a social as well as an academic level. Finally, I am indebted to Dr. Neurath for the suggestion of the title of this volume.

PROTEASES AND THE EVOLUTION OF PROTEIN CHEMISTRY[1]

Hans Neurath

Department of Biochemistry
University of Washington
Seattle, Washington

I am honored and pleased to have been asked to present the opening lecture of this distinguished International Symposium on Proteins, an occasion of timely significance and importance. While protein chemists and protein chemistry have contributed in many different ways to the spectacular developments in modern biology, particularly molecular biology, there have not been many international occasions of late for protein chemists to speak to each other and to review the latest developments in the field of their primary interests. The gathering this week is one such occasion, restricted by necessity to a discrete number of topics because protein chemistry has become so complex and pervasive that no single conference could deal with its entirety.

I have chosen for this lecture the topic of proteases and the evolution of protein chemistry, for two reasons: one is my long-standing interest in proteolytic enzymes and the other my good fortune of having witnessed the evolution of protein chemistry over a period of more than three decades. I also thought that it might be of interest to take a somewhat detached view of the field, even if it would be historically incomplete and colored by personal experiences and impressions -- to see where we came from, where we are, and where we are going. In short, I should like to talk about the evolution of our knowledge of proteins.

---

[1]Supported in part by the National Institutes of Health (GM-15731).

Research on proteases is intimately interwoven with the
fields of protein chemistry and enzymology, and some of the
earliest triumphs of protein chemistry have been scored in the
field of proteases. Proteases are historically associated
with digestive functions and the proteases of the gastro-
intestinal tract were among the first to be purified and
crystallized. A great deal of our understanding of enzyme
catalysis and of the functional role of specific amino acid
residues has emanated from the study of proteases. The modern
era of protein chemistry actually began with the painstaking
studies of Northrop, Kunitz, Herriott, and Anson (1) on
trypsin, chymotrypsin, pepsin and carboxypeptidase A, four
enzymes which served in the succeeding years as models in
studies of the relation between the structure, function, and
evolution of proteins. In fact, the entire history of protein
chemistry could be documented by following the literature of
any one of these enzymes. An early observation that each of
these enzymes is secreted as an inactive zymogen has added an
element of interest which today overshadows their purely
digestive functions. For, it is now recognized that the
zymogen-enzyme conversion by limited proteolysis constitutes
an important mechanism of physiological regulation (2,3). The
repertoire of proteolytic enzymes of well defined structure
and/or well established physiological function has grown
enormously since these early days, perhaps to several hundred
in number (4). Many of these can be grouped according to the
chemical nature of their active site or their evolutionary
relationships (5,6) but many more remain to be characterized
and assigned to existing or yet to be established mechanistic
classes.

One factor that has contributed significantly to the
expansion of our knowledge is the manner in which proteins are
selected for study. Whereas in earlier years proteins,
including the proteases, were selected on the basis of their
availability, crystallization of the protein being the major
criterion of the success of an isolation procedure, today any
enzyme that can be assayed can also be isolated and purified
using suitable ligands (substrates, cofactors, or inhibitors)
to form an affinity chromatographic system. Proteins are no
longer being studied simply because they are there but because
of their importance for a given biological problem or process.
Occasionally, by a historical accident, we have rediscovered a
protein because of its newly recognized role in a physiological
process, such as concanavalin A, originally described by
Sumner in 1938 (7), or tropomyosin, first described by Bailey
in 1946 (8). In other instances, e.g. proinsulin (9), we were
slow in recognizing the existence of an inactive precursor
form of a well known active protein.

Procedures for isolating proteins have evolved from relatively crude methods to seemingly simple but refined techniques, and so have the methods for judging the homogeneity of the final product (10). A shopping list for establishing a protein laboratory in the 1940's essentially included a refrigerated low-speed centrifuge, a Beckman DU spectrophotometer, an analytical ultracentrifuge, a Tiselius moving boundary apparatus, and a Lamm diffusion apparatus. These were the principal items of equipment which Frank Putnam and I used in 1946 for isolating bovine carboxypeptidase A and determinining its molecular properties (11). In 1978, the same problem can be solved with the same assurance that the protein is pure, by use of an affinity column (if the enzyme cannot be purchased), a handful of acrylamide, and a powerpack (12). This is not to say that the more complex physico-chemical methods of analysis, such as ultracentrifugation, or moving boundary electrophoresis, were unnecessary for the study of proteins; indeed, they were, and still are, essential for establishing the criteria of protein purity and for creating the knowledge which subsequently enabled us to develop the simpler methods of gel electrophoresis, molecular exclusion chromatography, and others.

A similar pattern of evolution pertains to the methods for determining the covalent structure of proteins (to be discussed in more detail by Dr. Walsh) (13). The elegant simplicity of Sanger's analysis of the sequence of the 51 amino acid residues of insulin (14) certainly stands out as the beginning of a new era of protein research. A bottle of fluorodinitrobenzene, filter paper, a high voltage electrophoresis system, and an almost infinite amount of time were all that was required in those pioneering days. Some 20 years later, a "state of the art" sequence laboratory requires one or more sequenators of the liquid and/or solid phase type, a computer-linked amino acid analyzer, and a high pressure liquid chromatography system to determine infinitely more rapidly and accurately sequences 10 or 20 times as long (15). Looking ahead into the 1980's one may anticipate that $25 worth of restriction enzymes and acrylamide gel would suffice to determine the sequence of the protein from that of the coding DNA or messenger RNA, at a rate of 1000 nucleotides per day, corresponding to some 300 amino acid residues of the protein (16). Again, we shall have gone through phases, starting with the simple and laborious, proceeding to the sophisticated and more efficient, and winding up with a conceptually sophisticated but experimentally simple and more efficient method of experimentation.

I could not speak with the same knowledge about an equally important major development in the field of protein chemistry, namely x-ray structure   analysis, because I have never been a practitioner of this highly specialized skill.   However, like all of us, I have been an admirer, and intellectually a beneficiary of its successes.   It is evident that the ever-increasing rate of resolving protein structure has required far-reaching improvements and simplifications in recording and interpreting electron density maps (17).   The detailed analysis of the structures of several proteases, such as trypsin, chymotrypsin, carboxypeptidase, elastase, trypsin-inhibitor complexes, and thermolysin, has established firm grounds for the postulated existence of families of proteases, for their homology and patterns of evolution, and for predicting the structure of one proteolytic enzyme from that of another (18).

Let me now turn to an illustration of the role that proteolytic enzymes have played not only in the development of experimental procedures, but perhaps more importantly, in the advancement of our knowledge of the detailed chemical structure and biological functions of proteins.   I should like to choose three particular examples: the active site of enzymes, the role of proteases in biological regulation, and biochemical evolution.

ACTIVE SITE OF ENZYMES

Our knowledge of the active site of proteases, as of any enzyme, has evolved by applying a combination of kinetic, chemical, and spectroscopic probes which in one form or another were specifically adapted to the study of proteolytic enzymes.   The historical path leading to the elucidation of the active site of the serine proteases started with the discovery of relatively simple peptide, ester, and amide substrates designed to delineate substrate specificities (19). This was followed by the fundamental observation that these proteases were inactivated by diisopropyl phosphorofluoridate which reacts stoichiometrically with a specific serine residue of the active site of these enzymes (20).   Inactivation by photooxidation (21) or by active-site directed chloromethyl-ketones (22) led to the implication of a histidine residue in the catalytic function of chymotrypsin and later on, a number of other tailor-made active site titrants and pseudosubstrates were described (23).   The application of conventional and fast reaction kinetics measurements together with spectroscopic evidence (24) led to the postulate of a tetrahedral acylenzyme intermediate which was subsequently confirmed by x-ray analysis of chymotrypsin (25), subtilisin (26), and a trypsin-inhibitor complex (27). Similar studies of other proteases, e.g. carboxypeptidase, pepsin, and thermolysin (see 17) generated important concepts such as those of "induced fit" (28), "reporter groups" (29), and "entatic site" (30). These should not simply be regarded as shorthand designations but as reminders that the three-dimensional structure of enzymes allows for interactions which cannot be readily predicted or derived from the study of simpler model compounds. In this regard, globular proteins, including the proteases, differ significantly from certain protein hormones, such as ACTH or beta lipotropin. The physiological functions or destinations of these particular hormones are inscribed in adjacent linear sequence segments of polypeptide chains whereas in insulin and most active proteins the functional amino acid residues are located in distant segments of the polypeptide chains and are only brought together by three-dimensional folding (31).

It is reassuring that with few exceptions our notions of the nature and configuration of the active site of enzymes were found to be consistent with the results of x-ray analysis of protein crystals, though uncertainties and discrepancies of details exist. For instance, there is little doubt that the charge-relay system, Asp...His...Ser, is an integral component of the active site of all serine proteases of both mammalian and microbial origin (25,26), although details of the inter-actions between these three residues are still being debated (32). In another instance, the side chain inter-actions at the active site of bovine carboxypeptidase A in solution seem to differ from those deduced from x-ray analysis of the protein crystal (33). Be this as it may, there is no doubt that the x-ray structure together with the amino acid sequence provide the basic information on the structure of a protein and its relation to biological function (17).

ZYMOGEN ACTIVATION

The zymogen-enzyme conversion has proven to be an excellent model system to study the last and most crucial step in the biosynthesis of an enzyme. Although the phenomenon has been known since the early days of enzymology (34) and protein chemistry (1), our knowledge of the detailed conformational changes responsible for the genesis of enzymatic activity is still incomplete. Initially, experimental studies focussed on the identification of the end groups created by tryptic activation of zymogens, notably trypsinogen and chymotryp-sinogen, and on the chemical identification of the activation

peptides (see 35). This approach, refined by extension to the
determination of partial amino acid sequences, still forms the
basis of contemporary studies of the structural characterization of the individual zymogen activation reactions
(cascades) of the blood coagulation system (36). Deeper
insights into the mechanism of zymogen activation have been
gained by analyzing specific zymogen/enzyme pairs by a variety
of kinetic and conformational probes (37-39), including the
direct comparison of the electron density maps of chymotrypsin
and trypsin with those of their respective zymogens (40,41).
As a result, our ideas of the nature of the activation process
have progressed from the somewhat primitive notion that it is
initiated by bringing the serine and histidine residues into
the proper juxtaposition (42), to the realization that in the
zymogen the charge relay system is essentially intact, whereas
the substrate binding site is distorted and therefore functionally impaired (43). Furthermore, zymogens should no
longer be considered to be enzymatically inert since some of
them (e.g. trypsinogen, chymotrypsinogen, pepsinogen,
procarboxypeptidase, Factor X) display a weak intrinsic
activity which becomes greatly enhanced upon activation (44).
While the physiological significance of the intrinsic activity
of zymogens is conjectural, the phenomenon itself has provided
a new approach to elucidate the conformational changes
attendant activation (45). Of greater significance is the
role which zymogen activation plays in a large variety of
physiological systems, e.g. the conversion of procollagen to
collagen, proinsulin to insulin, in macromolecular assembly,
immune reactions, development and differentiation, or in the
processing of certain secretory proteins (46). The study of
these and many other physiological systems has led to the
realization that zymogen activation by limited proteolysis
constitutes an important mechanism of physiological regulation.
It depends uniquely on the initiation by an activating protease, it is essentially irreversible, and thus is capable of
creating a permanent change in the molecular environment of
cell constituents. In its most general aspects, zymogen
activation exemplifies an even broader concept, namely that
many nascent proteins are capable of expressing sequentially
more than one message of information and are subsequently
processed, by proteolytic "restriction" enzymes, into fragments
serving different physiological functions. This editing of
superchains has been found to occur in certain protein
hormones (47), secretory proteins (48,49), immunoglobulins
(50), etc. The lifetime of a protein may, in fact, be
described as a series of essentially irreversible steps of
limited proteolysis. In the case of many secretory proteins
(e.g. those of the pancreas, the hen oviduct, endocrine
glands or lymphocytes) the first step is the removal of an

amino terminal extension (the so-called signal peptide or leader sequence) specifying the extracellular destination of the protein. In the next step, limited proteolysis of a zymogen, poised for activation, produces the physiologically active protein. In the last step, limited proteolysis is believed to prepare the active protein for ultimate degradation (51). Although the phenomenon of limited proteolysis was discovered by Linderstrøm-Lang and Ottesen over 25 years ago (52), a general theory that would predict the exact predisposition of a protein for enzymatic peptide bond cleavage surprisingly is still lacking.

EVOLUTION

Perhaps one of the most interesting developments of protein chemistry during the last decade is the study of the biological history of proteins themselves, i.e. their homology and evolution. Among the first clues of the evolutionary relationship among proteins of diverse species was the sequence homology of cytochromes C, ranging from yeast to man (53); and one of the first indications of the kinship of proteins of different physiological function was the conformational similarity of myoglobin and the subunits of hemoglobin (54). Evidence for the structural homology of proteases emerged from the sequence analysis of bovine pancreatic chymotrypsin by Hartley (55), followed shortly thereafter by the determination of the amino acid sequence of bovine trypsin by Kenneth Walsh and myself (56). While the similarity in amino acid composition of these two serine proteases initially suggested the possibility of a structural relationship, the actual proof of sequence homology had to await the placement in sequence of the last peptides containing the functional amino acid residues of trypsin, specifically serine 195 and histidine 57. X-ray analysis of these two serine proteases and of porcine elastase firmly proved the similarity of these three enzymes at the three-dimensional level and confirmed the correct prediction of the structure of one proteolytic enzyme from that of its homologous counterpart (57). Structural homology has also provided a physical basis for the classification of proteases, and of proteins in general, in terms of their mechanistic similarities and evolutionary history (58). These structural analyses have provided a rational explanation for biological differentiation and evolution of function, besides confirming prior notions of the evolution of the species. They have also brought to light the nature of the molecular lesions responsible for over a

hundred hemoglobinapathies (59), starting with the keen
perception of an altered electrophoretic migration of sickle
cell hemoglobin (60).   Such single point mutations are
expressed in various forms in allotypes [e.g. beta lactoglo-
bulin (61), carboxypeptidase A (62)], and isoenzymes (63).
They are an expression of hyper-variable segments in the
corresponding position of the gene, accounting in particular
for antibody diversity (64).

We have accepted as a basic tenet of protein evolution
that those amino acid residues which are important for
function or maintenance of structure resist evolutionary
pressures and will normally occur in corresponding positions
in proteins that have descended from a common ancestor.   These
invariant residues have aided in tracing the pattern of
evolution of the serine proteases such as the trypsins of
vertebrate, invertebrate or bacterial origin (65).   The
invariant residues of the active site have been guide posts in
establishing the structural relationship among pancreatic
trypsin, chymotrypsin, and elastase, and their relation to the
larger and more complex regulatory proteases of the blood
clotting system (66).   Comparison of the relative positions of
the functional residues in chymotrypsin and in subtilisin (26)
has clearly shown that these two serine proteases, although
mechanistically similar, could not have descended from a common
ancestor; nor could have thermolysin and carboxypeptidase (67),
two metalloenzymes which have similar configurations of their
active sites.   Detailed structural comparison of seemingly
unrelated proteins sometimes brings to light unsuspected
similarities in structure.   Who could have predicted that
mammalian alpha lactalbumin and avian lysozyme would be homologous proteins (68) or that proteins would acquire in the
course of evolution structural traits from other proteins and
incorporate them into their own structure, as in the case of
the Clq component of the complement system which contains
essentially a collagen-like sequence in the stem portion and
a unique sequence in the antibody binding portion (69).   In
yet other examples, evolution has tinkered with active site
residues of ovomucoid inhibitors in closely related species
(70), or has produced non-covalent partnerships, as in the
case of allosteric enzymes or in the macromolecular assembly
of viruses, phages or ribosomes.   None of these phenomena
could have been discovered without the analytical methods of
protein chemistry developed during the last decade or so.

OUTLOOK

There is no doubt that in protein chemistry, as in most
hybrid fields of science, the surges of progress have been
catalyzed by major advances in the parent fields, i.e.
chemistry, physics, molecular biology and biological
engineering. In the invention of the fraction collector, the
amino acid analyzer, the sequenators, and the x-ray diffracto-
meter, we have benefitted from the advances in engineering.
We have incorporated in our experimental armamentarium the
methods of n.m.r. spectrometry, fluorescence polarization,
laser Raman spectroscopy, and magnetic circular dichroism.
We use high speed computers to calculate protein conformations
and to display the final structures; we are extending our
inquiries into problems of membranes, receptor sites, antibody
diversity, cell transformation and many other areas of modern
molecular and cell biology. In short, while we have come a
long way during the last few decades I believe the exciting
years of protein chemistry are before and not behind us.
To be sure, the procedures for the isolation and characterization are more or less the same for most proteins but the
individuality of a newly discovered protein presents unique
challenges and poses new problems. The task ahead is
enormous: According to a recent estimate, approximately
100,000 amino acid residues have been placed in sequence,
representing approximately 800 polypeptide chains. Yet the
coding capacity of mammalian genomes corresponds to 2 billion
amino acid residues and while only a small fraction of the
nuclear DNA is responsible for coding proteins, it is evident
that the amino acid sequences of many key proteins yet remain
to be established (71). We have yet to learn how to determine
the sequences of proteins available only in microgram
quantities. The number of proteins of known three-dimensional
structure is miniscule when viewed in the same light, and no
satisfactory alternative is yet in sight to determine the
conformation of those proteins which for one reason or another
resist all attempts of crystallization. We are far from
understanding the principles of mutual recognition of proteins,
of hormones and hormone receptors, of antibodies and antigens
and generally of the likes and dislikes of protein surface
groups. We have only recently learned to appreciate that
proteins undergo internal vibrations, and can adjust their
conformation to the modulating pressures of their physiological
environment. Despite the recent triumphs of organic peptide
chemistry in synthesizing biologically active enzymes, the
chemical synthesis of protein hybrids by specific fusion of
peptide chains derived from two different proteins (e.g. the

light and heavy chains of different coagulation proteases) or the predetermined synthesis of genetic variants of the same protein still represent formidable problems. In yet another area, despite the enormous progress in our understanding of the mechanism and regulation of protein biosynthesis, our knowledge of the other half of the ledger of protein turnover is woefully inadequate and incomplete. While there is no doubt that proteases are involved in protein degradation and that some of the reactions are energy-dependent (72), we have at best an inkling of the existence of factors which regulate protein breakdown.

The scientific program to follow will bring us to the forefront of knowledge in key areas of protein chemistry: current techniques of impressive elegance and precision; the prediction of protein conformation; the structure, regulatory functions and evolution of enzymes; systems of interacting proteins; and finally more complex physiological processes and their molecular regulation. There is little doubt that today, as in the years past, protein chemistry offers unlimited challenges and opportunities to inquiring minds, and I hope that in the overture of this distinguished International Symposium on Protein Chemistry I have succeeded in sharing with you not only impressions of the past but also a feeling of the exciting developments of the future.

ACKNOWLEDGMENT

I am indebted to my friend and collaborator Dr. Kenneth A. Walsh for fruitful discussions and a critical review of this manuscript.

REFERENCES

1. Northrop, J.H., Kunitz, M., and Herriott, R., "Crystalline Enzymes," 2nd ed., Columbia University Press, New York, 1948.
2. Neurath, H., and Walsh, K.A., Proc. Nat. Acad. Sci. U.S.A. 73:3825 (1976).
3. "Proteases and Biological Control" (E. Reich, D.B. Rifkins, and E. Shaw, eds.), Cold Spring Harbor Laboratory, 1975.
4. "Methods in Enzymology," Vol. 40 (L. Lorand, ed.), Academic Press, New York, 1976.

5. Hartley, B.S., Ann. Rev. Biochem. 29:45 (1960).
6. de Haën, C., and Neurath, H., in "Biochemical and
    Biophysical Perspectives in Marine Biology," (D.C. Malms
    and J.R. Sargeant, eds.), Vol. 3, p. 1, Academic Press,
    New York, 1976.
7. Sumner, J.B., Gralen, N., and Eriksson-Quensel, I.-B.,
    J. Biol. Chem. 125:33 (1938).
8. Bailey, K., Nature 157:368 (1946).
9. Steiner, D.F., Cunningham, D.D., Spigelman, L., and
    Aten, B., Science 157:697 (1967).
10. "Methods in Enzymology," Vol. 22 (W.B. Jacoby, ed.),
    Academic Press, 1971.
11. Putnam, F.W., and Neurath, H., J. Biol. Chem. 166:603
    (1946).
12. Weber, K., and Osborn, M., in "The Proteins," 3rd ed.
    (H. Neurath and R.L. Hill, eds.), Vol. 1, p. 180,
    Academic Press, New York, 1975.
13. Walsh, K.A., Ericsson, L.H., and Titani, K., this Volume.
14. Sanger, F., Science 129:1340 (1959).
15. Konigsberg, W.H., and Steinman, H.M., in "The Proteins,"
    3rd ed. (H. Neurath and R.L. Hill, eds.) Vol. 3, p. 2,
    Academic Press, New York, 1977.
16. Sanger, F., Nicklen, S., and Coulson, A.R., Proc. Nat.
    Acad. Sci. U.S.A. 74:5463 (1977).
17. Matthews, B.W., in "The Proteins," 3rd ed. (H. Neurath
    and R.L. Hill, eds.) Vol. 3, p. 404, Academic Press, New
    York, 1977.
18. Shotton, D.M., and Watson, H.C., Phil. Trans. Royal Soc.
    London B257:111 (1969).
19. Neurath, H., and Schwert, G.W., Chem. Rev. 46:69 (1950).
20. Jansen, E.F., Nutting, M.D.F., and Balls, A.K., J. Biol.
    Chem. 109:189 (1949).
21. Weil, J., James, S., and Buchert, A.R., Arch. Biochem.
    Biophys. 46:266 (1953).
22. Schoellman, G., and Shaw, E., Biochemistry 2:252 (1963).
23. Shaw, E., in "The Enzymes," 3rd ed. (P. Boyer, ed.) Vol.
    1, p. 91, Academic Press, New York, 1970.
24. Bender, M.L., and Killheffer, J.V., C.R.C. Critical
    Reviews of Biochemistry 1:149 (1973).
25. Henderson, R., Wright, C.S., Hess, G.P., and Blow, D.M.,
    Cold Spring Harbor Symp. Quant. Biol. 36:63 (1971).
26. Robertus, J.D., Kraut, J., Alden, R.A., and Birktoft,
    J.J., Biochemistry 11:4293 (1972).
27. Huber, R., Kukla, D., Bode, W., Schwager, P., Bartels,
    K., Deisenhofer, J., and Steigemann, W., J. Mol. Biol.
    89:73 (1974).
28. Koshland, D.E. Jr., Proc. Nat. Acad. Sci. U.S.A. 44:98
    (1958).

29. Burr, M.E., and Koshland, D.E. Jr., Proc. Nat. Acad. Sci. U.S.A. 52:1017 (1964).

30. Vallee, B.L., and Williams, R.J.P., Proc. Nat. Acad. Sci. U.S.A. 59:498 (1968).

31. Roberts, J.L., and Herbert, E., Proc. Nat. Acad. Sci. U.S.A. 74:5300 (1977).

32. Mathews, D.A., Alden, R.A., Birktoft, J.J., Freer, S.T., and Kraut, J., J. Biol. Chem. 252:8875 (1977).

33. Johansen, J.T., and Vallee, B.L., Biochemistry 14:649 (1975).

34. Kühne, W.F., Verhandlungen des Heidelb.-Naturhist.-Med. Vereins, N.A. p. 2, 1876.

35. Neurath, H., Adv. Protein Chem. 12:320 (1957).

36. Davie, E.W., and Fujikawa, K., Ann. Rev. Biochem. 44:799 (1975).

37. Gertler, A., Walsh, K.A., and Neurath, H., Biochemistry 13:1302 (1974).

38. Kerr, M.A., Walsh, K.A., and Neurath, H., Biochemistry 14:5088 (1975).

39. Bazzone, T.J., and Vallee, B.L., Biochemistry 15:868 (1976).

40. Birktoft, J.J., Kraut, J., and Freer, S.T., Biochemistry 15:4481 (1976).

41. Fehlhammer, H., Bode, W., and Huber, R., J. Mol. Biol. 111:415 (1977).

42. Neurath, H., and Dixon, G.H., Federation Proc. 16:793 (1957).

43. Lonsdale-Eccles, J.D., Neurath, H., and Walsh, K.A., Biochemistry, in press.

44. Morgan, P.H., Robinson, N.C., Walsh, K.A., and Neurath, H., Proc. Nat. Acad. Sci. U.S.A. 69:3312 (1972).

45. Kerr, M.A., Walsh, K.A., and Neurath, H., Biochemistry 15:5566 (1976).

46. Neurath, H., and Walsh, K.A., in Proceedings 11th FEBS Meeting, "Regulatory Proteolytic Enzymes and Their Inhibitors," (S. Magnusson, M. Ottesen, B. Foltman, and H. Neurath, eds.), Pergamon Press, Oxford, in press.

47. Chan, S.J., Keim, P., and Steiner, D.F., Proc. Nat. Acad. Sci. U.S.A. 73:1964 (1976).

48. Devillers-Thiery, A., Kindt, T., Scheele, G., and Blobel, G., Proc. Nat. Acad. Sci. U.S.A. 72:5016 (1975).

49. Palmiter, R.D., Thibodeau, S.N., Gagnon, J., and Walsh, K.A., in Proceedings of the 11th FEBS Meeting, "Regulatory Proteolytic Enzymes and Their Inhibitors," (S. Magnusson, M. Ottesen, B. Foltman, and H. Neurath, eds.), Pergamon Press, Oxford, in press.

50. Burstein, Y., Zemell, R., Kantor, F., and Schechter, I., Proc. Nat. Acad. Sci. U.S.A. 74:3157 (1977).

51. Katunuma, N., in "Current Topics in Regulation," (B.L. Horecker and E.R. Stadtman, eds.), Vol. 7, p. 175, Academic Press, New York and London, 1973.
52. Linderstrøm-Lang, K.U., and Ottesen, M., C.R. Trav. Lab. Carlsberg 26:403 (1949).
53. Fitch, W.M., and Margoliash, E., Brookhaven Symp. Quant. Biol. 21:217 (1968).
54. Ingram, V.M., Nature 189:704 (1961).
55. Hartley, B.S., Nature 201:1284 (1964).
56. Walsh, K.A., and Neurath, H., Proc. Nat. Acad. Sci. U.S.A. 52:884 (1964).
57. Hartley, B.S., Phil. Trans. Royal Soc. London B257:77 (1969).
58. "Structure, Function and Evolution in Proteins," Brookhaven Symposia in Biology, No. 21 (1968).
59. "Treatment of Haemoglobinapathies and Allied Disorders," World Health Organization Technical Report Series 509, Geneva, 1972.
60. Pauling, L., Corey, R.B., Singer, S.J., and Wells, I.C., Science 110:543 (1949).
61. Aschaffenburg, R., and Drewry, J., Nature 180:376 (1955).
62. Walsh, K.A., Ericsson, L.H., and Neurath, H., Proc. Nat. Acad. Sci. U.S.A. 56:1339 (1966).
63. "Multiple Forms of Enzymes," Ann. N.Y. Acad. Sci. 151: 1-689 (1968).
64. Edelman, G.R., and Gall, W.E., Ann. Rev. Biochem. 38:415 (1969).
65. de Haën, C., Neurath, H., and Teller, D.C., J. Mol. Biol. 92:225 (1975).
66. Titani, K.A., Fujikawa, K., Enfield, D.E., Ericsson, L.H., Walsh, K.A., and Neurath, H., Proc. Nat. Acad. Sci. U.S.A. 72:3082 (1975).
67. Kester, W.R., and Matthews, B.W., J. Biol. Chem. 252:7704 (1977).
68. Brews, K., Vanaman, T.C., and Hill, R.L., J. Biol. Chem. 242:3747 (1967).
69. Reid, K.B.M., and Porter, R.R., Biochem. J. 155:19 (1976).
70. Kato, I., and Laskowski, M. Jr., in Proceedings of the 11th FEBS Meeting, "Regulatory Proteolytic Enzymes and Their Inhibitors," (S. Magnusson, M. Ottesen, B. Foltman and H. Neurath, eds.) Pergamon Press, Oxford, in press.
71. Edman, P., Carlsberg Res. Commun. 42:1 (1977).
72. Kowit, J.D., and Goldberg, A.L., J. Biol. Chem. 252:8350 (1977).

SOME REFLECTIONS UPON FUTURE DEVELOPMENTAL TRENDS IN PROTEIN SEPARATION METHODS

Jerker Porath

Institute of Biochemistry
University of Uppsala
Uppsala, Sweden

INTRODUCTION

To try to be a prophet and foretell the future is a risky task for a serious scientist and I am sure that much of what I have to say may not come true. I hope nevertheless that my guesses will make some sense. Thoughts about possible extensions and uses of unexplored knowledge are the cradles of invention and discovery, so I will reluctantly make an attempt to forecast the future of protein fractionation methods.

The first question that may arise is whether there is a need for further improvements and the introduction of new techniques for isolation and physical characterization of proteins. We can give an affirmative answer for many reasons. We need sophisticated large scale methods for the isolation of minute quantities of proteins extracted from samples weighing tons, thousands of tons or perhaps even more. As an example it is worth to mention that ton quantities of brain tissue were used for laboratory isolation of thyroid hormone releasing factor (TRF) from hypothalami. In the future there will certainly be an increasing demand for isolating proteins in much larger quantities than has been done so far from precious starting materials, such as tissue and physiological fluids of human origin,

which are of clinical importance as regards both
therapy and diagnosis. The same may be true for
proteins used in veterinary medicine. At the same
time I also feel that there will probably be an
increasing demand for subultramicro scale methods
which give high resolving power.

Large Scale Fractionation Processes

If we first consider large scale procedures
it is always the opinion of engineers that
continuous processes should be developed. We
may distinguish among those based on one-phase,
two-phase and multi-phase systems. One-phase
systems appear to be the easiest to develop for
industry. Separation methods which employ
electrical fields (e.g., electrophoresis and
electrodialysis) present practical problems,
especially in connection with effective removal
of the continuously generated joule heat. High
resolution electrophoresis or dialysis of protein
mixtures in kilogram quantities per hour cannot
yet be achieved. Electrodialysis of proteins is
now being investigated theoretically by two research groups headed by Dr. Archer Martin (1)
in England and by Professor Harry Rilbe (1) in
Sweden. Useful group fractionation may eventually
be practicable. Dr. Martin believes that displacement electrophoresis (isotachophoresis) should
be applicable on a fairly large scale. However,
no apparatus capable of handling kilogram quantities of proteins has yet been described. Gel
electrophoretical methods are not likely to be
developed for large scale separations simply
because the problems associated with the dissipa-
tion of the generated heat put a limit to its use.
High pressure ultrafiltration and extraction
with multiple liquid phase systems seem to be a
feasible approach. Polymer phase systems have been
studied by Per-Åke Albertsson (2). He has been
able to construct systems composed of as many as
18 phases in equilibrium with each other, and all
of them containing water as the main component.
Such multiple phase systems can be arranged one
above the other in order of increasing hydrophobicity. The differences in density between the
phases are quite small, but the separation of such

phases can be speeded up by centrifugation. For
two-phase systems continuous centrifugation may
presumably be feasible. There is, however, a
serious drawback associated with this method, *viz.*
the separation of the polymers on a large scale
from the purified proteins. Until such methods
have been developed it is thus not likely that more
sophisticated procedures such as affinity extraction can be used. Large scale protein chromatography is not easy to perform for the following
reasons: 1) uniform flow throughout the bed is
difficult to achieve, 2) bed compression must be
kept within narrow limits to maintain sufficiently
high flow rates, 3) the bed material should
be easy to regenerate, 4) if the proteins are to
be used clinically the sorbent must be able to
withstand sterilization *in situ* before the run
and 5) the cost must be reasonably low. The
second condition may set a practical limit for
molecular sieves and adsorbents based on hydrophilic gel materials. Such gels seem to be rather
soft mechanically, which means that in practice
shallow beds must be used.

Large composite columns may be prohibitively
expensive and are economically justifiable only
for the purification of very precious materials.
Adsorbents are more feasible in practice since
comparatively smaller columns will suffice. For
crude separations stream-lifted, suspended beds
may be used. Continuously operated columns
with countercurrent flow of adsorbent and solvent
can perhaps also be developed.

Some Theoretical Considerations

J. Calvin Giddings has made a theoretical
treatment of the basis of macromolecular chromatography (3). He starts out from the interdependence of the distribution constant, K, and
$\Delta\mu_o$, the change in chemical potential for the
transfer of one mole of solute from mobile to
stationary phase:

$$K = e^{-\dfrac{\Delta\mu_o}{RT}}$$

1)

where R and T are the gas constant and temperature,

respectively. He discusses the contributions made
by the constituent groups to $\Delta\mu_O$ in the same way
as described by Martin (4). The shortcomings of
conventional adsorption and partition chromatography reside in the vast range of $\Delta\mu^O$ for
proteins. Evaluation of $\Delta\mu^O$ provides a basis for
theoretical treatment of multipoint adsorption-
desorption processes in chromatography of macromolecules which contain a multitude of interacting functional groups exposed at the surface.
The key problem is thus to find a way to reduce
the effective range of $\Delta\mu^O$. Giddings arrives at
the conclusion that field-flow fractionation
processes should give solutions to this problem.
In these methods cross-field flow in the separation
channel forces the solute to stream in layers
parallel to the wall in concentration gradients
which depend on some property of the solutes,
e.g., diffusion coefficients.

If $\Delta\mu_O$ can be kept within the range of 1-4
kcal/mole corresponding to single point attachment or attachment at a few points, then linear
adsorption chromatography may be possible. The
problem is thus to produce adsorbents which
are directed towards specific side groups that are
exposed on the surfaces of the protein molecules
and which occur at low frequency.

Charge-transfer adsorbents, which are at
present under development in my laboratory, can
be produced to exhibit a high affinity for the
indole group of tryptophan residues (1,5). Metal
chelate adsorbents with high affinity for imidazole
and thiol groups can also be synthesized easily
(6). These adsorbents are likely to be helpful
tools for protein chemists whether they are used
in linear chromatography to achieve separations
at high resolution or in displacement chromatography for group fractionation.

Adsorption, extraction and precipitation
methods are also based on the surface characteristics of protein molecules. On the other hand
ultracentrifugation and exclusion methods make
use of molecular weight or size and shape. Electrophoresis in gels in some of its varied forms
is the method that at present offers the highest
resolution capability. No doubt it will be
developed further to better suit preparative
purposes.

Frequently, due to the large $\nabla\mu_O$, the equilibrium distribution of proteins between two phases will be one-sided. This fact might be used to advantage both in group separations and for extracting specific proteins from crude mixtures. Very popular among biochemists at present are the biospecific adsorption methods that are grouped together under the heading of biospecific affinity chromatography. Such one-sided material distributions do not require the use of chromatographic techniques. Batchwise adsorption and desorption can be more convenient on a large as well as on a small scale. The specific methods are here to stay and can be further improved and refined. They will not make other methods superfluous or outdated as is evident by the fact that they also accomplish group fractionation, $i.e.$, by separating out isoenzymes which in turn must be further fractionated by electrophoresis for example.

After this brief survey of separation methods that are well known already or have recently been introduced for fractionation of proteins I shall now outline some of my speculations as regards future trends in the development of methods for the separation of proteins. One conceivable trend is the utilization of unknown properties of proteins to achieve further fractionation and characterization. Would it thus be possible to devise separation methods based on some properties found at the interior regions of the molecules solely and to develop them in such a way that the operations will leave the proteins undamaged? By the use of artificial or natural gravitational fields proteins may be separated according to differences in density by isopycnic methods. These mthods exemplify the use of properties that depend on the overall composition of the protein molecules, $i.e.$, the interior as well as the surface characteristics.

Since the protein core seems inaccessible for manipulation and since unfolding should be avoided in the isolation or purification processes the problem may seem at first sight to be insurmountable. However, this is not always necessarily the case. Electrical and magnetic fields, stationary or oscillating, for example, penetrate the protein molecules and will be distorted as a result of the conducting properties of the interior molecular regions. After proteins are isolated using such methods, however, they are not unfolded or inactivated.

Alexander Kolin (1,7) has pointed out the possibility to use electromagnetic fields for fractionation of cell particles when they differ in conductivity from the surrounding fluid. Kolin has called the method *magnetoelectrophoresis*. The principle is the following: if a particle is subjected to crossed magnetic and electrical fields, a force, F, will act on it in a direction perpendicular to both fields. The separation parameter is the electrical conductivity of the particle, which is a function of its core composition. If the electrical conductivities of the particle and the surrounding fluid are σ' and σ'', respectively, η is the viscosity of the fluid and a is the diameter of the particle, the migration velocity, v, of the latter will be

$$\underline{v} = \underline{J} \times \underline{B} \cdot \frac{\sigma'-\sigma''}{\sigma'+2\sigma''} \cdot \frac{a^2}{3\eta} \qquad 2)$$

where $\underline{J}$ and $\underline{B}$ are the current and magentic field strength, respectively.

The direction of migration is given by the relative magnitudes of σ' and σ'' and the particle will apparently be at rest when $\sigma'=\sigma''$. Kolin suggests that isoperichoric focusing of particles should be possible in a manner analogous to isoelectric focusing. The environmental gradient parameter should in this case be the conductivity of the surrounding medium. The principle has been tested on model non-conducting polymer particles and cells. Rough calculations applying to Kolin formula have convinced me that there might be a future for magnetoelectrophoresis of proteins. Stronger magnetic fields are needed. The method might find applications for fractionation of proteins on a pico- or fentogram scale employing extremely thin wires or capillaries. The experimental difficulties to be encountered for analysis and fraction collecting might seem almost insurmountable at the present stage of its development. However, new methods for monitoring the movement of liquid on submicro scales might be envisaged making use of electrokinetic or electromagnetokinetic fluid displacement rather than hydraulic flow. At such a scale, electrophoresis, electromagnetophoresis or "chromatography" should be

applicable for the isolation of proteins from a
single cell or even from subcellular particles
such as chromosomes.

We may expect new problems to arise when known
principles are to be applied on a small scale. In
addition to the difficulties involved in the control of liquid displacement and the collection of
well-defined fractions there will be the action
of capillary forces, evaporation and undesired
adsorption which will not be easy to avoid,
suppress or counteract. However, when the need
for such methods arises ingenuity will be invoked
and the problems will be solved or circumvented
one way or another.

REFERENCES

1. Chromatographic Reviews, 1st Tiselius Sympo-
    sium 1977 (to be published).
2. Albertsson, P. A., Endeavour, Vol. 1, p. 69
    (1977).
3. Giddings, J. C., J. Gas Chromatogr., Aug. 1967
    p. 413.
4. Martin, A.J.P., Biochem. Symp., Cambridge,
    Univ. Press 3:4 (1949).
5. Porath, J., and Dahlgren Caldwell, K.,
    J. Chromatogr. 133:180 (1977).
6. Porath, J., Carlsson, J., Olsson, I., and
    Belfrage, G., Nature 258:598 (1975).
7. Kolin, A., Science 117:135 (1953).

FLUORESCENT METHODS FOR ISOLATION, CHARACTERIZATION AND ASSAY OF PEPTIDES AND PROTEINS

Sidney Udenfriend

Roche Institute of Molecular Biology
Nutley, New Jersey

INTRODUCTION

There are common problems relating to purification of peptides and proteins which are of importance in fields as diverse as virology, endocrinology, immunology and neurobiology. Current technology which includes U.V. detection, ninhydrin amino acid analyzers, and the automated sequenators are applicable to proteins and peptides that are present at relatively high concentration in tissues or can be readily concentrated from large amounts of tissue. However, many important proteins and peptides are present in tissues at concentrations lower than a microgram per gram, a range where the generally used methods are difficult to apply. For example, isolation of many of the known hormones from the pituitary gland, has been carried out using hundreds of thousands of glands obtained from animals killed at the slaughterhouse. While this approach has been highly successful it required an initial outlay for tissue alone approaching the million dollar range. The cost of solvents and chemicals is, of course, proportionally high. There are other problems associated with the use of slaughterhouse materials. Proteolysis can occur during collection, during transfer, prior to freezing, during thawing, and while working up the tissues. With slaughterhouse material it is difficult to control the first few steps, resulting in two types of artifacts, the appearance of peptides which are normally not present in the tissues and the disappearance of larger proteins and peptides. Finally, methods which require quantities of proteins and peptides in the micromole range and higher cannot be applied to the tissues of small laboratory animals which are generally used for biological experimentation.

Fluorescamine (Fluram[T.M.]) a fluorometric reagent for primary amines (1) has been applied to the assay of amino acids, peptides and proteins in the picomole range (2).　A related reagent, 2-methoxy-2,4-diphenyl-3(2H) furanone (MDPF) was subsequently introduced (3) (Fig. 1).

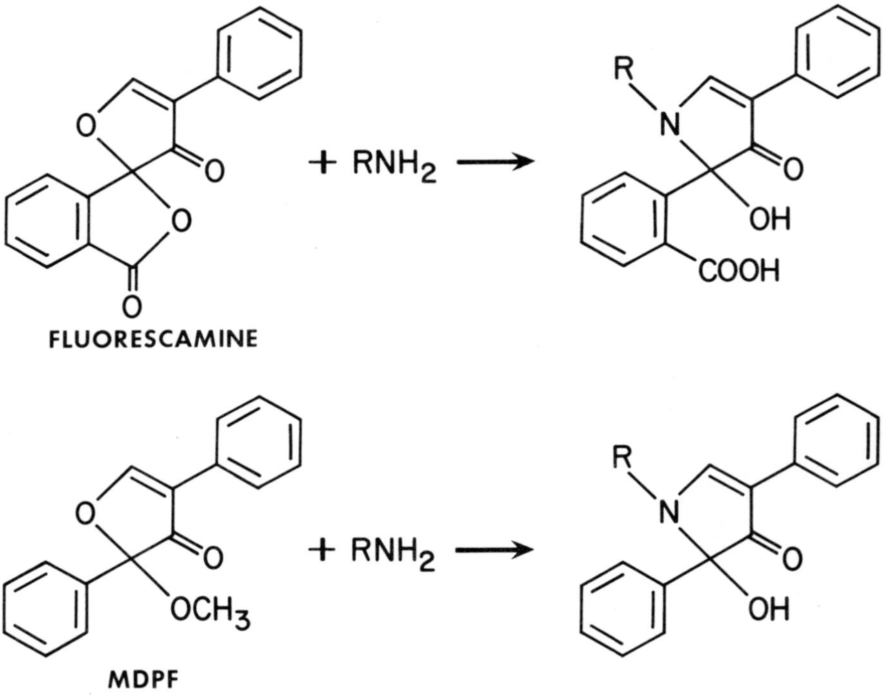

Fig. 1.　Reaction of fluorescamine and MDPF with primary amines.

These two reagents are now being used in all phases of protein and peptide chemistry including isolation, molecular weight determination, amino acid assay, fingerprinting and sequencing, as well as peptide synthesis and quantitative assay of tissue extracts.　We shall discuss each of these processes and present some applications for which fluorescamine or MDPF are uniquely suited.

AMINO ACID ASSAY

Fluorescamine is ideally suited for automated amino acid
analysis. Its reaction with amino acids is essentially in-
stantaneous at room temperature and destruction of excess reagent occurs rapidly. Fluorescamine and its hydrolysis products are non-fluorescent while the amino acid derivatives
are highly fluorescent and picomole quantities are sufficient
for assay. Another advantage of fluorescamine is that it is
relatively insensitive to ammonia. The imino acids proline
and hydroxyproline can be assayed after oxidation to the corresponding amino aldehydes (4). A fully automated amino acid
analyzer which utilizes fluorescamine for detection has been
developed (5). A standard mixture (250 pmol) run on the
fluorescamine amino acid analyzer is shown in Fig. 2.

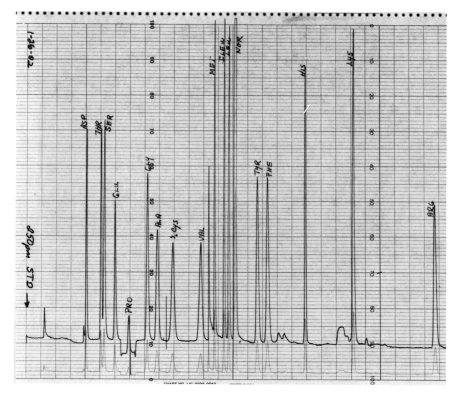

Fig. 2. Automated fluorescamine assay of an amino acid
mixture (250 pmol).

A unique application of the fluorometric methods was in the identification of a tissue protein which crossreacts with antibody to the enzyme prolyl hydroxylase (6). The enzyme and the crossreacting protein (CRP) are found in all tissues which produce collagen, with the latter always in excess. Microgram quantities of both were isolated from newborn rat skin. When subjected to SDS gel electrophoresis the enzyme, a tetramer of molecular weight 240,000, was resolved into its two component subunits, a larger one (S_L) about 64,000 daltons, and a smaller one (S_S) about 60,000 daltons. When CRP was subjected to SDS gel electrophoresis it migrated to the same position as (S_S) indicating that both were of the same molecular weight. Each of the Coomassie stained bands, obtained on electrophoresis containing microgram quantities of protein, were cut out, the gel segments were subjected to acid hydrolysis, and then to automated amino acid assay with fluorescamine (7). The amino acid compositions of CRP and S_S were found to be almost the same, but not identical. Fluorescamine and MDPF were then used along with high performance liquid chromatography (HPLC) for fingerprinting microgram quantities of the two 60,000 dalton peptides. Their great similarity was further confirmed when it was shown that only one or two of the approximately 70 peptides obtained on treatment with trypsin were different.

Previous attempts to determine the amino acid composition of protein bands cut out of polyacrylamide gels by ninhydrin assay (8) have been limited by the large amounts of ammonia which are released from gels during hydrolysis. o-Pthalaldehyde, which has also been used for fluorometric amino acid assay, at the picomole level (10), is also unsuitable for assay of gel slices.

PROTEIN AND PEPTIDE ISOLATION

The properties of fluorescamine are ideal for monitoring flow systems. Reaction with peptides is rapid and the reagent and its hydrolysis products are non-fluorescent. When applied to the quantitative assay of peptides by chromatography, the column effluent is directed into a fluorometric detection system, in the amino acid analyzer. Quantitative assay of the dipeptide, carnosine, has been carried out on extracts of nasal mucosa and olfactory bulb (10). As little as 5 pmole are sufficient for analysis (11).

With the fluorescamine detection system aliquots for assay do not contribute appreciably to the losses incurred during purification since an automatic stream-sampling valve

utilizes only a small portion of a column effluent for assay
(picomoles) while the remainder (nanomoles) is directed to a
fraction collector (12). This detection system has been used
for the chromatography of nanomole quantities of peptides
ranging in size from a few amino acid residues to large proteins. Some applications are described below.

The purity of peptides which are to serve as calibration
standards for various types of assays must be established
prior to use. Furthermore, peptides that are to serve as
antigenic determinants in the preparation of antibodies must
be homogeneous to yield antisera specific for the peptide.
Although synthetic peptides can now be made in high purity impure preparations are common (Fig. 3). Impure peptide preparations can be purified by chromatography on a high efficiency column (13). Radioactive peptides present another
problem. They are prepared in small amounts and are of high
specific activity. They are also expensive and laboratories
which prepare them must waste relatively large (and therefore
costly) amounts to detect and remove impurities. Since few
scientists who use them are equipped to monitor their purity
with the small amounts available, impure labelled peptides
are frequently used as prime standards in biological studies.
Fluorescamine methodology is ideally suited for the purification of labelled peptides without sacrificing much for assay.

Much of the research in our laboratory has been concerned
with biologically active peptides. We are, at present, particularly interested in the opioid peptides. To monitor these
peptides in column effluents we have developed a competitive
binding assay with neuroblastoma X glioma hybrid cells (14)
using tritiated leucine-enkephalin as the displaced ligand
(15). Both the bioassay and the fluorometric procedures are
sensitive at the picomole level making it possible to use
tissues from relatively few laboratory animals (rats and
guinea pigs) instead of slaughterhouse material (16). As
noted above, use of laboratory animals is important for several reasons. Among these are the ability to remove the tissues from the animals rapidly and to homogenize them under conditions where proteolytic degradation is minimal. Besides
generating superfluous peptides that can complicate isolation,
proteolysis can significantly lower the yield of the protein
or peptide under investigation. Both problems were encountered in our isolation of β-lipotropin from frozen glands obtained from a commercial source. Chromatography of the pituitary extract on Sephadex indicated that a substantial proportion of the larger proteins had been degraded to small peptides. Recovery of β-lipotropin was therefore markedly reduced and the smaller endorphins were correspondingly increased (16).

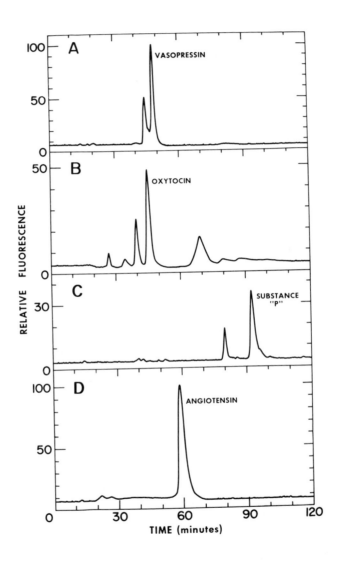

Fig. 3.   Evaluation of purity of commercial peptides by fluorescamine - HPLC.

Since the rat has traditionally been used in studies on
analgesia and behavior, studies on the opioid peptides will
have to be done in the rat. It was necessary, therefore, to
isolate and characterize the various biologically active peptides from rat tissues since there might be differences in
their primary structures compared to the corresponding peptides which were obtained from the larger animals. We have
isolated β-lipotropin from relatively few fresh rat anterior
pituitaries (16). An extract of 40 rat glands was fractionated on G-75, then on Partisil SCX (a strong cation ex-
changer), and finally on Lichrosorb RP-18 (reverse-phase)
(Fig. 4). Each step was monitored with the automated fluorescamine detection system. Aliquots of each fraction were
treated with trypsin to convert the β-lipotropin to a biologically active cleavage product, which was then determined
by the receptor binding assay. A yield of 4 nmol of β-lipotropin was obtained from the 300 mg of anterior pituitary
which was used. Less than 20% of the starting material was
consumed by the monitoring procedure. Some of the isolated
material was used to determine molecular weight and amino acid
composition by fluorescamine procedures. Similar procedures
were used to isolate and characterize rat β-endorphin (17)
and are being used in the isolation of opioid peptides from
other regions of the brain. The resolving power of the high
efficiency columns which we have utilized is demonstrated by
the separation of the hexadecapeptide α-endorphin from the
heptadecapeptide γ-endorphin, the latter differs only by an
additional leucine residue at the carboxy terminus (Fig. 5).

Utilizing similar procedures we demonstrated the presence
of a large protein in extracts of rat pituitary glands which
on treatment with trypsin yielded opioid active material (16),
presumably the precursor of the recognized opioid peptides.
Independently Mains and Eipper (18) showed that mouse pituicytes in culture formed a protein of M.W. ca 31,000, and presented evidence that it was a precursor to both corticotropin
and opioid peptides. Roberts and Herbert (19) recently cor-
roborated the latter finding in mouse pituicytes. We have
now shown (20) that a comparable ∿ 30K precursor common to
both corticotropin and opioid peptides is present not only
in the pituitary of rodents, but in other species as well
(Fig. 6). Furthermore, the product of trypsin digestion was
shown to be the nonapeptide β-lipotropin 61-69 which is also
the product obtained from β-lipotropin and β-endorphin by
trypsin. Mains and Eipper (18) have also identified the
radioactive nonapeptide as a trypsin product of the mouse
∿ 30K protein. We have suggested the name pro-opiocortin (20)

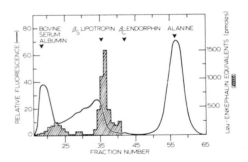

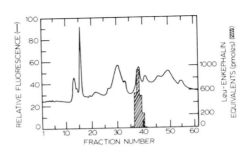

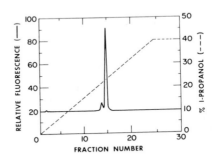

Fig. 4.   Chromatographic steps for the overall purification of rat β-lipotropin.

for this ∿ 30K protein to signify its precursor relation to the two different physiologicall active peptides.

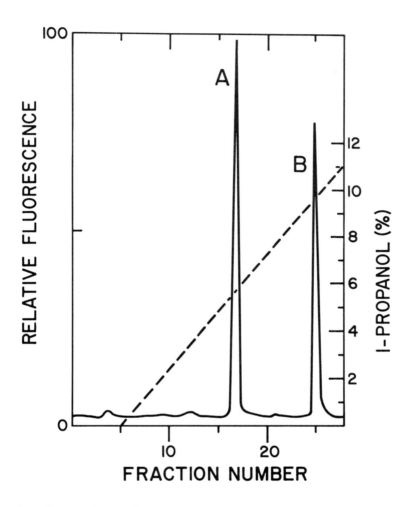

Fig. 5. High performance liquid chromatography of α-endorphin and γ-endorphin.

The smallest opioid peptides, the enkephalins, were first found in the brain (21) and there is substantial evidence that they are not derived from pituitary sources (22), (23). If this is so one might expect precursor(s) in brain tissue. In recent studies (24) we have found in striated tissue from rat, guinea pig and beef a protein which appears to be larger than pro-opiocortin which on treatment with trypsin yields

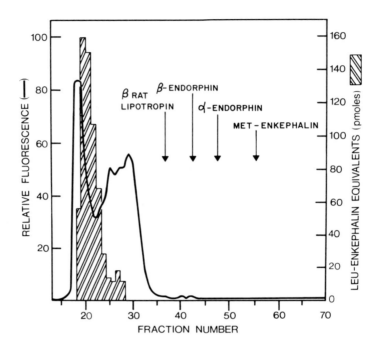

Fig. 6.   Rechromatography of pro-opiocortin on Sephadex
G-75.

opioid active material.   It is of interest that unlike the
pituitary, striatal tissue contains no intermediate sized
peptides comparable to β-lipotropin or β-endorphin.

PRIMARY STRUCTURE

The fluorometric procedures are also important for determining primary structure when only nanomole quantities are
available.   Picomole amounts have been used for molecular
weight determination utilizing comparative gel filtration
(16).   Fluorescence has also been used for fingerprinting by
a variety of procedures (24, 25, 26).   In our studies the
cleavage products obtained by enzymatic or chemical degradation of proteins or peptides have been resolved by HPLC using
the fluorescence monitoring system.   The identity of rat and
camel endorphin was established by such methods (17) showing
that they both yielded exactly the same pattern of tryptic

peptides on HPLC. Alternatively, separation of peptides can be achieved by disc gel electrophoresis followed by staining with fluorescamine. Another approach is to prelabel proteins and peptides with MDPF which makes them fluorescent and carry out electrophoresis in the presence of sodium dodecyl sulfate. Since MDPF products have no additional charge their electrophoretic patterns are unchanged. Application of the latter procedure to the tryptic peptides of aldolase is shown in Fig. 7. With this methodology the tedious job of staining and destaining is not required since the bands can be visualized with an ultraviolet lamp even during the electrophoresis.

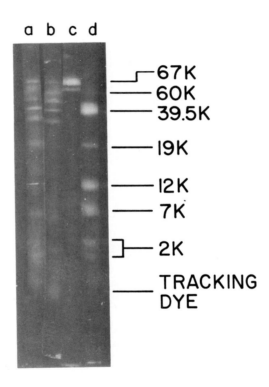

Fig. 7. Polyacrylamide gel electrophoresis in SDS of MDPF-aldolase (c) and MDPF-aldolase fragmented with cyanogen bromide (d).

Furthermore, because gels need not be fixed protein and peptide derivatives are recoverable in high yield. This prelabeling technique is especially useful with peptides below
a few thousand daltons, that cannot be precipitated in gels.
The MDPF fluorescent label on peptides has been found to be
stable even after cleavage at methionyl residues by cyanogen
bromide in formic acid (28).

Doctor C. Y. Lai in our Institute has been able to hydrolyze the thiazolinone derivatives formed in the Edman degradation and obtain high yields of the respective amino acids
(29). This procedure permits him to carry out sequence analysis utilizing as little as one nanomole of a tryptic peptide.

MONITORING PEPTIDE SYNTHESIS

At each step in solid state peptide synthesis it is necessary to check for completeness of reaction. Fluorescamine
can detect less than 0.5% of uncoupled peptide (30, 31). Alternatively, the growing peptide may be removed from a small
portion of the resin in order to check for homogeneity at intermediate steps. Larger peptides end up with accumulated
errors, requiring tedious purificaiton of the final product.
Felix et al. (32) have also used fluorescamine to abort incom-
pletely reacted chains at each step of the solid state synthesis.

QUANTITATIVE ASSAY

There are advantages to forming the fluorophors prior to
resolution of a peptide mixture by column chromatography, in
contrast with post-column labeling. Only one pump (for the
column) is required since reagent has already been added.
This makes for simpler instrumentation and a lower and more
stable baseline. The latter thereby extends the lower limit
of detection.

Assay of the nonapeptides oxytocin and vasopressin in
individual rat posterior pituitaries have been carried out by
the prelabeling technique (33). Each pituitary is homogenized in dilute acid and proteins are precipitated with trichloroacetic acid. A copper Sephadex column is used to remove α-amino acids and the resultant peptide fraction is
treated with fluorescamine. The fluorescent derivatives are
then resolved on a reverse-phase column with a fluorescent
monitoring system.

Specificity and accuracy of the assay was checked by the following procedures. Fluorophors of synthetic oxytocin and vasopressin were shown to co-chromatograph with the tissue-derived nonapeptides. The putative oxytocin and vasopressin fluorophors collected in the column effluent were hydrolyzed and subjected to amino acid analysis. The expected compositions were obtained. When aliquots of pituitary extracts were analyzed by a pressor assay specific for vasopressin, the results were in agreement with the fluorescence assay. The sensitivity of peptide assay by prelabeling is often at or below the picomole level.

Although fluorescamine has been used as the prelabeling agent in the earliest studies with oxytocin and vasopressin, MDPF has properties which may make it the reagent of choice for this methodology (34). Fluorophors derived from MDPF are more stable. They are also more hydrophobic which aids in the reverse-phase chromatography of small peptide fluorophors. Furthermore, the MDPF residue can be removed from many peptides by hydrazinolysis to release the intact native peptide (35).

DISCUSSION

Assay at the picomole level is really not much of a problem if one takes into consideration several important factors. First of all losses due to adsorption become readily apparent at these low levels. In our studies these have been minimized in many ways. The use of any kind of glassware as test tubes, pipettes, etc. is eliminated and polypropylene is used instead. Where glassware must be used it is treated with Sili-clad. Peptide standards and extracts are usually stored in dilute hydrochloric acid (0.01 M). Two other problems which are apparently increased in this range are oxidation of methionine and bacterial destruction. The antioxidant thiodiglycol and the antimicrobial agent pentachlorophenol have been found satisfactory when added in sufficient amount to all buffers used in chromatography. A problem that arises during purification of tissue extracts is losses due to traces of tissue proteases. Phenylmethylsulfonyl fluoride, or other protease inhibitors are always added during tissue homogenization, and are generally included in all solutions. Background contamination is minimized by distilling all volatile solvents and reagents, including pyridine, acetic acid, and hydrochloric acid. With such precautions we have not found it cumbersome to operate at the picomole level. In fact, there are many offsetting features that make the micromethodology far simpler than it appears.

Instrumentation for these procedures is only now becoming commercially available. The instruments which we now use are not unusually complicated. However, their construction and maintenance require a substantial part of the investigator's time. Commercial equipment of appropriate design should be even more reliable and have greater sensitivity. When commercial instruments become available, fluorescamine and MDPF, and other fluorescent reagents, will most likely become widely used, not only because of their inherently greater sensitivity and specificity, but for the resulting economy and other considerations cited above.

ACKNOWLEDGMENTS

The studies reported here were carried out collaboratively with Drs. Stanley Stein, Menachem Rubinstein and Randolph V. Lewis. I also wish to acknowledge the assistance of Mr. Larry Brink and Ms. Louise D. Gerber.

REFERENCES

1. Weigele, M., DeBernardo, S., Tengi, J. P., and Leimgruber, W., J. Amer. Chem. Soc. 94, 5927 (1972).
2. Udenfriend, S., Stein, S., Böhlen, P., Dairman, W., Leimgruber, W., and Weigele, M., Science 178, 871 (1972).
3. Weigele, M., DeBernardo, S., Leimgruber, W., Cleeland, R. and Grunberg, E., Biochem. Biophys. Res. Commun. 54, 899 (1973).
4. Weigele, M., DeBernardo, S., and Leimgruber, W., Biochem. Biophys. Res. Commun. 50, 352 (1973).
5. Stein, S., Böhlen, P., Stone, J., Dairman, W., and Udenfriend, S., Arch. Biochem. Biophys. 155, 203 (1973).
6. Chen-Kiang, S., Cardinale, G., and Udenfriend, S., Proc. Natl. Acad. Sci. U.S.A. 74, 4420 (1977).
7. Stein, S., Chang, C. H., Böhlen, P., Imai, K., and Udenfriend, S., Anal. Biochem. 60, 272 (1974).
8. Houston, L. L., Anal. Biochem. 44, 81 (1971).
9. Benson, J. R., and Hare, P. E., Proc. Natl. Acad. Sci. U.S.A. 72, 619 (1975).
10. Margolis, F. L., Science 184, 909 (1974).
11. Stein, S., in "The Peptides" (E. Gross, ed.), in press Academic Press, New York (1979).

12. Böhlen, P., Stein, S., Stone, J., and Udenfriend, S.,
    Anal. Biochem. 67, 438 (1975).
13. Radhakrishnan, A. N., Stein, S., Licht, A., Gruber, K. A.
    and Udenfriend, S., J. Chromatog. 132, 552 (1977).
14. Klee, W. A., and Nirenberg, M. Proc. Natl. Acad. Sci.
    U.S.A. 71, 3474 (1974).
15. Gerber, L. D., Stein, S., Rubinstein, M., Wideman, J.,
    and Udenfriend, S., Brain Research (in press).
16. Rubinstein, M., Stein, S., Gerber, L. D., and Udenfriend,
    S., Proc. Natl. Acad. Sci. U.S.A. 74, 3052 (1977).
17. Rubinstein, M., Stein, S., and Udenfriend, S., Proc.
    Natl. Acad. Sci. U.S.A. 74, 4969 (1977).
18. Mains, R. E., Eipper, B. S., and Ling, N., Proc. Natl.
    Acad. Sci. U.S.A. 74, 3014 (1977).
19. Roberts, J. L. and Herbert, E., Proc. Natl. Acad. Sci.
    U.S.A. 74, 5300 (1977).
20. Rubinstein, M., Stein, S., and Udenfriend, S., Proc.
    Natl. Acad. Sci. U.S.A. 75, 331 (1978).
21. Hughes, J., Smith, T. W., Kosterlitz, H. W., Fothergill,
    L. A., Morgan, B. A., and Morris, H. R., Nature 258,
    577 (1975).
22. Rossier, J., Vargo, T. M., Minick, S., Ling, N., Bloom,
    F. E., and Guillemin, R., Proc. Natl. Acad. Sci. U.S.A.
    74, 5162 (1977).
23. Yang, H.-Y. T., Hong, J. S., Fratta, W., and Costa, E.,
    Advan. Biochem. Psychopharm. (in press).
24. Lewis, R. V., Stein, S., Rubinstein, M., Gerber, L. D.,
    and Udenfriend, S. (in preparation).
25. Nakai, N., Lai, C. Y., and Horecker, B. L., Anal. Biochem.
    58, 563 (1974).
26. Furlan, M., and Beck, E. A. J. Chromatog. 101, 244
    (1974).
27. Mendez, E., and Lai, C. Y. Anal. Biochem. 65, 281 (1975).
28. Chen-Kiang, S., "Studies on the Biosynthesis of Collagen"
    P.D. Thesis, Columbia University, New York (1977).
29. Mendez, E., and Lai, C. Y., Anal. Biochem. 58, 563 (1975).
30. Felix, A. M., and Jimenez, M. H. Anal. Biochem. 52, 377
    (1973).
31. Tometsko, A. M., and Vogelstein, E. Anal. Biochem. 64,
    438 (1975).
32. Felix, A. M., Jimenez, M. H., Vergona, R., and Cohen,
    M. R., Intl. J. Peptide Protein Res. 7, 11 (1975).
33. Gruber, K. A., Stein, S., Radhakrishnan, A. N., Brink,
    L., and Udenfriend, S., Proc. Natl. Acad. Sci. U.S.A.
    73, 1314 (1976).
34. Wideman, J., Brink, L., and Stein, S., Anal. Biochem.
    (in press).
35. Weigele, M., and DeBernardo, S., Personal communication.

STRATEGIES OF AMINO ACID SEQUENCE ANALYSIS

Kenneth A. Walsh[1], Lowell H. Ericsson and Koiti Titani

Department of Biochemistry
University of Washington
Seattle, Washington

The amino acid sequence of a protein can be thought of as a sentence of cell biology written in a universal twenty-letter alphabet. The information coded in the sequence dictates the three-dimensional folding and the biological function of the protein. Some proteins contain segments of information specifying the ultimate destination of that protein (1), others control the timing of their expression (2), and still others respond to metabolic controls (3); all protein sequences contain traits of their ancestry. Fifteen years ago it was a major challenge to attempt to undertake the sequence analysis of a 20,000 dalton protein. Today, such a project is a relatively routine matter and polypeptide chains of 40,000 to 100,000 daltons have been analyzed.

Questions are being raised whether sequence analysis is best attempted on the protein or on its polynucleotide counterpart. The latter has the potential of providing not only primary sequence data but also untranslated transcriptional control regions of the parent DNA (4). However, current work is revealing surprising deviations from the previously assumed co-linearity between a protein and its gene and processing events are found along the pathway of transcription, messenger secretion, translation, and during the post-translational period (5). Hence, it becomes important to compare the structure of the gene with those of the corresponding nuclear RNA, mRNA, and protein products of both in vivo and in vitro

[1]Supported by NIH Grant GM-15731

translation. But in the final analysis, it is the product of
translation which is the functioning form and structural
questions will continue to focus on the phenotype even if the
genotypic origin is fully established.

During the past few years, new methods have been developed
to improve the strategy of sequence analysis, and to minia-
turize the experimental tactics. As a result it is now possible to explore at the molecular level physiological systems
previously beyond reach and to correlate the results with
mechanisms of gene expression, processing events, folding
patterns, questions of metabolic control, mechanisms of enzyme
action and protein evolution.

The earliest methods of sequence analysis in Sanger's
laboratory (6) involved the cleavage of a protein into overlapping sets of small peptides, analysis of the partial
sequence of amino acids in each peptide, and deduction of the
alignment of the peptides. The principle of this general
strategy remains the same today, but more precise methods were
introduced in the laboratory of Moore and Stein (7), and a
stepwise degradation procedure developed by Edman (8) was
automated in liquid and in solid phase modes (9,10). Contin-
uing refinement of these techniques has improved the efficiency of sequence analysis by nearly two orders of magnitude,
as judged by the progressively smaller number of peptides
which has been necessary to deduce the complete sequence of a
protein (Table I).

TABLE I. Progress in the Efficiency of Sequence Analysis[a]

	Number of Residues	Approx. Number of Peptides Isolated	Relative Efficiency (mean residues proven/peptide isolated)	Year
Insulin	51	152	0.3	1952
Ribonuclease	124	95	1.3	1962
Trypsin (bovine)	229	200	1.1	1964
Carboxypeptidase A	307	199	1.5	1969
Thermolysin	316	173	1.8	1972
Trypsin (porcine)	223	11	20.2	1972
Amyloid Protein A	76	2	38.0	1972
Group Specific Protease	224	12	18.7	1978

[a]These examples are taken from references (6, 11-17).

In principle the strategy has evolved from one involving limited stepwise degradation of a large number of small peptides to one of extensive degradation of a small number of large peptides. An early example appeared in 1972 when Hermodson <u>et</u> <u>al</u>. (16) isolated only two peptides to deduce the complete sequence of a 76-residue protein (Amyloid protein A). Edman degradation of the intact protein placed 37 residues, including the only two methionyl residues at positions 17 and 24 (Fig. 1). Cleavage with cyanogen bromide yielded a polypeptide fragment representing residues 25–76, analysis of which extended the sequence to residue 70, including all sites of potential tryptic cleavage. Isolation of the carboxyl-terminal peptide from a tryptic digest yielded the remainder of the structure. More traditional methods of analysis would have involved the laborious isolation and analysis of about 40 more peptides. This example serves to direct our attention towards the principal problems of structural analysis, namely:

A) How to limit specific cleavage of proteins and generate large fragments.
B) How to isolate these large, and relatively insoluble, denatured fragments.
C) How to obtain the maximal stepwise degradation of each isolated fragment.

These will now be considered in turn, with illustrations largely taken from our sequence analysis of rabbit muscle glycogen phosphorylase, an enzyme containing 841 residues in a single polypeptide chain (19).

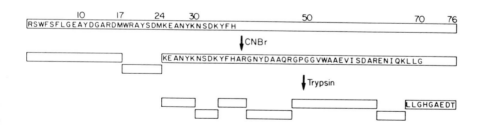

FIGURE 1. Schematic diagram of the strategy of sequence analysis of Amyloid protein A (16). The length of each bar is proportional to the length of the protein or fragment. A one-letter code (18) is used to designate the amino acid residues placed in sequence in the intact protein (top) and in two derived fragments (below).

A. Limited Cleavage Techniques

Ideally, one would like to use a series of different
cleavage methods so specific that each would cleave only one
peptide bond in the original protein. Then, once the amino-
terminal sequence of the native protein has been established,
the α-amino group could be blocked (e.g., by succinylation),
the specific cleavage method applied, and a single sequence
from the point of cleavage derived by Edman degradation of the
mixture of the two cleaved products. This principle has been
used both with proteins and certain fragments as illustrated
by analysis (Fig. 2) of a segment of phosphorylase (representing residues 17-264 of the molecule). Succinylation, followed by specific cleavage at the single Asp-Pro bond, yielded
by Edman degradation a unique sequence of 17 residues (20)
which happened to overlap two other peptides generated by
cleavage of methionyl bonds. The rare occurrence of Asp-Pro
and Asn-Gly sequences (approximately once per 200 residues)
makes them suitable for this approach. The method will be

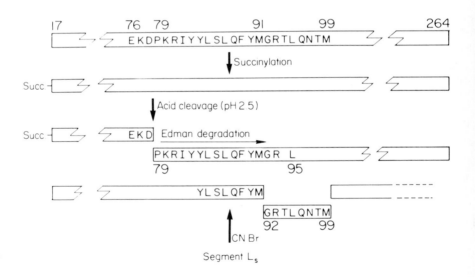

FIGURE 2. Strategy of analysis of a portion of Segment
L_s of phosphorylase (20). The residue numbers indicate the
length of the fragments. Sequenator analysis of the whole
reaction mixture after succinylation and cleavage of Asp-Pro
(D-P) identified residues 79-95, overlapping two other fragments separately derived from a CNBr digest.

successful even if the cleavage is not quantitative.

When initiating sequence analysis of a new protein it is useful to test the susceptibility of the protein to acid cleavage of Asp-Pro bonds or to NH_2OH cleavage of Asn-Gly bonds. These reactions can be monitored by SDS gel analysis, or more advantageously by placing the whole, desalted, reaction mixture in a sequenator and observing directly the number of new sequences (Fig. 3).

Most other cleavage techniques (Table II) produce, characteristically, many fragments and the principal problem becomes one of their separation and purification. In general, the two most useful techniques are cleavage of methionyl bonds by cyanogen bromide and of argininyl bonds by tryptic digestion of an N^ε-acylated protein. Sometimes the amino acid composition reveals potential cleavage loci in suitably low frequency. At other times it is possible to reduce the number of susceptible cleavage sites, for example, by selective oxidation of exposed methionyl residues in native proteins (30). In yet other cases, it may be advantageous to promote incomplete cleavage so that the reaction mixture would contain not only the expected cleavage products, but also overlapping peptides which can be separately analyzed.

One of the most selective cleavage techniques is seldom

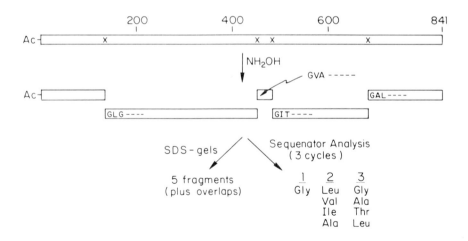

FIGURE 3. Schematic illustration of methods used to detect the number of Asn-Gly cleavage loci (X) in phosphorylase. Sequenator analysis of the whole reaction mixture revealed 4 different fragments. The amino-terminus of the protein is acetylated.

TABLE II.   Agents Useful for Limited and Specific Cleavage[a]

Residue		Cleavage Agent	Ref.
n	n + 1		
Met	– X	CNBr	(21)
Asn	– Gly	NH$_2$OH	(22)
Asp	– Pro	Acid	(23)
Trp	– X	BNPS-Skatole	(24)
Arg	– X	Trypsin + Substrate with blocked Lys.	(25)
Arg	– X	Clostripain	(26)
Lys	– X	Trypsin + Substrate with blocked Arg.	(27)
X	– Lys	Myxobacter protease	(28)
Glu	– X	Staphylococcus protease	(29)
Various		Limited Proteolysis	(Table III)

[a]Cleavage is specifically directed between residues n and
n + 1.   X indicates any residue.

applied. It takes advantage of the intrinsic resistance of
the compact structure of a native protein to proteolysis. It
has been noted that proteases of broad specificity, such as
subtilisin, act in limited and specific ways on large proteins
if the time of treatment is short and the native structure of
the substrate is preserved (Table III). In fact, subtilisin
was discovered at the Carlsberg laboratory by the serendipi-
dous observation that it specifically nicks ovalbumin to form
plakalbumin (31). Subtilisin has since been used to specifically fragment ribonuclease (32) and phosphorylase (33, 19),
and in general to produce large fragments useful for sequence
analysis. In the case of phosphorylase, cleavage by subtilisin generated two very large segments which, by their very
nature, determined the subsequent strategy of analysis of the
whole molecule. Prior to this finding, we were confronted
with a molecule containing 841 amino acid residues including
21 methionines, 3 Asp-Pro bonds, 4 Asn-Gly bonds, and a much
larger number of each other cleavage point. On the one hand,
it was not possible to separate all 22 products of methionine
cleavage (39), whereas on the other, the incomplete cleavage
of Asp-Pro or Asn-Gly bonds yielded complex mixtures not
amenable to separation. However after phosphorylase was first
separated into the two major segments, subsequent cleavage of
these segments by other techniques yielded simpler mixtures of
fragments which could be separated from each other. While it

TABLE III. Some Examples of Limited Proteolysis[a] in Sequence Analysis[b].

		Ref.
γ-Globulin [224 + 446] $\xrightarrow{papain\ or\ trypsin}$	F$_{ab}$ [214 + 222], F$_c$ [224]	(34,35)
Ribonuclease [124] $\xrightarrow{subtilisin}$	S-protein [104], S-peptide [20]	(32)
Proinsulin [81] $\xrightarrow{in\ vivo}$	A chain [21], B chain [30], C-peptide [26]	(36)
Porcine trypsin [223] $\xrightarrow{autolysis}$	α-N [125], α-C [98]	(15)
Carboxypeptidase B [306] $\xrightarrow[pancreatic\ juice]{autoactivated}$	Light chain [86], Heavy chain [217]	(37)
Ovomucoid [188] $\xrightarrow{Staph.\ protease}$	Domains I and II [131], Domain III [57]	(38)
Phosphorylase [841] $\xrightarrow{subtilisin}$	Light segment [248], Heavy segment [577]	(19,33)

[a]Numbers in brackets following proteins or fragments refer to the chain lengths.

[b]This list does not include many examples of limited proteolysis during zymogen activation (2).

may seem unsatisfactory to apply an empirical cleavage procedure, the conversion of phosphorylase from one exceptionally
long chain or two of manageable length (one, the size of
chymotrypsinogen, the other equivalent to prothrombin) fully
justified this seemingly arbitrary approach.

Recently, Cleveland et al. have described an interesting
variant of the technique. It involves limited proteolysis of
proteins in the presence of sodium dodecyl sulfate (40) where
the specific conformation of each SDS-protein complex appears
to restrict proteolysis to a limited number of sites.

B. Isolation of Large, Denatured Fragments

Isolation of suitable fragments has always been a major
problem in sequence analysis. While powerful methods have
been routinely applied to resolve native proteins or mixtures
of small peptides, neither set of methods is directly applicable to mixtures of large, denatured polypeptide fragments
which tend to aggregate and precipitate. In general these
difficulties are alleviated by gel filtration in denaturing
solvents at a pH as far removed as possible from the isoelectric point. High concentrations of formic, acetic or
propionic acids often achieve this goal, but fragments of
acidic proteins are generally more soluble at high pH. Urea
or guanidine hydrochloride are alternative solutions but urea
can lead to carbamylation of amino-terminal residues at high
pH and guanidine hydrochloride precludes ion exchange chromatography.

Since current tactics of limited cleavage have led to
relatively simple mixtures of a small number of large fragments, separation techniques of high resolution are not
necessarily required. In many cases it is sufficient to
separate fragments solely on the basis of size. In fact,
complete resolution of fragments from each other may not even
be necessary. Separate analysis of peptides from the leading
and trailing edges of a seemingly Gaussian peak may well yield
two different single sequences, admittedly with less than
quantitative recovery. If the sequences are unique and the
fragments occur in significant concentration, it may not be
worth the time and effort to resolve the two poorly separated
components.

The disadvantage of the insolubility of large fragments
can sometimes be turned into an advantage by selective precipitation of the least soluble component in a mixture. For
example, in our analysis of the sequence of phosphorylase a
single set of fractions from gel filtration of products of

cyanogen bromide cleavage yielded a mixture of five peptides
ranging in length from 17 to 37 residues (Fig. 4). After one
of these (CB21) was selectively precipitated, simply by ad-
justing to pH 4.0, the other four could be separated by alternative chromatographic techniques.

Besides separation on the basis of size, examples of high
resolution on the basis of charge include those of Fowler and
Zabin (41) with β-galactosidase fragments separated on carboxymethyl cellulose in 8 M urea at pH 5, and ours of phosphorylase fragments separated on SP-Sephadex in urea (20).

Methods producing cleavage in low yield (e.g., at tryptophan) are usually unsatisfactory because the reaction mixture
contains both the primary cleavage products and overlapping
fragments. If there is only one cleavage point, the two
products are usually separable from each other and from

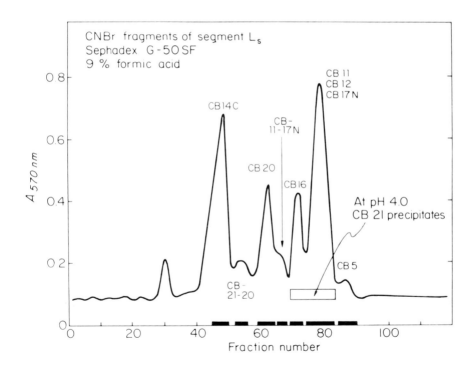

FIGURE 4. Separation of fragments generated by CNBr
cleavage of segment L_s of phosphorylase (20). Of the 5 fragments in fractions 70-85, only one (CB21) precipitated at pH
4.0, facilitating not only its isolation but also that of
fragments CB16, CB11, CB12 and CB17N.

uncleaved protein. However, if, for example, there are three
cleavage points, a maximum of ten possible components could be
present in the mixture and complete resolution is unlikely.
Occasionally a desired peptide is isolated from such a mixture in spite of its complexity. For example, incomplete
cleavage of 3 tryptophanyl bonds in fragment CB18 of phosphorylase (Fig. 5) yielded a fragment (residues 38 to 78)
which was of favorable size and concentration (42).

Other strategies of specific peptide isolation are the
"diagonal" methods reviewed by Hartley (43), and residue-
specific covalent attachment to solid supports. A particularly useful tactic has been developed by Horn and Laursen
(44) which takes advantage of the unique reactivity of homoserine lactone by selectively attaching homoseryl peptides to
immobilized supports. Figure 6 illustrates the application of
this procedure to analysis of fragment CB24 of phosphorylase

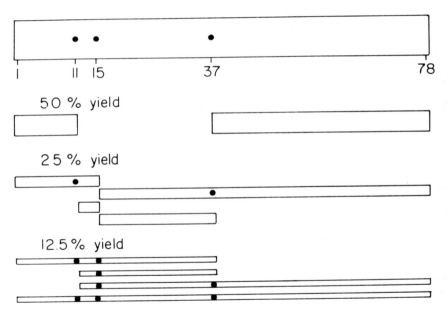

FIGURE 5. Fragments generated by incomplete cleavage at
the three tryptophanyl residues (●) of fragment CB18 of phosphorylase (42). The length of each bar is proportional to the
length of a fragment. The height of each bar is proportional
to the yield expected if 50% cleavage takes place in a random
manner at each tryptophanyl residue. The upper bar represents
intact CB18. The fragment representing residues 38-78 was
easily separated from this complex mixture.

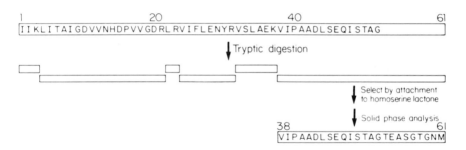

FIGURE 6. Strategy of sequence analysis of fragment CB24
of phosphorylase (45). The carboxyl-terminal methionyl residue (M) was in the homoserine form. Tryptic cleavage took
place at arginyl (R) and lysyl (K) residues. Only two series
of Edman degradations (indicated in one-letter codes) were
necessary to place 61 residues.

where sequenator analysis had placed 52 of its 61 residues,
including all five lysyl and arginyl residues. The whole
fragment was then cleaved with trypsin, the single carboxyl-
terminal homoseryl residue converted to the lactone, and the
mixture of 6 peptides added to an aminated resin. Only the
lactone-bearing peptide (residues 38-61) became attached to
the resin whereas the rest were washed away. Solid phase
Edman degradation of this peptide completed the sequence of
this fragment. Thus, the purification was a by-product of the
preparation for degradation.

C. Automated Edman Degradation

While other methods are also used for analysis of an isolated fragment (e.g., mass spectrometry (46)), the most widely
adopted techniques employ the Edman automated degradation in
a spinning cup (9) or on a solid support (10). As previously
indicated, a solid phase method may offer advantages in the
purification strategy, and has the potential for degradation
through the carboxyl-terminal amino acid of the fragment.
However, problems in peptide attachment persist and the spinning cup method has found wider application. Both modes of
operation allow complete control of the experimental conditions, including the ability to preclude oxidative side reactions. Initially, the spinning cup method was limited by
progressive loss of peptide during extraction of reagents and
products. Peptides were then modified to decrease their

solubility in extractants (47). Now, by using more volatile
reagents, less concentrated buffers, less polar extractants,
better drying protocols and more gentle extraction, the methodology can be applied to many small peptides as well as proteins (48-51). Recently a polycationic carrier (polybrene)
has been introduced to stabilize the protein film and improve
retention of small peptides (52).

Early attempts to improve the instrumentation were directed at enhancing the stepwise yield at each degradation
cycle because an increase of one percent in stepwise yield
results in about 10 to 20 additional residue identifications.
However yields decrease at residues of proline or glutamine
and at the sequence Asn-Gly (53). Few reports describe more
than 60 consecutive identifications. It appears that the
problem is not so much one of the stepwise yield itself but
of the gradual acid-catalyzed hydrolysis of peptide bonds in
the interior of the fragment during degradation. Typically,
the unambiguous identifications of PTH-amino acids in early
cycles of the degradation gradually become less clear in later
cycles as the yield decreases, and new amino-terminal residues
are generated by uncontrolled acid-catalyzed cleavage. The
problem is particularly severe in proteins with a high content
of hydroxylated amino acids.

A second problem, which limits the extent of productive
degradations, is the instability of serine and threonine residues in the anhydrous carboxylic acids used to cleave the PTH-
amino acid from the protein. Tarr has discussed alternative
reagents for this phase of the degradation cycle (54) but it
remains to be seen whether these alternatives are compatible
with the automated instrumentation.

Occasionally, one is presented with a mixture of peptides
which differ in length by only one or two amino acids at their
amino-termini. For example, treatment with CNBr of a peptide
having the sequence A-B-Met-Met-X-Y-Z---, yields several
products including homoseryl-X-Y-Z--- and X-Y-Z---. These are
likely to be so similar that they will be inseparable.
Degradation of the mixture is nevertheless interpretable, once
the pattern is recognized, and the results serve almost as
duplicate analyses (55). Another source of these "ragged
ends" is cleavage by limited proteolysis which may well not be
absolutely specific. For example, subtilisin cleaved phosphorylase at three alternative loci within four residues of
each other (Fig. 7). The three chains to the right of the
cleavage points thus differed in their amino-termini, but were
each so similar and large (approximately 70,000 daltons) that
they were inseparable. Nonetheless, we were able to deduce
simultaneously the sequence of all three chains by semi-quan-
titative analysis of the yields of the three PTH-amino acids
in each cycle because the fragments were present in

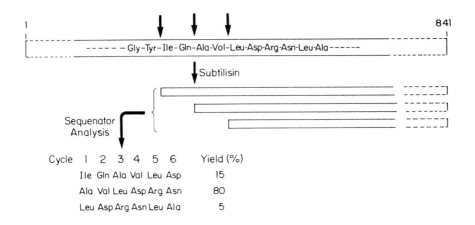

FIGURE 7. A "ragged end" generated by limited proteolysis
of phosphorylase b (841 residues) by subtilisin (20).
Sequenator analysis of a mixture of the large segments (from
the points of cleavage to the carboxyl-terminus) yielded
multiple products indicating the 3 points of cleavage indicated by the arrows at the top.

sufficiently different proportions [80:15:5]. Since a glutaminyl residue was observed in one of the minor chains at
cycle 2, we interrupted a duplicate degradation after removal
of the first residue and promoted spontaneous cyclization of
the amino-terminal glutamine to a blocked pyroglutamyl residue. From that point on we observed only two chains, simpli-
fying interpretation of the data from the three-chain mixture.

Each cycle of the Edman degradation yields a PTH-amino
acid and the final step is that of identification. Various
chromatographic techniques are employed before or after back
hydrolysis to the free amino acid. The method of choice
appears to be high performance liquid chromatography of the
PTH-amino acids themselves because it is sensitive, relatively
quantitative, rapid, and non-destructive (56,57). In addition,
it is easily adapted to collection of the pure PTH-amino acid
for mass spectrometry or analysis of radioactivity, techniques
which have proven to be valuable for the identification of
modified amino acids or for microanalytical purposes (e.g.,
58,59).

In several cases, X-ray crystallographic analyses have
yielded partial sequences of proteins (60), although only 50-
75 percent of the residues were correctly identified in those

cases where electron density maps were of the highest quality.
While this method is powerful for aligning fragments derived by
chemical methods (61), it cannot become the method of choice
for sequence analysis because of the problems of local disorder in the crystals and the uncertainties associated with
isosteric residues (62).

D. Occurrence of Covalently Modified Residues

Although proteins are almost entirely composed of the 20
conventional L-amino acids, many proteins are found to contain
one or more covalently modified residues which are crucial to
their functions (5). Natural processing events have been
described which cause attachment of large prosthetic groups or
oligosaccharides, and others which introduce small chemical
groups (hydroxyl-, methyl, acetyl, iodo-, carboxyl-, etc.).
Proteolytic processing events alter the length of the polypeptide chain during control phenomena associated with secretion (1) or zymogen activation (2). Some of these events
occur during translation, others are post-translational
phenomena (Table IV). In either case, the sequence analyst

TABLE IV. Some Protein Processing Events[a]

During Translation	After Translation
Acetylation, Formylation	γ-carboxylation, Methylation, Phosphorylation
Hydroxylation (HO-Pro, HO-Lys)	
Glycosylation	Adenylylation, ADP-ribosylation
	Attachment of: biotin, heme, lipoyl
	Crosslinks: desmosine, bi-tyrosine, Nϵ(γ-Glu)-Lys
	Glycosylation
---------------- Involving Peptide Bond Cleavage -------------	
Removal of initiator methionine	Zymogen activation
Removal of signal peptide	

[a]See review by Uy and Wold (5)

must be alert to the possibility of such anomalies because the biological purpose of the modification is usually a clue to the function of the protein. The modifications are easy to overlook both during amino acid analysis of the protein and during sequence analysis, particularly if the linkage is acid-labile. Furthermore, a modified residue, may be missed if its phenylthiohydantoin (e.g., of a glycosylated amino acid) is not extracted from the reaction mixture in the routine mode of Edman degradation. Furthermore, the modified amino acid may resemble another conventional amino acid in the analytical system employed. As an example, two laboratories have recently reported that the amino-terminal residues of two different pili proteins are α-N-methyl-phenylalanine (58,63). In one of these cases (58), this residue was first misidentified as tyrosine because its silylated phenylthiohydantoin had the precise mobility of silylated PTH-tyrosine on gas/liquid chromatography. The anomaly was recognized, however, when the phenylthiohydantoin was characterized by high performance liquid chromatography which ruled out tyrosine. Mass spectrometry of this phenylthiohydantoin and chemical synthesis ultimately identified the derivative. This illustration draws attention to the importance of examining the products of Edman degradation by more than one technique.

A particularly vexing problem in sequence analysis is presented by proteins with blocked α-amino groups. The common blocks are N-formyl-methionyl residues (in prokaryotes), N-acetyl groups or pyroglutamyl residues (5), but other anomalies have been reported, e.g., the spontaneous blocking in acid of a protein with an amino-terminal tryptophanyl residue (64). Each of these groups prevents direct sequence analysis of the intact protein and unblocking procedures are not always effective. Palmiter (65) has described a method to prevent acetylation during in vitro translation by diverting acetyl-coenzyme A to citrate synthesis. Acidic treatments have been reported which remove formyl groups (66,67). Doolittle has isolated enzymes which remove pyroglutamyl residues from some protein substrates (68).

E. Microanalytical Procedures

In general, modern techniques of sequence analysis require about 20 to 100 n moles of protein or fragment for each Edman degradation. Since this quantity is difficult to obtain in many cases of great biological interest, several laboratories have undertaken the development of more sensitive techniques. Moderate improvements are achieved by increasing the

sensitivity of phenylthiohydantoin detection (e.g., 56,57), by
optimizing polypeptide retention in the spinning cup (48–51),
or by covalent retention using solid phase techniques (10).
Increases of 3 to 6 orders of magnitude have been achieved by
appropriate application of radiolabelling techniques, either
by treating labelled protein with cold phenylisothiocyanate
(PITC) or by treating cold protein with labelled PITC (Fig. 8).
The procedures with cold proteins have been successfully
applied (69,70) by avoiding the obvious hazard of trace contamination by foreign peptides. The procedures with labelled
proteins require methods to introduce specifically labelled
amino acids into the proteins and to purify the labelled
proteins on a microscale (e.g., 59,71). Some proteins can be
labelled in cell or tissue culture, others are obtained by
mRNA translation in vitro. In each case immunological selection of a labelled protein can provide a product suitable for
analysis. In our laboratory we have analyzed the labelled
products of both in vivo and in vitro synthesis at the 0.1 to
1.0 picomole level (59). These specific studies were directed
towards a comparison of the amino-terminal sequences of the
nascent chains and the completed chains of certain secretory
proteins. We introduced up to 15 different labelled amino

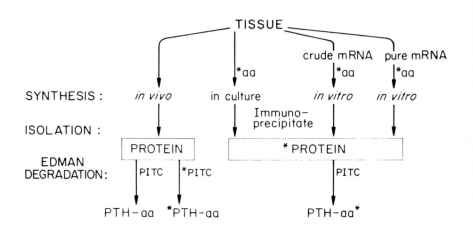

FIGURE 8.   Approaches to microanalysis of amino acid
sequences. Labelled phenylthiohydantoins (PTH-aa) are produced either from proteins containing labelled amino acids
(*aa) or from cold proteins and labelled phenylisothiocyanate
(*PITC).   Identification of the PTH-aa requires techniques
for their separation and detection.

acids into a single synthetic product and separated the labeled PTH-amino acids from a single degradative analysis. Since the data were unambiguous, it appears that subpicomole analysis is technically possible. Developmental work in several other laboratories shows similar promise of yielding ultrasensitive techniques (71-73).

SUMMARY

This report describes in general terms some of the strategical and tactical considerations which guide amino acid sequence analysis. It is not intended to be either comprehensive or detailed and does not include many important techniques used and developed in other laboratories. It does attempt to give some perspective on the accelerating productivity of the field and on the tactics which have contributed to the strategy of analysis in our laboratory.

ACKNOWLEDGMENTS

We are grateful for the active collaboration of many co-workers in our laboratory over the years who have made these studies possible, in particular, Dr. Hans Neurath, Dr. Mark Hermodson, Dr. Atsushi Koide, Dr. Jacques Hermann, Dr. Jean Gagnon, Dr. Richard Woodbury, Stephen Thibodeau, Santosh Kumar, Richard Granberg and Roger Wade.

REFERENCES

1. Blobel, G., and Dobberstein, B. J. Cell Biol. 67, 852 (1975).
2. Neurath, H., and Walsh, K.A. Proc. Natl. Acad. Sci. (USA) 73, 3825 (1976).
3. Stadtman, E.R., in "The Enzymes", Third Edition, Volume I (P.D. Boyer, ed.) p. 398, Academic Press, New York, 1970.
4. Shine, J., Seeburg, P.H., Martial, J.A., Baxter, J.D., and Goodman, H.M. Nature (London) 270, 494 (1977).
5. Uy, R., and Wold, F. Science 198, 890 (1977).
6. Sanger, F., and Thompson, E.O.P. Biochem. J. 53, 353 (1953).

7. Spackman, D.H., Stein, W.H., and Moore, S. Anal. Chem.
    30, 1190 (1958).
8. Edman, P. Acta Chem. Scand. 4, 283 (1950).
9. Edman, P., and Begg, G. Eur. J. Biochem. 1, 80 (1967).
10. Laursen, R.A., Bonner, A.G., and Horn, M.J., in "Instru-
    mentation in Amino Acid Sequence Analysis" (R.N. Perham,
    ed.) p. 73, Academic Press, 1975.
11. Hirs, C.H.W., Moore, S., and Stein, W.H. J. Biol. Chem.
    235, 633 (1960).
12. Walsh, K.A., and Neurath, H. Proc. Natl. Acad. Sci.
    (USA) 52, 884 (1964).
13. Bradshaw, R.A., Ericsson, L.H., Walsh, K.A., and Neurath,
    H. Proc. Natl. Acad. Sci. (USA) 63, 1389 (1969).
14. Titani, K., Hermodson, M.A., Ericsson, L.H., Walsh, K.A.,
    and Neurath, H. Nature New Biology 238, 35 (1972).
15. Hermodson, M.A., Ericsson, L.H., Neurath, H., and Walsh,
    K.A. Biochemistry 12, 3146 (1973).
16. Hermodson, M.A., Kuhn, R.W., Walsh, K.A., Neurath, H.,
    Eriksen, N., and Benditt, E.P. Biochemistry 11, 2934
    (1972).
17. Woodbury, R.G., Katunuma, N., Kobayashi, K., Titani, K.,
    and Neurath, H. Biochemistry, in press.
18. Dayhoff, M.O. "Atlas of Protein Sequence and Structure",
    Volume 5, Nat. Biomedical Research Foundation, Silver
    Spring, Md., 1972.
19. Titani, K., Koide, A., Hermann, J., Ericsson, L.H., Kumar,
    S., Wade, R.W., Walsh, K.A., Neurath, H., and Fischer,
    E.H. Proc. Natl. Acad. Sci. (USA) 74, 4762 (1977).
20. Koide, A., Titani, K., Ericsson, L.H., Kumar, S., Walsh,
    K.A., and Neurath, H. In preparation.
21. Gross, E. Methods Enzymol. 11, 238 (1967).
22. Bornstein, P., and Balian, G. Methods Enzymol. 47, 132
    (1977).
23. Landon, M. Methods Enzymol. 47, 145 (1977).
24. Omenn, G.S., Fontana, A., and Anfinsen, C.B. J. Biol.
    Chem. 245, 1895 (1970).
25. Yaoi, Y., Titani, K., and Narita, K. J. Biochem. Japan
    56, 222 (1964).
26. Mitchell, W.M. Methods Enzymol. 47, 165 (1977).
27. Smith, E.L. Methods Enzymol. 47, 156 (1977).
28. Wingard, M., Matsueda, G., and Wolfe, R.S. J. Bact. 112,
    940 (1972).
29. Drapeau, G.R. Methods Enzymol. 47, 189 (1977).
30. Schechter, Y., Burstein, Y., and Patchornik, A. Bio-
    chemistry 14, 4497 (1975).
31. Linderstrøm-Lang, K.U., and Ottesen, M. C.R. Trav. Lab.
    Carlsberg 26, 403 (1949).

32. Richards, F.M., and Vithayathil, P.J. J. Biol. Chem. 234, 1459 (1959).
33. Raibaud, O., and Goldberg, M.E. Biochemistry 12, 5154 (1973).
34. Porter, R.R. Biochem. J. 73, 119 (1959).
35. Edelman, G.M., Cunningham, B.A., Gall, W.E., Gottlieb, P.D., Rutishauser, U., and Waxdal, M.J. Proc. Natl. Acad. Sci.(USA) 63, 78 (1969).
36. Chance, R.E., Ellis, R.M., and Bromer, W.W. Science 161, 165 (1968).
37. Reeck, G.R., Walsh, K.A., Hermodson, M.A., and Neurath, H. Proc. Natl. Acad. Sci. (USA) 68, 1226 (1971).
38. Kato, I., Kohr, W.J., and Laskowski, M., Jr. Federation Proc. 36, 764 (1977).
39. Saari, J.C., and Fischer, E.H. Biochemistry 12, 5225 (1973).
40. Cleveland, D.W., Fischer, S.G., Kirschner, M.W., and Laemmli, U.K. J. Biol. Chem. 252, 1102 (1977).
41. Fowler, A.V., and Zabin, I. Proc. Natl. Acad. Sci. (USA) 74, 1507 (1977).
42. Hermann, J., Titani, K., Ericsson, L.H., Walsh, K.A., and Neurath, H. In preparation.
43. Hartley, B.S. Biochem. J. 119, 805 (1970).
44. Horn, M.J., and Laursen, R.A. FEBS Lett. 36, 285 (1973).
45. Titani, K., Koide, A., Ericsson, L.H., Kumar, K., Wade, R.W., Walsh, K.A., and Neurath, H. In preparation.
46. Falter, H., in "Advanced Methods in Protein Sequence Determination" (S.B. Needleman, ed.) p. 123, Springer-Verlag Press, Berlin, 1977.
47. Braunitzer, G., Chen, R., Schrank, B., and Stangl, A. Hoppe-Seyler's Z. Physiol. Chem. 353, 832 (1972).
48. Niall, H.D. Methods Enzymol. 27, 942 (1973).
49. Hermodson, M.A., Schmer, G., and Kurachi, K. J. Biol. Chem. 252, 6276 (1977).
50. Crewther, W.G., and Inglis, A.S. Analyt. Biochem. 68, 572 (1975).
51. Brauer, A.W., Margolies, M.N., and Haber, E. Biochemistry 14, 3029 (1975).
52. Tarr, G.E., Beecher, J.F., Bell, M., and McKean, D.J. Analyt. Biochem. In press.
53. Hermodson, M.A., Ericsson, L.H., Titani, K., Neurath, H., and Walsh, K.A. Biochemistry 11, 4493 (1972).
54. Tarr, G.E. Methods Enzymol. 47, 335 (1977).
55. Titani, K., Ericsson, L.H., Walsh, K.A., and Neurath, H. Proc. Natl. Acad. Sci. (USA) 72, 1666 (1975).
56. Bridgen, P.J., Cross, G.A.M., and Bridgen, J. Nature (London) 263, 613 (1976).

57. Zimmerman, C.L., Appella, E., and Pisano, J.J. Analyt.
     Biochem. 77, 569 (1977).
58. Hermodson, M.A., Chen, K.C-S., and Buchanan, T.M. Bio-
     chemistry. In press.
59. Palmiter, R.D., Gagnon, J., Ericsson, L.H., and Walsh,
     K.A. J. Biol. Chem. 252, 6386 (1977).
60. Kannan, K.K., in "Advanced Methods in Protein Sequence
     Determination" (S.B. Needleman, ed.) p. 75, Springer-
     Verlag Press, Berlin, 1977.
61. Matthews, B.W., Colman, P.M., Jansonius, J.N., Titani,
     K., Walsh, K.A., and Neurath, H. Nature New Biology 238,
     41 (1972).
62. Lipscomb, W.N., Hartsuck, J.A., Quiocho, F.A., Reeke,
     G.N., Jr. Proc. Natl. Acad. Sci. (USA) 64, 28 (1969).
63. Frost, L.S., Carpenter, M., and Paranchych, W. Nature
     (London) 271, 87 (1978).
64. Fujikawa, K., Legaz, M.E., and Davie, E.W. Biochemistry
     11, 4882 (1972).
65. Palmiter, R.D. J. Biol. Chem. In press.
66. Sarges, R., and Witkop, B. J. Am. Chem. Soc. 87, 2011
     (1965).
67. Palmiter, R.D., Gagnon, J., and Walsh, K.A. Proc. Natl.
     Acad. Sci. (USA) 75, 94 (1978).
68. Doolittle, R.F. Methods Enzymol. 25, 231 (1972).
69. Bridgen, J. Methods Enzymol. 47, 321 (1977).
70. Jacobs, J.W., and Niall, H.D. J. Biol. Chem. 250, 3629
     (1975).
71. Ballou, B., McKean, D.J., Freedlender, E.F., and Smithies,
     O. Proc. Natl. Acad. Sci. (USA) 73, 4487 (1976).
72. Walker, J.E., Shaw, D.C., Northrup, F.D., and Horsnell,
     T., in "Solid Phase Methods in Protein Sequence Analysis"
     Volume 5 (A. Previero and M.-A. Coletti-Previero, eds.),
     Elsevier/North Holland Press, 1978.
73. Silver, J., and Hood, L. Proc. Natl. Acad. Sci. (USA)
     73, 599 (1976).

SOLUTION VS SOLID-PHASE METHODS IN
PEPTIDE CHEMISTRY

Theodor Wieland

Max-Planck-Institut für medizinische
Forschung, Abteilung Naturstoff-Chemie
Heidelberg, West Germany

I. INTRODUCTION

Since the early fifties, peptide chemistry,
formerly a stepchild of organic chemistry, has developed increasingly. In the early sixties R.B.
Merrifield, with the introduction of his solid phase
synthesis, stimulated the progress in this area to
an extent which opened a reasonable chance to synthesize proteins in not too far future. In the first
half of the seventies these hopes were inspired: By
conventional synthesis in solution as well as on a
solid phase many homogenous peptides of about 50
amino acids were synthesized and several laboratories were busy with building up polypeptides in
the protein range, i.e. of 100 amino acids and more,
mainly on solid supports. In several cases products
with distinct and specific biological activities
were obtained, such as RNAase- or lysozyme functions,
acyl carrier function, growth promoting potency (for
references see the reviews 1 - 4). Never has it been
claimed that these polypeptides consist of only one
homogenous sort of molecules; the proteins obtained
by solid phase synthesis are mixtures of extremely
similar polypeptides. The fact of their displaying
considerable activity e.g. as enzymes is evidence
that the specificity is due not only to one primary
structure but that there exist thousands of mini-
mally altered analogues exhibiting similar specificity
(immune specificity included), and biological activity.

59

This is a result of great value although not intended at the beginning. Nevertheless efforts are being
made in several laboratories to obtain synthetically
proteins of the same homogeneity as it has been attained at polypeptides of the order of insulin
(20 + 31 amino acids), human big gastrin (5) (34
amino acids) or calcitonin (32 amino acids). The
steeply rising curve of results from which one could
extrapolate to the achievement of these goals, apparently did not continue its strong upward trend. It
appears that a sort of barrier blocks the way further. In the following I shall try to understand
some of the reasons responsible for the present
situation. For this purpose we need repeat briefly
the fundamentals of peptide synthesis.

A. Principles of Peptide Synthesis

In a chemical sense, the synthesis of a peptide
seems to be a rather simple and monotonous matter
for it consists essentially in repeatedly removing
a molecule of water from between the carboxylic
group of one and the amino group of another amino
acid. Experimentally, however, differences in the
ease of formation of a given peptide bond can be
tremendous, depending on the space filling properties
of the amino acids concerned and on the neighbouring
amino acid sequences i.e. the conformations of the
reaction components. Difficulties in polypeptide
synthesis arise also largely because a number of the
amino acids are polyfunctional molecules whose functional groups can not safely be protected even at
present and because even simple amino acids such as
alanine, valine etc.have a "personality" of its own.
Each synthesis must, therefore, be planned to ac-
commodate differences in reactivity of amino acids,
danger of racemization and the complexity of the
target molecule. Nevertheless will the experimentor
not be charmed against bitter disappointments when
in the late stages of his synthesis a coupling reaction, proved quite well in similar cases, will not
be successful.
    The basic scheme generally followed in peptide
synthesis proceeds through three stages (Figure 1).
In the first stage the amino group of one amino acid
(or peptide), the acylating component, is protected
by a reversibly removable residue, and the carboxylic group of the amine component to be acylated is

Protection

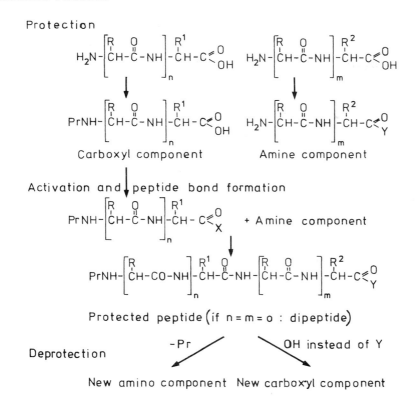

Carboxyl component   Amine component

Activation and peptide bond formation

Protected peptide (if n = m = o : dipeptide)

Deprotection   -Pr   OH instead of Y

New amino component   New carboxyl component

FIGURE 1.   General scheme of a peptide synthesis.

likewise blocked. The actual peptide bond formation (second stage) is accomplished by activation of the carboxylic group in the acylating component followed by the reaction with the amine component to form a protected peptide. Partial deprotection in the third stage affords peptide components ready for further condensations, complete deprotection provides the desired end product of the synthesis.

II. THE MOST USED COUPLING METHODS

Out of several dozens, four efficient methods for activation of the carboxyl component have found wide and general application during the past two decades.

A. Azide Method

Still at present the historically first method to form an activated carboxyl component, the formation of an azide (Th. Curtius, 1902), is one of those processes. Reaction of an alkylester with hydrazine affords the hydrazide which by reaction with an nitrosylating agent is transformed to the rather unstable azide. N-protected hydrazides (e.g. benzyloxycarbonylhydrazides), obtained from amino acids by a "peptide" synthesis with derivatives of hydrazine, are sometimes employed through various steps of a peptide synthesis. They are protected forms, as it were, which in the right moment can be converted into reactive species as shown in Figure 2. A side reaction with azides is the Curtius-rearrangement yielding isocyanates (and further derivatives). Although the reaction rate of the nitrosylating agent is much greater with the hydrazide moiety, minimal side reactions with nitrogen atoms of the peptide bond and with imino groups of the peptide itself (e.g. proline) can not be excluded. Yet the azide method causes least racemization, although its credit as entirely free from racemization has been disproved in the past years.

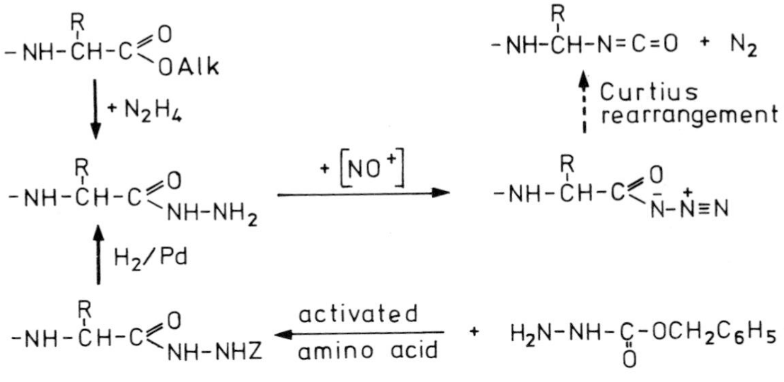

FIGURE 2. Formation of an amino acid azide.

B. Mixed Anhydride Method

The mixed anhydride method (MA) worked out 1951
independently by R. Boissonnas, J.R. Vaughan and in
our laboratory, uses in the present fashion the anhydrides of the carboxylic group with sec.-butyl-
carbonate prepared from the N-methylmorpholine salt
of the amino acid and sec.butyloxycarbonylchloride
at -15°C within several minutes, and reacted in situ
shortly thereafter with the amine component, usually
for 2 - 4 h. In the case of sterically hindered
amino groups (prolonged reaction time), attack at
the wrong anhydride moiety can become noticeable
leading to unwanted alkyloxycarbonyl derivatives.
Racemization which is minimal under normal circumstances will become significant, too. Nevertheless,
the method is in our hands the most covenient one
and proved effective as unique among all others
tried in cases where the activation of the carboxyl
had to be attained in the presence of an intramolecular (lactonizing) γ-hydroxyl group without exclusively forming the lactone (6). The yields are quite
satisfactory particularly when an excess of the
mixed anhydride is employed from which the protected
carboxyl component easily can be recovered by treatment of the reaction mixture with cold aqueous bi-
carbonate [M.A.Tilak (7)]. Special purification of
the peptides formed seems not to be necessary, the
product of each peptide formation can be obtained by
precipitation with water, deblocked after drying,
and subjected to a next elongation reaction. Up to
16 coupling steps have been carried out in a "repe-
titive" synthesis by vanZon and Beyermann (8). A
further simplification of this method came from the
introduction of the α,α-dimethyl-3,5-dimethoxybenzyl-
oxycarbonyl group (Ddz) for N-protection by Birr (9).
This group possesses a considerably increased acid
lability as compared to the frequently used tert.-
butyloxycarbonyl residue (Boc), but is only half as
acid sensitive as the biphenylisopropyloxycarbonyl
function (Bpoc) when measured in 80 % acetic acid.
The adamantyloxycarbonyl residue (Adoc) is rather
more stable than Boc and therefore lends itself to
permanent protection of side chain amino groups. By
the way, Ddz is also photosensitive and, consequently, can be removed by irradiation with a high pressure mercury lamp (9).

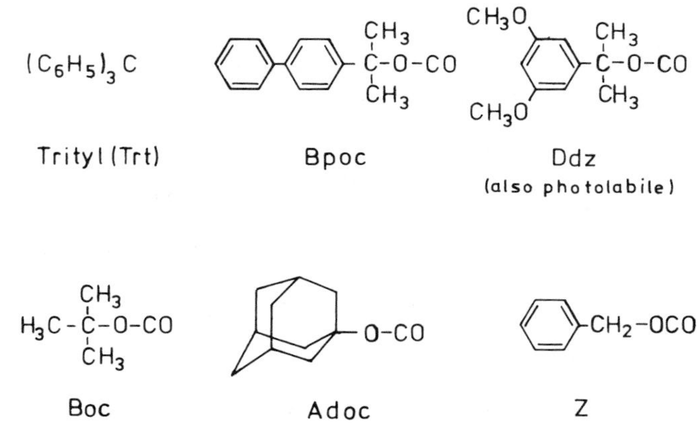

FIGURE 3.   Some α-amino protecting groups
with increasing acid stability.

Due to their high acid lability Ddz-peptides can
be deblocked by 5 % trichloroacetic acid in dichloromethane within only 15 minutes at 20°C. For
further reaction in a incremental (stepwise) chain
elongation the deblocked product needs not to be
isolated as in the afore mentioned process, but is
accessible to the next coupling reaction just after
neutralization of the acid by a tertiary base like
N-methylmorpholine (NMM). For each coupling product,
we recommend purification by chromatography on
Sephadex LH-20 with methanol as a solvent. This procedure running overnight provides a clean Ddz-
peptide ester which is introduced in the deblocking
process whilst simultanously the mixed anhydride of
the next Ddz-amino acid is being prepared. The time
schedule for one step of this MA-method has been represented in Figure 4. Since selective cleavage of
the Ddz residue from the amino function is feasible
without touching any tert.butyl group tert.butyl-
esters of amino acids can likewise be employed in
the synthesis described.

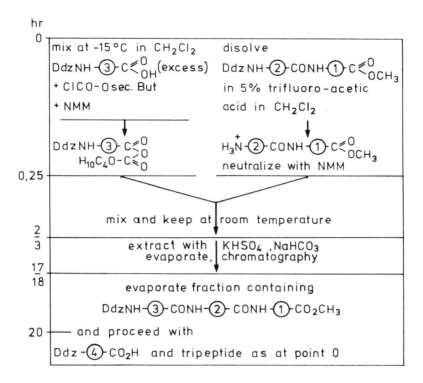

FIGURE 4. Cyclus of a repeating peptide synthesis using an (1.5 - 2 M) excess of mixed anhydrides of Ddz-amino acids.

Another very powerful and simple mixed anhydride method described in 1956 but nearly fallen into oblivion is that using phosphorus oxytrichloride (10). The N-protected carboxyl component together with the amine component is dissolved in an inert solvent like tetrahydrofurane and admixed at -15°C with one equivalent of $POCl_3$. The method affords very satisfactory yields of products also from sterically hindered di- and tripeptides, but, that is true, has not yet been proved by more difficult problems. The use of symmetric anhydrides of N-protected amino acids in modern peptide synthesis (11) may also be mentioned here.

   We have mostly discussed stepwise chain elongation reactions which offer no problems of racemization as long as protecting groups of the urethane
type are employed. Such problems, however, become
prominent when the condensation of two peptides is
to be carried out where the carboxyl group to be
activated is part of an L-amino acid acylated by
a second amino acyl residue and therefore prone to
racemization. C-terminal glycine or proline, of
course, cannot be racemized. One can state that the
problem of avoiding this very cumbersome reaction
has been overcome in principle at the time being.
The conditions for the racemization "free" MA
method of coupling fragments are short reaction time
and low temperature, e.g. Z-glycyl-L-phenylalanine
in tetrahydrofurane activated with sec. butyl-
chloroformate in the presence of exactly one equiv.
of NMM at -15°C followed 1 minute later by 1 equiv.
of ethylglycinate and allowed to stand for 3 minutes
at 22°C. In this case 98 % tripeptide with a content
of 0.2 % DL-compound were obtained (12). At unfavorable conditions, which demand longer reaction times
(sterical hindrance by bulky side chains at the
termini of the reactants or by conformation), the
extent of racemization may be higher.

                    Anderson Test

                          $CH_2-C_6H_5$
                          |
ZNH-CH$_2$-CONH-CH - CO$_2$H   +   H$_2$N-CH$_2$-CO$_2$C$_2$H$_5$

                    L

                               $CH_2C_6H_5$
                               |
    $\xrightarrow{MA}$ ZNH-CH$_2$-CONH-CH - CONH-CH$_2$-CO$_2$C$_2$H$_5$

                          L
                       (and D)

D,L-compound very sparingly soluble in ethanol

C. Carbodiimide Coupling

The coupling reagent most widely used up to now, at least in peptide syntheses on a solid phase, is dicyclohexylcarbodiimide (DCC) (13). Because of considerable racemization the method in its original form cannot be used for the racemization-free assembly of peptides from acylpeptides containing carboxyl-terminal L-amino acids. However, addition of N-hydroxysuccinimide (14) or 1-hydroxybenztriazole (15) appears to provide a relatively safe procedure for coupling of peptides.

The DCC method serves not only to make peptide bonds, but also other derivatives from (N-protected) amino acids like active esters. We have to assume that the N-hydroxy compounds mentioned afore will be incorporated by DCC into the carboxylic group to form active N-hydroxy esters, e.g.

$$\text{PrNH-CH-C}\underset{\text{OH}}{\overset{O}{\lessgtr}}^{R} + \text{HO-N} \quad \xrightarrow{\text{(DCC)}} \quad \text{PrNH-CH-C}\underset{O-N}{\overset{O}{\lessgtr}}^{R}$$

D. Active Esters

Active esters are the fourth abundant group of peptide forming reagents. A compound of this type, although not a true ester, has been formulated just above. Active esters are the tools of Nature's protein and peptide synthesis: The building stones of the proteins, as you know, are activated by esterification with an OH-group of the ribose moiety of the terminal adenylic acid of the transfer ribonucleic acids. For a non-ribosomal synthesis of peptides like the gramicidins micro-organisms employ amino acids activated by linkage to thiol groups. Prior to this realization by Lipmanns group in 1969 (16), thioesters of amino acids have been prepared as the first active esters in our laboratory.

1. Thiophenyl Esters. As early as in 1951 thiophenyl esters of N-protected amino acids and peptides were synthesized by condensation via mixed anhydrides (17), and unprotected S-aminoacyl derivatives followed in 1952 (18). A prominent feature of aminoacyl mercaptanes is their surprising, high stability against hydrolysis; this makes them ideal acylation agents in water containing alkaline solvents in

which also alkali salts of the amine components
(amino acids or peptides) are soluble. Since the
solubility problem is one of the most troublesome
obstacles in synthesis of higher peptides the
possibility of working with water and water miscible
solvents may be of advantage in future attempts of
protein synthesis.

The unprotected aminoacyl thiophenols were introduced also as amine components in peptide syntheses with protected amino acids via mixed anhydrides
so yielding dipeptide thiophenylesters which in turn
could serve via their free amino group as acyl acceptor components in a subsequent synthesis of a tripeptide (19). This principle, recreated in another
laboratory a little later using p-nitrophenylesters
(20), has been named "backing-off procedure".

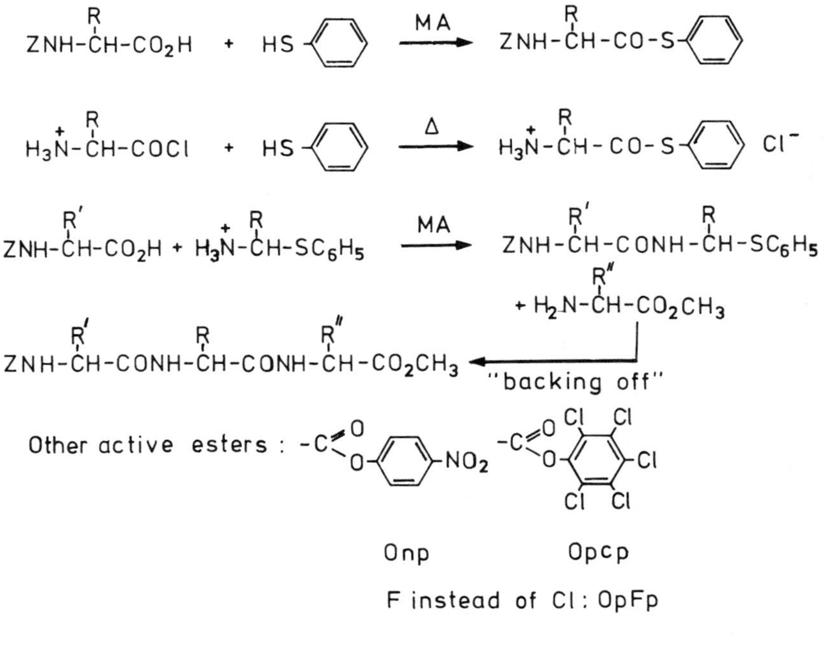

FIGURE 5.   N-Protected and N-free amino acid
            thiophenyl esters in peptide
            synthesis; other active esters.

2. o-Nitrophenyl Esters and Others. p-Nitro-
phenyl esters (21) (Onp) are probably the most widely used reagents of this class, followed by esters
of more or less highly halogenated phenols from
which the pentafluorophenylesters (22) are the most
reactive compounds. The active esters of phenols
substituted by less electron attracting groups are
relatively slow reacting aminoacyl donors. This
shortcoming, however, is compensated by the
"mildness" which generates fewer byproducts than a
reaction with highly ("super") activated carboxyl
components.

E. Stepwise Approach vs Fragment Condensation

1. Stepwise Approach. Active esters are very
useful reagents for the stepwise approach of peptide
synthesis. Apparently, the longest biologically
active peptide ever synthesized was the intestinal
hormone secretin (27 amino acids) which was prepared
by using p-nitrophenylesters of the N-protected
building stones by Bodansky et al. (23). Since the
purified end product showed a biological activity
comparable to that of natural porcine secretin the
authors state that their stepwise synthesis "may be
applicable to even more ambitious endeavors". In a
critical evaluation (2), however, the homogeneity of
the synthetic peptide is questioned by arguments
like inadequate analytical controls and non-optimal
isolation procedures of the intermediates. In the
opinion of the present author, too, the above state-
ment concerning "ambitious endeavors" sounds too
optimistic. Inherently in a stepwise approach the
separation of the product (n+1) from its predecessor
(n) becomes the more difficult the greater the number n is. Mere precipitation, as in the foregoing
case will not safely lead to absolutely pure intermediates, though thin layer chromatography would
seem to disprove the presence of (similarly migrat-
ing) contaminants. Therefore it is suggested to
employ a stepwise approach only to synthesize peptides up to 10 amino acids or so. In order to facilitate the separation of the end product the last
amino acids to be coupled to the fragment should be
a characteristic one which makes the peptide significantly different from its precursor peptides.

2. Fragment Condensation. In fragment condensation purification is facilitated, for in most instances the products exhibit physical and chemical properties different from those of the starting peptides. Therefore, this methods provide the most promising route to homogenous polypeptides or even proteins. Successful condensations of fragments imply racemization-free coupling methods, which could be elaborated in various laboratories up to date. Difficulties of the fragment method are the frequently observed low coupling yields, particularly, when large fragments are to be connected and the sparing solubility of some protected or partially protected peptides. For chromatographic and as reaction solvents besides dimethylsulfoxide, dimethylformamide and hexamethylphosphotriamide, such exotic liquids as N-methylpyrrolidone or trifluoroethanol are being utilized. For coupling of the fragments, particularly of the large ones, the DCC-hydroxysuccinimide procedure (see p. 9) is the method of choice. For side chain protection - and this applies in particular to syntheses on a solid phase - residues must adhere as permenently as possible: e.g. for protecting the ε-amino group of lysine the Adoc group was preferred to the generally used Boc group by Kenner et al. in their synthetic approach towards a lysozyme analogue (129 amino acids). This group also introduced phenyl esters into peptide chemistry, for these compounds are very mildly hydrolyzed, by oxidation with hydrogenperoxide anion at pH 10.5. An excellent example for modern strategy and tactics in protein synthesis by fragment assembly techniques is given by Kenner in his Bakerian Lecture (24).

a. Difficulties of a protein synthesis. Why, then, inspite of all progress made in the last two decades, is the synthesis of homogenous proteins of 100 and more amino acids so extremely difficult, if not still infeasible? One reason is the expenditure of time and of material needed. Let's assume a team of 6 chemists is to synthesize a hecatopeptide (100 amino acids) via decapeptide fragments. One person is full-time engaged in preparing the (side chain) protected amino acids, the other 5 will be synthesizing the decapeptides producing one peptide bond in the average of 3 days, what is a optimistic rather than realistic guess. Ten decapeptides (90 peptide bonds) could so be obtained in, say, 270 working days, i.e. about one year, but for this work

would be done in parallel by 5 chemists, every person would be engaged for 2 - 3 months. For the coupling of 5 decapeptides each to yield two pentaconta-
peptides let's assume another 1 - 2 months, so that
these two large fragments would be available within
half a year from the beginning of the synthesis. At
an average yield of 50 % (what is usual, if purification is included) starting from 0.1 moles of amino
acids one would obtain the two big fragments only
in the one-gram range. Now let the crucial final
condensation 50 + 50 = 100, yield only 10 percent,
what is not unusual in peptide chemistry, and let
the ultimate deprotection reactions damage a greater
part of the protected synthetic protein. As a way
out of such frustrations the <u>solid phase</u> peptide
synthesis can be helpful.

III. SOLID PHASE PEPTIDE SYNTHESIS

This brilliant invention of R.B.Merrifield in
1962 (3) ought not to be analyzed in detail here.
The basic features of the procedure involve phase
separation of the peptide from reagents and side
products. The growing peptide chain is kept covalently attached to an entirely insoluble support
throughout all stages of the synthesis. For multi-
stage syntheses of large polypeptides 100 % quantitative incorporation of each amino acid residue
must be accomplished. Finally the peptide should be
fully removable from the support without undergoing
any changes of the backbone and side chains. The
procedure offers many advantages over solution syntheses: 1) avoiding of losses which occur during
purification of intermediates; 2) reaction may be
driven to completion by using high excesses of
reactants; 3) no solubility problems arise (as
mentioned before limited solubility may constitute
the most serious obstacle for solution synthesis);
4) omission of the ordinary laboratory operations
such as evaporation, crystallization, precipitation,
extraction etc. allows speeding up of the synthesis
and makes the procedure suitable for automation.
If all these conditions would be fulfilled by
100 percent, the method surely would be the optimum.
However, nobody is perfect!

A. Polystyrene Support

As a support, a majority of workers use a chlo-
romethylated polystyrene resin crosslinked by co-
polymerization of styrene with p-divinylbenzene. In
a 2 percent crosslinked resin as it is commonly
employed, only a minor but not negligible part of
the gel would be inaccessible because of steric
hindrance close to the crosslinking areas. The chlo-
romethylation to an extent of 0.5 - 1.0 mequiv. of
chloromethyl groups per gram of resin yields a support in which one chloromethyl group meets 10 to 20
phenyl residues. The distribution pattern of these
residues is not known, but it is feasible that the
"free hanging" functions are a little more reactive
than those neighbouring to the crosslinks. For this
reason alone it is desirable to reduce the number of
crosslinks to a minimum, i.e. to use resins cross-
linked as weakly as possible. By this the swelling
factor of the particles will also increase. The
resin beads are not a surface support, but freely
permeable to reagents. Nearly all of the functional
groups are located in the interior. It is clear that
the reactivity will be enhanced by loosening of the
matrix, by reducing the places where the dissolved
reagents have less access. 0.5 % crosslinked polystyrene which swells to a tenfold of its dry volume,
e.g. in dichloromethane, happens to be too fragile
for use in a stirred or shaken reaction vessel, but
is conveniently applied in a centrifugal flow reactor designed by Chr. Birr in our laboratory (25).

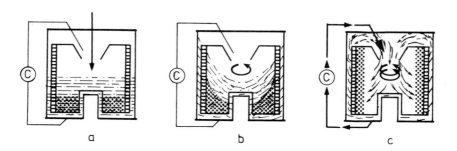

a   b   c

FIGURE 6.   The centrifugal reactor for solid-phase
            peptide synthesis, a) before rotation,
            b) spinning at low speed, c) in function
            solution circulating (25)

1. The Centrifugal Reactor. The resin is contained in a rapidly magnetically rotated Teflon beaker, the wall of which consists of porous quartz and which is mounted inside a slightly larger glass housing (Figure 6). The solution introduced through the cap top, is driven by centrifugal forces through the resin adhering along the porous wall to the inner wall of the housing which it climbs and plunges back from there into the rotating basket. The device, additionally, operates as a rotary pump that can provide an external spectroscopic flow cell with a circulating current of reagent solutions or solvents.

a.   Monitoring of the reaction cycles. An equipment consisting of the synthesizer, a flow photometer and a continuous line recorder (26) allows monitoring of increasing or decreasing amounts of uv-absorbing compounds, in this case at 230 and 280 nm the chromophoric 3,5-dimethoxyphenyl moiety of the Ddz-group (p. 5) and 3,5-dimethoxy-α-methyl-styrene and 3,5-dimethoxyphenyldimethylcarbinol, the products of its fission. Figure 7 shows a diagram of continous recording and control of deblocking, washing, deprotonation, and coupling of a Ddz-amino acid to the amino component in the Merrifield process.

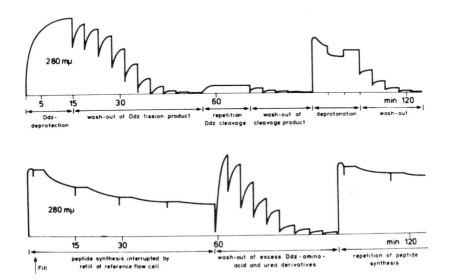

FIGURE 7.   Diagram of one reaction cycle of a solid-phase peptide synthesis monitored at 280 nm

Deblocking with 5 percent trifluoroacetic acid in
dichloromethylene creates a strong absorption in
the current liquid which gradually is washed out
until the recorder attains zero. After repetition of
the Ddz cleavage and deprotonation, peptide synthesis is indicated by a slow vanishing of the Ddz-
amino acid by its binding to the resin bound amine
component. In this way also a quantitative determination of the reagents and - most importantly - the
end of coupling and deblocking reactions is possible.
I need not stress again how crucial the attaining of
100 percent at each step of the solid phase operation is. Unfinished (truncated) and left out (deleted) peptide chains will generate a microhetero-
genous mixture which puts extremely high demands
on the fractionation power of purification methods.
A stepwise synthesis of an eicosapeptide (20 amino
acids) by an average yield of 99 % of every elongation will give a mixture of ~ 84 % of the target
with 16 % of all possible peptides mainly of one
less constituent. As it is apparent from the diagram
(Figure 7) cleavage of the Ddz residue is nearly
complete after 15 minutes, several washing out
processes of the fission products will appear in the
solution. The coupling reaction (DCC) comes to
completion not before 90 minutes and needs two
charges of Ddz amino acids. This clearly demonstrates that in order to be sure a double coupling (with
each an excess of at least 5 moles per mole of
peptide chain) is required.

    b. Blocking of truncated sequences. For sake
of security several reactions have been proposed to
blocking unreacted amino groups in order to prevent
further elongation at later stages. Our reagent,
3-nitrophthalic acid anhydride, energetically reacts
with amino groups to give 2-carboxy-3-nitrobenzoyl-
amides, which after splitting off from the resin
together with the desired end product can be easily
removed from the mixture by ion exchange chromatography due to their strong acidic nature (27).

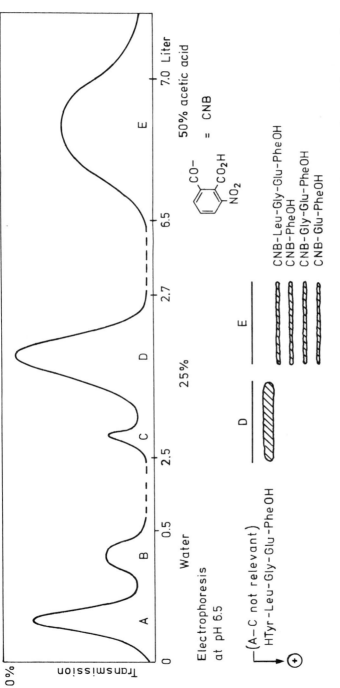

FIGURE 8. Chromatographic separation (Dowex 1x2) and paper electrophoretic analysis of a deliberately defective solid phase synthesis of H-Tyr-Leu-Gly-Glu-Phe-OH with blocking after every condensation step unreacted chains by 3-nitrophthalic acid anhydride [H. Wissenbach (28)].

The effectiveness of this method is demonstrated
by a paper electrophoretic analysis of the products
of a solid phase synthesis of a pentapeptide. In
order to simulate unfavorable reaction conditions
whereby considerable amounts of peptides escape
chain elongation, not the usual 3fold excess but
only equimolecular amounts of each Boc-amino acid
were applied in the DCC-condensation steps. The
reaction mixture, after cleavage from the resin, was
subjected to ion exchange chromatography on Dowex
1 x 2 with water-acetic acid as an eluent. Figure 8
displays the clean separation of the desired pentapeptide from the blocked truncated sequences (28).

c. Analysis by Edman-degradation on the support.
As a synthesis of a nonapeptide in which three times
each a 5 fold excess of Ddz-amino acids was employed
with additional application of 3-nitrophthalic anhydride, Chr.Birr and R.Frank analyzed the still
resin bound product by a modified Edman degradation
on the solid phase (29). The analysis of the PTH-
products by sequential high pressure liquid chromatography revealed that minimal amounts of failure
sequences were present inspite of all precautions
taken. Nevertheless, a complete purification of this
peptide could easily be accomplished and this should
be true also for longer peptides, say up to penta-
decapeptides generated by the same technique.

B. Some Other Links and Supports

Further improvement offers the change from the
generally used benzylester link of the peptide to
the resin to a phenacyl ester type (30). Following
synthesis of a peptide by the usual stepwise proce-

$$\text{—} \overset{|}{\bigcirc} \text{—} \underset{O}{\overset{||}{C}} \text{—} CH_2 \text{—} O \text{—} \underset{O}{\overset{||}{C}} \text{—peptide}$$

dure, the phenacyl ester link can be cleaved with
sodium thiophenoxide, 0.5 M aqueous sodium hydroxide,
ammonia or hydrazine, conditions which are much
milder than the strong acidic ones employed for the
cleavage of the benzyl esters.

Worth improving is also the chemistry of the
support. Sites of the polystyrene resin accessible
to solvents and solutes at the first steps of a peptide synthesis may become inaccessible with growing
length and, in parallel, decreasing swellability in
the solvents commonly used. Therefore, Sheppard et
al.(31) designed a polyamide support and a modification of linking of the first amino acid which allowed
working in highly polar organic media in which both
the peptide and polymer chains were fully solvated.
This system proved superior to the conventional
solid phase in the synthesis of a notoriously difficult decapeptide sequence using symmetrical Boc-
amino acid anhydrides or p-nitrophenylesters. No
swelling problems offers the utilization of a soluble
polymer as a support for synthesis and elongation of
peptide chains. This concept, outlined in Shemyakins
laboratory (32) has been further developed and
promisingly employed using polyethylene glycol as
carrier by E.Bayer's group (33).

C. Evaluation of Solid Phase Methods

Solid phase peptide synthesis, in the present
state, is a very valuable method for preparing in
relatively short time oligopeptides of such a purity
that further purification (by chromatography,
counter current distribution) easily yields uniform
products of molecular homogeneity. The upper limit
of the number of amino acids, depending on their
complexity, is in the order of ten to fifteen. If
lower demands are made to the quality of the preparations, e.g. for half quantitative biological
screening, also longer peptide chains may rapidly
and conveniently be built up on an automatic peptide
synthesizer. Peptides larger than eicosapeptides can
hardly be obtained in uniform state by the present
experimental techniques. This will require still improved synthetic procedures and more powerful fractionation methods. Whether homogenous proteins ever
will be obtained by stepwise solid phase synthesis
seems very doubtful. At the time being the power of
the method lies in the rapid provision of fragments
which in turn could be assembled by solution
techniques. Since side chain protection has to be
maintained during this process, specific removal of
the α-N-protecting residues is important, and hereby
the Ddz group (26) proves very useful.

IV. COMPARISON AND OUTLOOK

In this review, I tried to describe - from my
personal standpoint - the efficiency of modern peptide chemistry by looking into the formerly opposing
laboratories of both solution and solid-phase
chemists. We have seen, that each of these techniques
has its advantages and disadvantages. For an evaluation a table, borrowed from J.Meienhofers review (1)
is represented.

Solid-phase synthesis	Solution synthesis
Favorable features:	**Favorable features:**
Speed; convenience; no solubility problem; automation possible;	Homogenous product; adequate analysis; purification of intermediates
Unfavorable features:	**Unfavorable features:**
High excess of amino acid needed; product inhomogenous by less than 100 % N-deprotection and coupling yield; degradation danger on removal from solid-phase; extensice product purification required; no adequate fine analysis	Slow, laborious; solubility problems; failing of fragment condensation; racemization danger

The comparison shows that the strength of solid-
phase synthesis lies in operational efficiency, the
strength of solution methods in product quality. For
achieving the declared goal of peptide chemists,
synthesis of proteins, a combination of the existing
efficient procedures and further methodological improvements will be necessary. Approaches to using
water as a solvent for peptide syntheses by con-
structing water-solubilizing protecting groups (34)
and utilization of peptide forming enzymes (35) can
be taken into consideration. An encouraging experiment was the pepsin-catalyzed condensation of N-di-
methylphenylalanylleucylglycylglutamylphenylalanin
with tyrosylleucylglycylglutamylphenylalanine which
yielded the corresponding decapeptide by nearly
80 percent (28).

ACKNOWLEDGMENTS

The author is deeply indebted to all collabora-
tors, who have contributed to the results in peptide
chemistry from the early beginning 1948 until today,
30 years later. Several of them are mentioned in the
references, the anonymous ones cordially included.

REFERENCES

1. Meienhofer, J., in "Hormonal Proteins and Pep-
    tides" (Choh Hao Li, ed.), p. 45. Academic
    Press, New York and London, 1973.
2. Finn, F.M. and Hofmann, K., in "The Proteins"
    (Neurath, H. and Hill, R.L., eds)., Vol. II,
    p. 105. Academic Press, New York, San Francisco,
    London, 1976.
3. Erickson, B.W. and Merrifield, R.B., in "The
    Proteins" (Neurath, H. and Hill, R.L., eds).
    Vol. II, p. 255. Academic Press, New York,
    San Francisco, London, 1976.
4. Wünsch, E., in "Methoden der organischen Chemie"
    (E.Müller and O.Bayer, eds), Vol.XV, 1 and 2.
    Georg Thieme Verlag, Stuttgart, 1974.
5. Choudhury, A.M., Kenner, G.W., Moore, S.,
    Ramage, R., Richards, P.M., Thorpe, W.D.,
    Moroder, L., Wendlberger, G. and Wünsch, E., in
    "Peptides 1976" (A.Loffet, ed.), p. 257.
    Editions de l'Université de Bruxelles, 1976.
6. Munekata, E., Faulstich, H. and Wieland, Th.,
    Angew.Chem.Int.Ed.Engl. 16:267 (1977);
    Munekata, E., Faulstich, H. and Wieland, Th.,
    J.Amer.Chem.Soc. 99:6141 (1977).
7. Tilak, M.A., Tetrahedron Lett. (1970), 849.
8. vanZon, A. and Beyermann, H.C., Helv.chim.Acta
    56:1729 (1973).
9. Birr, Ch., Lochinger, W., Stahnke, G. and Lang,
    P., Justus Liebigs Ann.Chem. 763:162 (1972).
10. Wieland, Th. and Heinke, B., Justus Liebigs
    Ann.Chem. 599:70 (1956).
11. Flor, F., Birr, Ch. and Wieland, Th., Justus
    Liebigs Ann.Chem. (1973), 1601.
12. Kemp, D.S., Bernstein, Z. and Regek, jr. J.,
    J.Amer.Chem.Soc. 92:4756 (1970).
13. Sheehan, J.C. and Hess, G.P., J.Amer.Chem.Soc.
    77:1967 (1955).

14. Weygand, F., Hoffmann, D. and Wünsch, E.,
    Z.Naturforsch. B. 21:426 (1966).
15. König, W. and Geiger, R., Chem.Ber. 103:2024
    (1970).
16. Lipman, F., Accounts Chem.Res. 6:361 (1973).
17. Wieland, Th., Schäfer, W., and Bokelmann, E.,
    Justus Liebigs Ann.Chem. 573:99 (1951).
18. Wieland, Th. and Schäfer, W., Justus Liebigs
    Ann.Chem. 576:104 (1952).
19. Wieland, Th. and Heinke, B., Justus Liebigs
    Ann.Chem. 615:184 (1958).
20. Goodman, M. and Stueben, K., J.Amer.Chem.Soc.
    81:3980 (1959).
21. Bodansky, M., Nature (London), 175:685 (1955).
22. Kisfaludy, L., Roberts, J.E., Johnson, R.H.,
    Mayers, G.L. and Kovacs, J., J.Org.Chem.
    35:3563 (1970).
23. Bodansky, M., Ondetti, M.A., Levine, S.D. and
    Williams, N.J., J.Amer.Chem.Soc. 89:6753 (1967).
24. Kenner, G.W., Proc.R.Soc.Lond.A., 353:441 (1977).
25. Birr, Ch. and Lochinger, W., Synthesis (1971),
    319.
26. Birr, Ch., in "Peptides 1974" (Y.Wolman, ed.),
    p.117. John Wiley and Sons, New York, Toronto,
    1974.
27. Wieland, Th., Birr, Ch. and Wissenbach, H.,
    Angew.Chem.Int.Ed.Engl. 8:764 (1969).
28. Wissenbach, H., Doctoral thesis, University
    Heidelberg, 1971.
29. Birr, Ch. and Frank, R., FEBS Letters, 55:68
    (1975).
30. Weygand, F., in "Peptides 1968" (E.Bricas, ed.),
    p. 183. North Holland Publ., Amsterdam, 1968.
31. Atherton, E., Clive, D.L.J. and Sheppard, R.C.,
    J.Amer.Chem.Soc. 97:6584 (1975).
32. Shemyakin, M.M., Ovchinnikov, Yu.A., Kiryushkin,
    A.A. and Koznevnikova, J.V., Tetrahedron Lett.
    (1965), 2323.
33. Mutter, M., Hagenmaier, H. and Bayer, E.,
    Angew.Chem.Int.Ed.Engl. 10:811 (1971).
34. Wieland, Th., in "Proc.VIIth Europ. Peptide
    Symposium" (V.Bruckner and K.Medzihradszky,
    eds.) Acta Chim. Hung. 44:5 (1965).
35. Wieland, Th., Determann, H. and Albrecht, E.,
    Justus Liebigs Ann.Chem. 633:185 (1960);
    Determann, H., Heuer, J. and Jaworek, D.,
    Justus Liebigs Ann.Chem. 690:189 (1965) and
    papers on "plastein" formation, cited in the
    last reference.

IMPROVED PROCEDURES FOR THE SOLID-PHASE
SYNTHESIS AND PURIFICATION OF PEPTIDES[1]

Donald Yamashiro
James Blake

Hormone Research Laboratory
University of California
San Francisco, California

Solid-phase peptide synthesis introduced by
Merrifield in 1963 (1) has become a widely accepted
method. Its speed and simplicity are most fully
realized when the stepwise strategy is employed.
However, imperfections in the execution of this
strategy are cumulative since there is currently
no way that impurities which develop on the support
can be removed as they are formed. Thus, every
effort is made to reduce these imperfections.

We have directed our efforts toward minimizing
these imperfections by (a) improvements in side-
chain protection and coupling, and (b) improvements
in purification of the product. These two avenues
of investigation are related in that the choice of
synthetic strategies is frequently dictated by the
capabilities and shortcomings of purification
methods.

A set of side-chain protecting groups has been
developed for use with N^α-Boc protection and their
final removal in liquid HF (2) as summarized in
Table I. Acid stabilities of most of these were

---

[1]This work was supported in part by U.S. Public
Health Service Grant GM-2907 to Professor C. H. Li.

TABLE I. Stabilities of Side-chain Protecting Groups Removable in Liquid HF

Amino Acid	Side-chain Protection	Time in 50% TFA/CH$_2$Cl$_2$ (hr)	Loss of Protection (%)	Results in Liquid HF	References
Asp	Bzl	23	4	a	(3,4)
	4-BrBzl	71	2.5	a	(5)
Thr	Bzl	23	5	a	(3,4)
	4-ClBzl	71	2.5	a	(5)
Ser	Bzl	23	3	a	(3,4)
	4-BrBzl	71	1.3	a	(5)
Glu	Bzl	23	2	a	(3,4)
	4-BrBzl	b	b	a	(6)
Cys	4-MeOBzl	23	27	a	(3,4)
	3,4-Me$_2$Bzl	23	0.2	a	(3,6)
Met	unprotected	--	--	t-butylation	(7)
Tyr	Bzl	21	55	a; side reaction	(3,4,8)

82

	2,6-Cl$_2$Bzl	21	1.4	a; 9% side reaction[c]	(3,4,9,10)
	Z	b	b	a	(11)
	2-BrZ	24	1	a; trace side reaction[c]	(10)
Lys	Z	20	42	a	(3,4)
	2-BrZ	20	0.7	a	(3,9)
His	Boc	unstable	--	--	(12)
	Z	16[d]	ca. 8	removed	(13)
Arg	Tosyl	b	b	a	(12)

[a] Removed quantitatively in 10 min at 0°.
[b] Not determined
[c] Determined in synthesis of H-Phe-Lys-Gln-Thr-Tyr-Ser-Lys-Phe-OH.
[d] Determined on Acetyl-Ala-His(Z)-Phe-Polymer.

measured in solution phase on acetylamide derivatives. The advantages of 2-BrZ protection for
tyrosine over that of Bzl and 2,6-Cl$_2$ Bzl have been
pointed out for circumventing side reactions in
liquid HF. Although 2-BrZ protection is preferred
when exposure to acid is extensive, Z protection is
satisfactory if tyrosine is near or at the N-
terminus as in the case of β-endorphins. Since
use of Boc protection for histidine leads to its
immediate loss in the next deprotection step, the
more stable Z group would appear to offer advantages. Recent improvements in the reduction of
methionine sulfoxide to methionine (14) make
sulfoxide protection (15) a more attractive tactic.
However, the need to perform an additional deblocking step always poses a new set of problems as in
the case of formyl protection of tryptophan (16)
where side reactions can occur during deformylation
under basic conditions (17).

For the critical coupling reaction, the symmetrical anhydride method (18,19) has been our
choice as a means of reducing chain termination
observed in earlier work on the β-lipotropin sequence (20). This, along with the aforementioned
improvements, have been applied to the synthesis
of biologically active peptides of moderate size.
This is exemplified by the synthesis of human β-
endorphin in 32% overall yield (21). Such efforts
have provided facile access to synthetic analogs
of ACTH (22) and β-endorphin (11).

The most critical attention in solid-phase
peptide synthesis has been directed at the coupling
problem -- the necessity, and our apparent inability, to effect quantitative coupling in every
cycle between the peptide chain and the incoming
protected amino acid. Even under conditions where
the peptide polymer is swollen, and presumably
solvated beyond its usual state, this reaction
falls short of 100% completion (23). The beneficial
effects of 2,2,2-trifluoroethanol (TFE) in the
coupling reaction was demonstrated in the synthesis
of H-Lys$_5$-Glu$_3$-Leu$_2$-Trp(Nps)-Phe-OH where coupling
conducted in 20% TFE in CH$_2$Cl$_2$ gave a product containing only 1.6% of the deletion sequence H-Lys$_4$-
Glu$_3$-Leu$_2$-Trp(Nps)-Phe-OH. Coupling in CH$_2$Cl$_2$
alone gave 7.7% of the deletion sequence. It was
noted that TFE promoted swelling of the peptide
resin. Recent evidence indicates that this swelling

does not occur with all sequences and coupling efficiency is correspondingly unimproved. There is clearly a need for further investigations, particularly with longer sequences. The major consequence of imperfections in the coupling reactions is that deletion sequences can be expected to be present in the crude product. The possibility of removing these impurities, especially those where only a single amino acid residue is missing, merits some attention.

The method of preparative partition chromatography on Sephadex G-25 (24) has been extended in recent years to the more porous supports Sephadex G-50 (9) and agarose (25). In liquid-liquid partition methods, all portions of a molecule exposed to the solvent environment can serve as "handles" for the separation process. The method may be formulated in quantitative terms by application of the theory of A.J.P. Martin (26). If a peptide A is transferred from the stationary aqueous phase of a partition column to the mobile organic phase, the partial molar free energy change in the standard state is given by the equation

$$\Delta\mu_A^o = RT \ln \alpha_A' \qquad (1)$$

where the reference state is that of infinite dilution and α_A' is the ratio of the mole fraction of A in the stationary phase to its mole fraction in the mobile phase. If peptide B differs from A in a structural alteration X and is likewise transferred

$$\Delta\mu_B^o = RT \ln \alpha_B' \qquad (2)$$

The free energy change for transfer corresponding to X is

$$\Delta\mu_X^o = RT \ln(\frac{\alpha_B'}{\alpha_A'}) \qquad (3)$$

In the classical Martin-Synge equation for partition chromatography

$$\alpha = (\frac{V_m}{V_s}) \; (\frac{1-R_f}{R_f}) \qquad (4)$$

the distribution constant α is based on concentrations per unit volume and V_m and V_s are mobile and stationary phase volumes, respectively. The

Bate-Smith and Westall (27) concept of R_M may be expressed in terms of the natural logarithm.

$$R_M = \ln\left(\frac{1-R_f}{R_f}\right) \tag{5}$$

For peptides A and B equation 4 may therefore take the form

$$\ln \alpha_A = \ln\left(\frac{V_m}{V_s}\right)_A + R_{MA} \tag{6}$$

$$\ln \alpha_B = \ln\left(\frac{V_m}{V_s}\right)_B + R_{MB} \tag{7}$$

If peptides A and B differ only slightly in structure so that they show no differences due to molecular sieving or adsorption, we may assume

$$\ln\left(\frac{V_m}{V_s}\right)_A = \ln\left(\frac{V_m}{V_s}\right)_B \tag{8}$$

Subtracting equation 6 from 7, we have

$$\ln\left(\frac{\alpha_B}{\alpha_A}\right) = R_{MB} - R_{MA} = \Delta R_{MX} \tag{9}$$

where ΔR_{MX} is the change in R_M associated with the structural alteration X. In dilute solution, the ratio of distribution constants in equation 9 approximately equals the ratio in equation 3. Therefore, combination of the two equations gives

$$\Delta R_{MX} = \frac{\Delta \mu^{\circ} X}{RT} \tag{10}$$

which relates the chromatographic behavior of peptides A and B to the free energy of transfer associated with alteration X.

   An attempt is now made to apply these formulations to the problem of purifying peptides. The peptide chosen for the test was camel β-endorphin (28) ($β_C$-EP) whose structure is shown in Figure 1.

H-Tyr-Gly-Gly-Phe-Met-Thr-Ser-Glu-Lys-Ser-
            5   10
Gln-Thr-Pro-Leu-Val-Thr-Leu-Phe-Lys-Asn-
            15   20
Ala-Ile-Ile-Lys-Asn-Ala-His-Lys-Lys-Gly-Gln-OH
       25   30

FIGURE 1.   Amino acid sequence of camel β-
endorphin.   Amino acid residues underlined indicate
those omitted in synthesis of omission analogs.

Fifteen different analogs were synthesized by the
solid-phase method, each missing only a single
amino acid in the entire sequence.   The amino acids
omitted in these syntheses are underlined in
Figure 1, and these analogs are referred to here
as omission analogs.   All the omission analogs
along with β_C-endorphin were examined under dilute
conditions (1-5 mg each on a 1.18 x 60 cm column
of Sephadex G-50, 20-43 μ dry particle size) in the
solvent system 1-butanol-pyridine-0.6 M NH_4OA_C
(5:3:10).   From the observed R_f values, the
corresponding R_M values were calculated according
to equation 5.   The ΔR_{MX} for insertion of any
residue is given by

$$\Delta R_{MX} = R_M(\beta_C\text{-endorphin}) - R_M(\text{analog}) \quad (11)$$

The value for each insertion is listed in Table II.
The entire range of these values is covered by
partition chromatography of a mixture of β_C-EP and
five of the omission analogs (Figure 2).   A close
separation is shown by the partition chromatography
of a mixture of β_C-EP and Des-Gly[2]-β_C-EP (Figure 3).
The separations shown in Figures 2 and 3 were
achieved in a column exhibiting about 500-600
theoretical plates as calculated by the Glueckauf
equation (29).   Under these conditions the only
analogs that cannot be expected to be cleanly
separated from β_C-EP are those with ΔR_M (residue)
values less than 0.20.   However, preliminary experiments show that even the difficult separation of
β_C-EP from Des-Ser[7]-β_C-EP can be partially effected
by recycling chromatography.

TABLE II. Hydrophobicity Scale for Amino Acid Residues and their Side-Chains in β_C-Endorphin[a]

Amino Acid Omitted in β_C-EP	ΔR_M for residue insertion[b]	ΔR_M for side-chain[c]	$\Delta\mu°$ (cal/mole)[d]
Tyr[1]	-1.54	-1.77	-2300
Phe[4]	-1.19	-1.42	-2500
Leu[14]	-1.05	-1.28	-1800
Met[5]	-0.64	-0.87	-1300
Val[15]	-0.55	-0.78	-1500
Ile[22]	-0.35	-0.58	
Ser[7]	+0.08	-0.15	+300
Thr[6]	+0.08	-0.15	-400
Thr[12]	+0.12	-0.11	
Pro[13]	+0.15	-0.08	
Ala[21]	+0.19	-0.04	-500
Ser[10]	+0.21	-0.02	
Gly[2]	+0.23	---	+895[e]
Gln[11]	+0.29	+0.06	+260
Asn[20]	+0.52	+0.29	+150

[a]Solvent system: 1-butanol-pyridine-0.6M NH$_4$OAc (5:3:10).

[b]ΔR_M (residue) = $R_M(\beta_C$-EP)-R_M (analog)
[c]ΔR_M (side-chain)= ΔR_M (residue)-ΔR_M(Gly)

[d]For transfer of side-chains in amino acids from water to ethanol and/or dioxane; taken from Nozaki and Tanford (30).

[e]Residue value measured as difference between H-Gly-Gly-OH and H-Gly-Gly-Gly-OH.

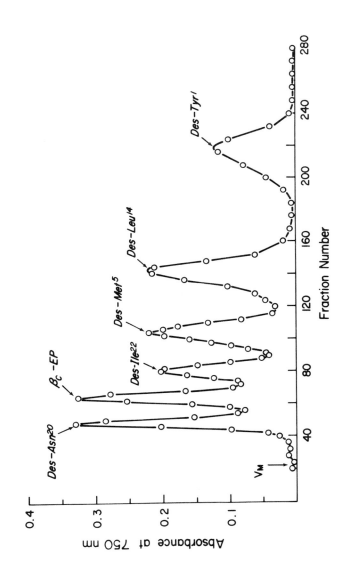

FIGURE 2. Partition chromatography of a mixture of β_C-endorphin and five omission analogs on Sephadex G-50; V_m, mobile phase volume; detection, Folin-Lowry method.

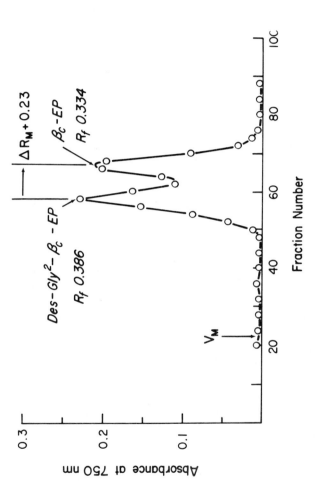

FIGURE 3. Partition chromatography of a mixture of β_c-endorphin and Des-Gly2-β_c-EP on Sephadex G-50; V_m, mobile phase volume; detection, Folin-Lowry method.

The separability of several of the omission analogs from β_C-EP on thin-layer chromatography was also tested (Figure 4). Detectable separations were observed only for those analogs missing Ile[22], Met[5], Leu[14], Phe[4], and Tyr[1]. In general, absence of a bulky hydrophobic side-chain was required for separation to be possible on thin-layer chromatography.

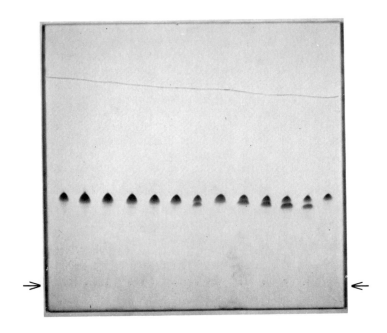

FIGURE 4. Thin-layer chromatography of mixtures of β_C-EP and omission analogs on silica gel in 1-butanol-pyridine-acetic acid-water (5:5:1:4); ninhydrin detection. Channels at extreme left and right run with β_C-EP alone. All other channels run with mixtures of β_C-EP and analogs with following omissions (left to right): Asn[20], Gln[11], Gly[2], Ser[10], Thr[12], Ile[22], Val[15], Met[5], Leu[14], Phe[4], and Tyr[1]. Arrows point to origin. Line toward top marks solvent front.

The additivity principle of Martin was applied
to the amino acid residues in Table II, thus:

$$\Delta R_M(\text{side-chain}) = \Delta R_M(\text{residue}) - \Delta R_M(\text{Gly}) \quad (12)$$

The ΔR_M values for each of the amino acid side-
chains are tabulated in Table II. From equation 10
it is evident that these are measures of the free
energy of transfer of the side-chains from the
aqueous phase to the organic phase. Similar values
for the transfer of side-chains in amino acids from
water to ethanol and/or dioxane have been reported
by Nozaki and Tanford (30) from measurements of the
solubilities of amino acids. These values are also
listed in Table II. A plot of ΔR_M (side-chain) for
amino acid side-chains in β_C-EP against $\Delta \mu°$ values
reported by Nozaki and Tanford is shown in Figure 5.
The ΔR_M (side-chain) value for Tyr[1] may be slightly
inaccurate since the "end effect" resulting from
its omission and appearance of a glycyl N-terminus
was not taken into account here. Nevertheless, an
approximate correlation of hydrophobicities is evident in spite of the very divergent conditions
under which the two sets of values were obtained.
Consequently, as a first approximation, relative
hydrophobicities in partition chromatography may be
expected to be independent of sequence. This type
of conclusion, of course, is the heart of the
Martin hypothesis.

The very large free energies of transfer associated with bulky aromatic side-chains are noteworthy. It may be seen in Table I that virtually
all the side-chain protecting groups possess a
similar structural feature. This would strongly
suggest that the presence of one such protecting
group not removed in the deblocking step would
cause such a substantial change in the R_M value of
β_C-EP that separation of such a by-product would
occur with ease.

A test of the hydrophobicity scale can be
illustrated with analogs of human β-endorphin.
The ΔR_M corresponding to replacement of one amino
acid by another can be calculated from Table II.
The calculated and observed values for two such
replacements are shown in Table III.

It is likely that ΔR_M for a given change will
vary within limits as a function of structure and
solvent environment. A variation that is of

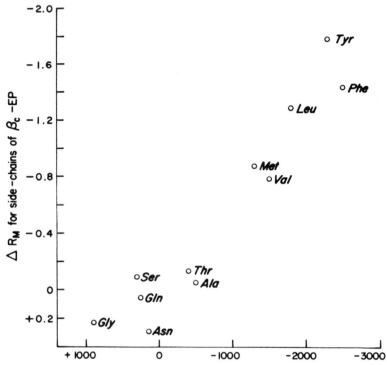

FIGURE 5. Plot of ΔR_M (side-chain) values in
β_C-EP from partition chromatography against
$\Delta\mu°$ (side-chain) values in amino acids. The ΔR_M
values for Ser and Thr are average values. The
values for Gly are the residue values.

TABLE III. Observed and Predicted ΔR_M Values for Analogs of β_h-Endorphin.

Peptide	R_f[a]	R_M	ΔR_M[b]	Predicted from table II
β_h-EP	0.40	0.405		
[Pro^5]-β_h-EP	0.23	1.21	+0.81	+0.79
[Leu^5]-β_h-EP	0.51	-0.04	-0.45	-0.41

[a]Values from references 21 and 31. Solvent system same as in Table II.
[b]$\Delta R_M = R_M$(Analog) $- R_M(\beta_h$-EP)

current interest is that associated with chain length. Data on this is very scant. The ΔR_M associated with the structural alteration of a methionine residue to a methionine sulfoxide residue has been measured as a function of chain length as shown in Table IV. Although the same solvent system could not be used in this series, the result is suggestive. The implication is that accessibility of an amino acid residue to the solvent environment decreases with increasing chain length in solvent systems of this type. Obviously, additional data would be helpful. Such quantitative measurements may afford a rational basis for assessing problems in the purification of large polypeptides and for design of synthetic strategies.

TABLE IV. Variation of ΔR_M as a Function of Chain Length

Chain Length	Peptide Pairs Separated	Support	Solvent System[a]	ΔR_M for -Met- to -Met(O)-
7	H-Leu-Gly-Arg-Leu-Gly-Met-Phe-OH H-Leu-Gly-Arg-Leu-Gly-Met(d-O)-Phe-OH	Sephadex G-50	A	-0.99
31	β_h-endorphin [Met(O)^5]-β_h-endorphin	Sephadex G-50	B	-0.74
39[b]	[Trp(Nps)^9]-α_s-ACTH [Met(O)^4,Trp(Nps)^9]-α_s-ACTH	Sephadex G-50	A	-0.50
91	β_s-LPH [Met(O)]-β_s-LPH	Agarose A-1.5m	C	-0.21

[a] A, 1-butanol-pyridine-0.1%HOAC (5:3:11); B, 1-butanol-pyridine-0.1N NH4OH containing 0.1% HOAc (2:1:3); C, 2-butanol-0.5N HOAc containing 0.5N NaCl and 0.03N TCA (8:11).

[b] Data taken from E. Canova-Davis and J. Ramachandran, Biochemistry 15:921 (1976).

ACKNOWLEDGMENTS

The authors thank Professor Choh Hao Li for his encouragement and support during the course of this work. The skilled technical assistance of Mr. W. F. Hain and Mr. Kenway Hoey is acknowledged.

REFERENCES

1. Merrifield, R. B., J. Am. Chem. Soc. 85:2149 (1963).
2. Sakakibara, S., Shimonishi, Y., Kishida, Y., Okada, M., and Sugihara, H., Bull. Chem. Soc. Jpn. 40:2164 (1967).
3. Yamashiro, D., Noble, R. L., and Li, C. H., in "Chemistry and Biology of Peptides, Proceedings of the Third American Peptide Symposium," (J. Meienhofer, ed.) p. 197. Ann Arbor Science Publishers, Ann Arbor, Mich., 1972.
4. Erickson, B. W., and Merrifield, R. B., in ref. 3, p. 191.
5. Yamashiro, D., J. Org. Chem. 42:523 (1977).
6. Yamashiro, D., Noble, R. L., and Li, C. H., J. Org. Chem. 38:3561 (1973).
7. Noble, R. L., Yamashiro, D., and Li, C. H., J. Am. Chem. Soc. 98:2324 (1976).
8. Iselin, B., Helv. Chim. Acta 45:1510 (1962).
9. Yamashiro, D., and Li, C. H., J. Am. Chem. Soc. 95:1310 (1973).
10. Yamashiro, D., and Li, C. H., J. Org. Chem. 38:591 (1973).
11. Yamashiro, D., Tseng, L-F., Doneen, B. A., Loh, H. H., and Li, C. H., Int. J. Pept. Prot. Res. 10:159 (1977).
12. Yamashiro, D., Blake, J., and Li, C. H., J. Am. Chem. Soc., 94:2855 (1972).
13. Blake, J., and Li, C. H., Int. J. Pept. Prot. Res., in press.
14. Houghten, R. A., and Li, C. H., in "Peptides, Proceedings of the Fifth American Peptide Symposium," (M. Goodman and J. Meienhofer, eds.) p. 458. John Wiley and Sons, New York, 1977.

15. Iselin, B., Helv. Chim. Acta 44:61 (1961).
16. Ohno, M., Tsukamoto, S., Makisumi, S., and
     Izumiya, N., Bull. Chem. Soc. Jpn. 45:2852
     (1972).
17. Yamashiro, D., and Li, C. H., J. Org. Chem.
     38:2594 (1973).
18. Wieland, T., Flor, F., and Birr, C., Liebigs
     Ann. Chem. 1595 (1973).
19. Hagenmaier, H., and Franck, H., Hoppe-Seyler's
     Z. Physiol. Chem. 353:1973 (1972).
20. Yamashiro, D., and Li, C. H., Proc. Nat. Acad.
     Sci. USA 71:4945 (1974).
21. Li, C. H., Yamashiro, D., Tseng, L-F., and
     Loh, H. H., J. Med. Chem. 20:325 (1977).
22. Lemaire, S., Yamashiro, D., Behrens, C., and
     Li, C. H., J. Am. Chem. Soc. 99:1577 (1977).
23. Yamashiro, D., Blake, J., and Li, C. H.,
     Tetrahedron Lett. 18:1469 (1976).
24. Yamashiro, D., Nature 201:76 (1964).
25. Yamashiro, D., and Li, C. H., Biochim. Biophys.
     Acta 451:124 (1976).
26. Martin, A.J.P., Biochem. Soc. Symp. 3:4 (1950).
27. Bate-Smith, E. C., and Westall, R. G.. Biochim.
     Biophys, Acta 4:427 (1950).
28. Li, C. H., and Chung, D., Proc. Nat. Acad. Sci.
     USA 73:1145 (1976).
29. Glueckauf, E., Trans. Far. Soc. 51:34 (1955).
30. Nozaki, Y., and Tanford, C., J. Biol. Chem.
     246:2211 (1971).
31. Yamashiro, D., Li, C. H., Tseng, L-F., and
     Loh, H. H., Int. J. Pept. Prot. Res., in press.

SURFACTANT-INDUCED CONFORMATION OF SOME
OLIGOPEPTIDES, POLYPEPTIDES AND PROTEINS:
HELIX- AND BETA-FORMING POTENTIAL OF AMINO ACID
SEQUENCE IN A PROTEINACEOUS ENVIRONMENT[1]

Jen Tsi Yang
Chuen-Shang C. Wu

Department of Biochemistry and Biophysics
and Cardiovascular Research Institute
University of California
San Francisco, California

I.   INTRODUCTION

Surfactants produce many diverse effects on proteins and other biological systems (1). They can denature proteins but also stabilize them against other denaturants. The ability to disperse, precipitate, and form complexes with proteins depends on experimental conditions. A unique feature of surfactants is their drastic effect on protein conformation at remarkedly low reagent concentrations such as 1 mM or less (2). It suggests a strong affinity between protein and surfactant molecules. Unlike guanidine hydrochloride or urea that unfolds the protein molecule, a surfactant can enhance or reduce the ordered structure in a protein molecule, the reason for which is still not clear.

Since 1940s the interaction between proteins and anionic surfactants has been extensively studied. Bovine serum albumin seemed to be a most studied protein. One of the arguments then involved the so-called all-or-none versus statistical binding (see, for example, Ref. 3). Currently we are interested in the conformation of proteins in surfactant solutions. We report here the binding of alkyl sulfates to synthetic

[1]This work was supported by the USPHS grants GM-10880 and HL-06285 (Program Project).

99

polypeptides as model compounds and also the effect of surfactants on some naturally-occurring oligo- and polypeptides.
The complex formation between cationic homopolypeptides and
anionic surfactants is highly cooperative and is independent
of any conformational transition. The induced conformation
of the sequential peptides in surfactant solution is related
to the structure-forming potential of their amino acid sequence.

II. HOMOPOLYPEPTIDES

A. Conformation of the Polypeptide-Surfactant Complex

The very simplicity of synthetic homopolypeptides makes
it ideal to study their conformations in surfactant solutions,
although the results are often more complicated than have been
expected. To illustrate, $(Lys)_n$ and its homologs with the
side group $R = -(CH_2)_mNH_2$ have the amino groups protonated in
neutral solution and their conformation is unordered. Raising
the pH to above 12 converts the unchanged $(Lys)_n$ ($m = 4$) into
a helical conformation, but $(Orn)_n$ ($m = 3$) to only a partial
helix (about 20%) (4). The only difference between the two
polypeptides is one methylene group in the side chain. Thus,
electrostatic interaction alone cannot account for the instability of the $(Orn)_n$ helix. More intriguing is the effect of
$NaDodSO_4$[2] on these cationic polypeptides. At ambient temperature $(Lys)_n$ adopts the β-form (5) and $(Orn)_n$ the helical conformation (6) in $NaDodSO_4$ solution.
    Figure 1 shows the changes in CD spectra of $(Lys)_n$ and
$(Orn)_n$ with increasing $NaDodSO_4$ concentration in neutral solution (7). The CD of an α-helix has a characteristic double
minimum at 222 and 208-210 nm and a maximum near 191-194 nm.
In contrast, the β-form has a single CD minimum near 217-218
nm and a maximum at about 197 nm, whereas the so-called
"random coil" has a strong negative band near 197 nm. The
surfactant solutions are well below the critical micelle concentration except for curves 10 (left) and 8 (right). Thus,
the molar ratio, R, of surfactant to peptide (residue) rather
than the critical micelle concentration determines the conformation of the polypeptide-surfactant complexes. At R
slightly greater than one the complexes precipitate, but
potentiometric titration indicates a stoichiometric binding

---

[2]Abbreviations used: $NaDodSO_4$, sodium dodecyl sulfate;
$NaDecSO_4$, sodium decyl sulfate; $DodNH_3Cl$, dodecylammonium
chloride; CD, circular dichroism.

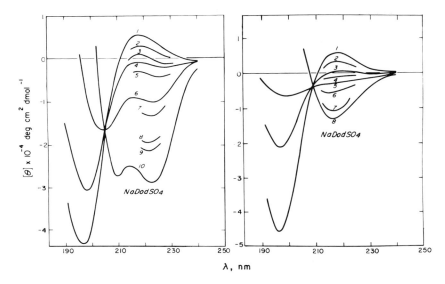

FIGURE1. The molar residue ellipticities of poly(L-
ornithine) (left) and poly(L-lysine) (right) in neutral
NaDodSO$_4$ solutions at 25°C. (Orn)$_n$: 0.108 mM. Molar ratio
(NaDodSO$_4$/residue) for curves 1 to 10: 0, 0.23, 0.31, 0.45,
0.55, 0.94, 1.10, 1.42, 1.57 and 240. (Lyn)$_n$: 0.117 mM.
Molar ratio for curves 1 to 8: 0, 0.22, 0.44, 0.60, 0.76,
0.97, 1.45 and 220.

isotherm. At higher NaDodSO$_4$ concentrations the complexes
redissolve and their conformation remains the same regardless
of the amount of excess surfactant.

   Figure 2 shows the CD spectra of (Lys)$_n$ and its four
homologs (8). The three higher homologs, poly(L-α, ω-diamino-
heptanoic acid) (m = 5), poly(L-α, ω-diaminooctanoic acid)
(m = 6) and poly(L-α, ω-diaminononanoic acid) (m = 7), all become helical in NaDodSO$_4$ solution. Why only the (Lys)$_n$-DodSO$_4$
complex behaves differently is difficult to explain. In this
respect, positively-charged (Arg)$_n$ also adopts the helical
conformation and (His)$_n$ the β-conformation in neutral NaDodSO$_4$
solution (9).

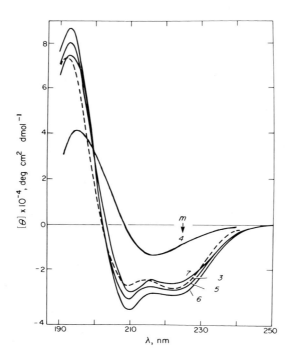

FIGURE 2. The molar residue ellipticities of poly(L-lysine) and its homologs in neutral NaDodSO$_4$ solution at 25°C. Polypeptide, 0.10-0.12 mM (residue); NaDodSO$_4$, 25-26 mM.

B.   Phase Diagram

The conformation of $(Lys)_n$ in NaDodSO$_4$ solution depends on the temperature and pH used (Fig. 3) (10). At constant temperature the complex undergoes a sharp β-to-helix transition with respect to pH. At higher temperature the middle point of the transition is shifted toward lower pH; the transition is cooperative and reversible. A phase diagram can be contructed from the data in Fig. 3. The shaded area in Fig. 4 represents the β-helix transition region (10). Raising the temperature above 55°C causes the β-form to precipitate. But the helical complex at, say, pH 11 indicates only a partial breaking-up of the conformation even after 1-hr heating at 70°C. This markedly contrasts the behavior of $(Lys)_n$ in aqueous solution without NaDodSO$_4$, which undergoes a helix-to-β transition at pH above 11 after about 10-min heating at 50°C (11). Similar helix-β transition (temperature unspecified)

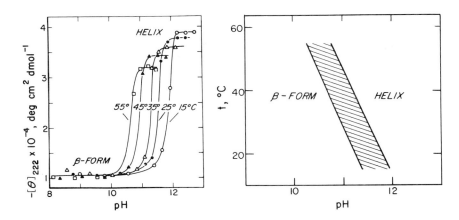

FIGURE 3 (left). The pH dependence of the molar residue ellipticity of poly(L-lysine) in NaDodSO$_4$ solution at various temperatures.
FIGURE 4 (right). The conformational phase diagram of poly(L-lysine) in NaDodSO$_4$ solution.

has been observed by the Russian school (12, 13). We postulate that the OH$^-$ and DodSO$_4$$^-$ ions compete for the binding sites, -(CH$_2$)$_4$NH$_3$$^+$. Because the binding constants are of the same order of magnitude, raising the (OH$^-$) concentration at high pH favors the formation of -(CH$_2$)$_4$NH$_3$$^+OH^-$ and therefore separates the bound DodSO$_4$ from the essentially uncharged polypeptide. Lowering the pH of the solution to neutral immediately re- verses the helix to the β-form for the (Lys)$_n$-DodSO$_4$ complex. On the other hand, deprotonated (Orn)$_n$ at pH above 12 has less than 40% helix even in the presence of NaDodSO$_4$ (K. Shirahama, unpublished data).

C. Binding Isotherm

Figure 5 shows the binding isotherm of sodium decyl sulfate (NaDecSO$_4$) to (Lys)$_n$, (Orn)$_n$ and (D,L-Orn)$_n$, noting that the racemic polypeptide does not have any conformational transition (14). At neutral pH the reaction can be written as M$^+$ + R$^-$ = M$^+$R$^-$, where M$^+$ is the polypeptide residue (monomer) and R$^-$ the free surfactant. The apparent binding constant, K$_a$, is simply

$$K_a = x/(1 - x)C_f \qquad (1)$$

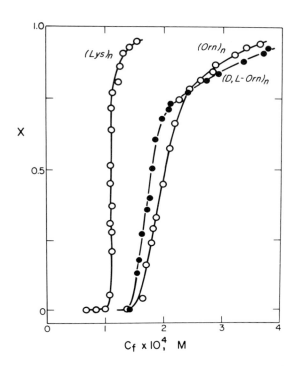

FIGURE 5. The binding isotherms of sodium decyl sulfate
to poly(L-lysine), poly(L-ornithine) and poly(D,L-ornithine)
in neutral solution at 25°C.

Here x is the degree of binding of the surfactant by the polypeptide residue. The free surfactant concentration, C_f, is
determined potentiometrically in a cell with a liquide membrane. The latter consists of dodecyl trimethylammonium
decyl sulfate $((Dod(CH_3)_3N^+)(DecSO_4^-))$ that is selectively
permeable to the $DecSO_4^-$ ions (14). We use $NaDecSO_4$ instead
of $NaDodSO_4$ because C_f of the latter is too low to insure
accurate measurements.
    The binding of $NaDecSO_4$ to the three polypeptides is
highly cooperative for $(Lys)_n$ and, to a lesser extent, for the
two polyornithines. The latter have quite similar profiles,
even though the L-polypeptide is helical and the D,L-poly-
peptide is unordered. The apparent binding constant increases
rather than decreases with increasing degree of binding (Fig.
6) (14). This is exactly the opposite of hydrogen ion titration of poly(L-glutamic acid) (15) and poly(L-lysine) (16),
for which the apparent association constant decreases with

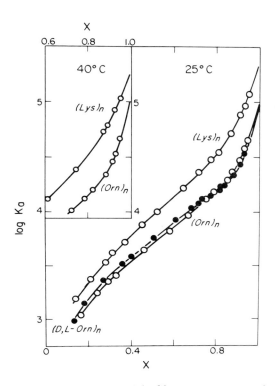

FIGURE 6.   The apparent binding constant at various
degrees of binding at 25°C and 40°C (inset).

increasing degree of protonation. Unlike the helix-coil transition of $(Glu)_n$ and $(Lys)_n$, which is also highly cooperative,
the hydrophobic interaction among the bound surfactant molecules enhances the successive binding of $DecSO_4^-$ ions onto the
cationic polypeptides. This hydrophobic interaction also
overshadows the coulombic repulsion among the charged side
groups of the polypeptide chain regardless of any conformational change that may have occurred (cf. $(Orn)_n$ versus $(D,L-Orn)_n$).
      By extrapolating x to unity (Fig. 6), the cooperative
binding constant of NaDecSO4 to the three polypeptides is of
the order of 10^5. The standard free energy change, ΔG^o, at
25°C was found to be -6.9 kcal/mol (residue) for $(Orn)_n$ and
$(D,L-Orn)_n$ and -7.4 kcal/mol (residue) for $(Lys)_n$. The
binding constants of each polypeptide at 40°C are not so different from those at 25°C. Thus, the interaction between the
surfactant and polypeptide is largely due to entropy changes.

D.   Theoretical Analysis

The Zimm-Bragg theory for helix-coil transition (17) can be adapted to analyze the cooperative binding isotherm, provided that the thermodynamic contribution of any conformational change accompanying the binding process is comparatively small.   Let the digital 0 be a vacant site and 1 a surfactant-bound site along the side groups of a polypeptide chain.   We can then write a hypothetical sequence as

000111000011.....

Zimm and Bragg define two parameters, an equilibrium constant s and initiation factor σ.   By denoting the concentration in brackets, we have

$$\sigma s = [01]/[00] \tag{2}$$

and

$$s = [11]/[10] \tag{3}$$

In our case (cf. Eq. (1)),

$$\sigma s = KC_f \tag{4}$$

Here K is the equilibrium binding constant for a surfactant molecule onto a site having two unoccupied nearest neighbors. It also depends on the electrical potential, ψ_o, on the surface of the polypeptide molecule; K is assumed to be constant unless the size of successive vacant sites becomes extremely small.   Thus, K in Eq. (4) is virtually independent of the degree of binding, x, up to a certain limit.

According to the Zimm-Bragg theory,

$$x = d\ln\lambda_o/d\ln s \tag{5}$$

with

$$\lambda_o = \{1 + s + [(1 - s)^2 + 4\sigma s]^{\frac{1}{2}}\}/2 \tag{6}$$

It follows that

$$x = 1/2 + (s - 1)/2[(1 - s)^2 + 4\sigma s]^{\frac{1}{2}} \tag{7}$$

Because s at x = 1/2 equals one, we have

$$x = C_f/(C_f)_{x=\frac{1}{2}} \tag{8}$$

Differentiating x in Eq. (7) with respect to C_f leads to

$$(dx/d\ln C_f)_{x=\frac{1}{2}} = 1/4\sigma^{\frac{1}{2}} \qquad (9)$$

We found $\sigma^{-1} = 77$, 161 and $\geq 10^4$ for $(Orn)_n$, $(D,L-Orn)_n$ and $(Lys)_n$, respectively. The agreement between the theory and experiments is good to about x = 0.7 (Fig. 7) (14). As x approaches one, the assumption of a constant ψ_o is no longer valid, thus accounting for the observed deviations in Fig. 7.

The average number of the cluster size of bound surfactant molecules, $\overline{m}$, can be shown as

$$\overline{m} = 2x[(1 - \sigma)/\sigma]/\{[4x(1 - x)(1 - \sigma)/\sigma + 1]^{\frac{1}{2}} - 1\} \qquad (10)$$

The $\overline{m}$ value increases with x; it is much larger for the complex of $DecSO_4^-$ with $(Lys)_n$ than with the two polyornithines. For instance, m is about 10, 14 and 101 at x = 0.5 for $(Orn)_n$, $(D,L-Orn)_n$ and $(Lys)_n$, respectively (14).

For the β-form of the $(Lys)_n$ complexes, the side group and surfactant anion form an ion pair and the alkyl chains of the bound surfactants are laterally associated. Another layer of

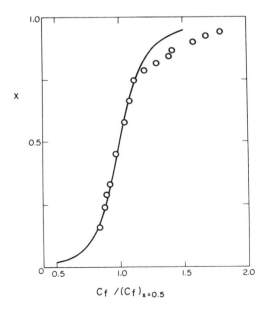

FIGURE 7. Comparison of observed and calculated binding isotherm of poly(L-ornithine) with $NaDecSO_4$ at 25°C. Points, experimental; solid line, calculated with $\sigma^{-1} = 77$.

surfactant anions having the $-SO_4^-$ heads exposed to the aqueous medium may form a double layer with the bound surfactants. Alternately, two stacked β-sheets may have a bilayer of the bound surfactants sandwiched between two polypeptide chains. In either case, the bound surfactants resemble the double layer of lipids in a membrane. The complex formation between $(Orn)_n$ and surfactant anions is also initiated through coulombic attraction, followed by a cooperative binding of successive surfactant molecules. These clusters provide a proteinaceous environment that stabilizes the helical turns. The "necklace" model has been proposed by Shirahama and Takagi (18, 19). Additional surfactant molecules can form micelle-like structures on the surface of the helix such that the non-polar alkyl chains will be shielded from the aqueous medium.

III. NATURALLY-OCCURRING OLIGOPEPTIDES, POLYPEPTIDES AND PROTEINS[3]

The real protein molecules, be they globular or fibrous, are unlike homopolypeptides. The side groups having positive and negative charges are distributed on the surface of a protein molecule in a "random" fashion. In many cases these charges may be too far apart to have a cooperative binding of the ionic surfactants. The so-called statistical binding of the first 10 to 12 surfactant anions by bovine serum albumin is just one such example (3). We also do not know how surfactants will bind to nonionic polar side groups, although such association has been reported for methyl cellulose and polyvinyl alcohol with $NaDodSO_4$ (20) and for polyvinylpyrrolidone with sodium alkyl sulfates (21).

Recently, it has become popular to predict the structural elements of a protein molecule from its amino acid sequence. For globular proteins the conformational parameters are computed from the probability of an amino acid residue located in a particular conformation according to x-ray diffraction results. Several proposed methods appear to have varying degrees of success (see, for example, Ref. 22). These empirical methods are only applicable to compact, rigid proteins, but not to oligo- and polypeptides that are flexible. We propose a working hypothesis that surfactants, be they ionic or nonionic, can provide a proteinaceous environment which offsets the

---

[3]Sleep peptide was purchased from Calbiochem; ACTH 4-11 and renin substrate were obtained from Pennisula Lab; insulin and glucagon were from Elanco; angiotensin I and II and Leu[15] human gastrin I were from Beckman.

hydrophilic interactions between the peptide backbone and water in aqueous solution. Lowering or raising the pH can further minimize any electrostatic repulsion among charges of the
same sign on the side groups, and between side groups and surfactant ions. Each of the 20 amino acid residues has a structure-forming potential for the helix, β-form, β-turn or unordered form. A sequential oligopeptide possessing such a
potential can adopt a particular conformation under suitable
conditions such as in a surfactant solution. Of course, long-
range interactions among amino acid residues that enhance the
stability of the protein structure is absent in our consideration. We test our idea on several naturally-occurring peptides with reasonable success. The induced conformation of
these peptides in surfactant solutions appears to be related
to their amino acid sequence. We use the Chou-Fasman method
(23, 24) for its simplicity, although other sequence-predictive
methods can equally be employed.

A. Sleep peptide

This nonapeptide has three glycine residues. If it were
a segment of a globular protein, the sequence-predictive method would indicate no ordered structure. Addition of NaDodSO$_4$
does not promote any helix or β-form (Fig. 8). Neither does
addition of DodNH$_3$Cl, even though Asp 5 and Glu 9 can attract
the surfactant cations. Thus, the proteinaceous environment
provided by the surfactants does not induce any ordered structure in this case. L. Peller (private communication) suggests
an alternate explanation that the surfactant micelles may not
cluster onto the peptide chain and therefore have no effect on
the peptide conformation.

In the same category is an octapeptide fragment of
adrenocorticotrophic hormone, ACTH 4-11, whose CD spectrum
indicates the absence of any helix or β-form.

B. Insulin B23-29

Insulin in ditute solution can dimerize through the formation of an antiparallel β-form between two B chains at the
C-terminal residues 24 and 29; Phe 24 of one molecule forms
two hydrogen bonds with Tyr 26 of the other (25). Thus, the
heptapeptide of insulin B23-29 (prepared by tryptic digestion)
has the β-forming potential, but in dilute solution (0.04 mg/
ml) it lacks any ordered conformation (Fig. 9). The effect of
NaDodSO$_4$ on this peptide depends on the molar ratio, R, of
surfactant to peptide (mean residue). At R = 5, a CD minimum
at 211 nm and a maximum at 195 nm suggest the presence of some

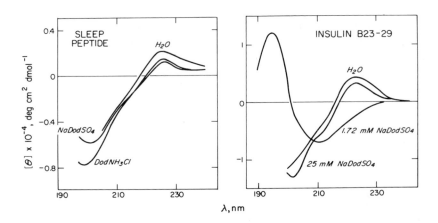

FIGURE 8 (left). The CD of sleep peptide. Concentrations: 0.89 mM peptide (mean residue); 25 mM NaDodSO$_4$; 20 mM DodNH$_3$Cl. Sequence: Trp-Ala-Gly-Gly-Asp-Ala-Ser-Gly-Glu.

FIGURE 9 (right). The CD of insulin B23-29. Concentration: 0.32 mM peptide (mean residue). Sequence: Gly-Phe-Phe-Tyr-Thr-Pro-Lys.

β-form. The peptide-NaDodSO$_4$ solution appears slightly turbid probably because of aggregation. However, the solution is clear when excess NaDodSO$_4$ is used, for instance at R = 150, but the CD spectrum reverses to that of the unordered form. Except Lys 7, this peptide has side groups that are largely hydrophobic. Conceivably, the oligopeptide in its unordered monomer state is stable inside the surfactant micelles.

Similarly, angiotensin I and II also adopt the β-form at R < 10, but they retain the CD profile with reduced magnitude at R = 100.

C. Glucagon 19-29

Glucagon is mainly helical in crystalline state (26). Residues 10 to 25 are in the α-helical form, which extends at both ends to residues 6 and 29 in a less regular helical structure; thus, the total helical content amounts to about 83%. But the CD spectrum of unaggregated glucagon in dilute acidic or alkaline solution (1 mg/ml or less) shows a helical content of less than 20% (27, 28). The glucagon conformation is also concentration dependent (29). Tryptic digestion breaks this polypeptide into three fragments, residues 1-12, 13-17 and 19-29, all of which appear to be unordered, as judged from their

CD spectra. But addition of NaDodSO$_4$ to glucagon 19-29 at pH
2 raises the helical content to about one half of the molecule
(Fig. 10), even though this undecapeptide has no Lys, His and
Arg residues. Glucagon 1-12 and 13-17 were found to have
little, if any, ordered structure in NaDodSO$_4$ solution. The
cationic surfactant DodNH$_3$Cl is almost as effective as NaDodSO$_4$
on glucagon 19-29. Similar results were observed for the
whole glucagon molecule in surfactant solutions. Unlike the
β-form of the insulin B23-29, the helical conformation of glucagon and its fragment 19-29 are independent of the surfactant
concentration, provided that the molar ratio, R, is greater
than one.

In contrast to the x-ray results, Chou and Fasman had proposed the conformational flexibility of glucagon and predicted

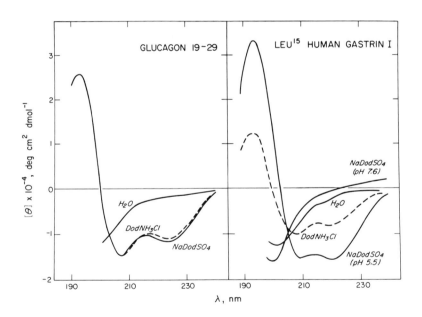

FIGURE 10 (left). The CD of glucagon 19-29 at pH 2 and
ionic strength 0.1. Concentrations: 0.36 mM peptide (mean
residue); 25 mM NaDodSO$_4$; 20 mM DodNH$_3$Cl. Sequence: Ala-Gln-
Asp-Phe-Val-Gln-Trp-Leu-Met-Asn-Thr.
FIGURE 11 (right). The CD of Leu[15] human gastrin I. Concentrations: 0.31 mM peptide (mean residue); 25 mM NaDodSO$_4$;
20 mM DodNH$_3$Cl. Sequence: pGlu-Gly-Pro-Trp-Leu-Glu-Glu-Glu-
Glu-Glu-Ala-Tyr-Gly-Trp-Leu-Asp-Phe-NH$_2$.

that residues 5-10 formed a β-sheet and residues 19-27 could
be either helical or in the β-conformation (30). The very
nature of any empirical method, be it the CD analysis or
sequence-predictive model, should be used with reservation.

D. Leu^{15} Human Gastrin I

This heptadecapeptide has one special feature: residues
6-10 consist of five Glu's in addition to Glu 1 and Asp 16,
which greatly contribute to the coulombic repulsion among the
negative charges of the side groups. Thus, addition of
DodNH$_3$Cl does induce the helical conformation (Fig. 11). But
the unordered form of this peptide at pH 5 or lower, where
Glu's are protonated, can also be converted into the helical
form in the presence of NaDodSO$_4$, which has no effect at neutral pH. The anionic surfactant appears to be more effective
than the cationic one in promoting the helical conformation.

E. Renin Substrate

The effect of NaDodSO$_4$ on this tetradecapeptide is difficult to determine. At low surfactant concentration (R = 4)
the CD spectrum with a minimum between 210 and 220 nm and a
maximum near 200 nm suggests the presence of a β-form (Fig.
12). The peptide-surfactant solution is slightly turbid probably because of aggregation. In concentrated NaDodSO$_4$ solution
(e.g. at R = 60) the CD spectrum shows three bands with a double minimum at 220 and 210 nm and maximum near 200 nm. This
seems to resemble the formation of a helix, but the magnitudes
of these bands are too small to warrant such a conclusion
(even after considering the chain-length dependence of CD of a
helix; see Ref. 31). The Chou-Fasman predictive method is
also equivocal in this case; the peptide has both the helix-
and β-forming potential. We should again emphasize the problems inherent in all these empirical proposals, including our
working hypothesis.

F. β-Endorphin and β-Lipotropin

Camel β$_c$-endorphin has 31 amino acid residues; its sequence
is identical with the C-terminal residues of sheep β$_s$-lipo-
tropin (91 residues) (32, 33). Judged from the CD spectra
(Fig. 13), both β$_c$-endorphin and β$_s$-lipotropin have little, if
any, secondary structure in aqueous solution (34). Hydrodynamic measurements also suggest that the molecules are neither
compact nor rigid. But addition of NaDodSO$_4$ of the order of

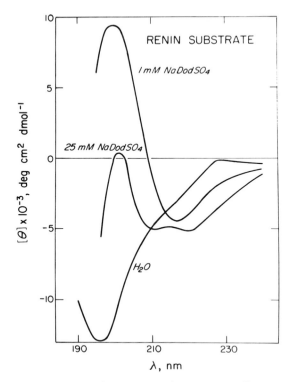

FIGURE 12.   The CD of renin substrate.   Concentration:
0.39 mM peptide (mean residue).   Sequence: Asp-Arg-Val-Tyr-
Ile-His-Pro-Phe-His-Leu-Leu-Val-Thr-Ser.

1 mM induces the helical conformation to the extent as much as
one half of each polypeptide molecule.   Lowering the temperature to 4°C precipitated most of the excess NaDodSO₄, but the
supernatant solution at 25°C gave essentially the same spectra
as shown in Fig. 13.   In contrast, methanol, which also pro-
motes the helical structure, must be raised to more than 50%
(v/v) in order to produce the same effect as NaDodSO₄.
     The helix-inducing effect of NaDodSO₄ on β-endorphin may
help us to understand the opiate-like activity of this hormone
polypeptide.   Celebroside sulfate, which like NaDodSO₄ is
amphipathic, is believed to be an active component of the opiate receptor in brain tissue. Probably the binding site of β-
endorphin to celebroside sulfate is stabilized by the induced
helical conformation of this polypeptide.

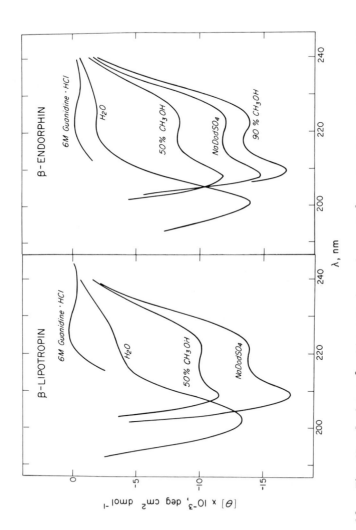

FIGURE 13. The CD of sheep β_S-lipotropin and camel β_C-endorphin. Concentrations: 0.5 mM peptide (mean residue); 3 mM NaDodSO₄.

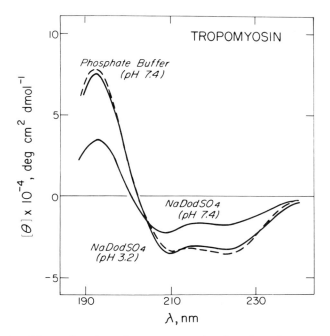

FIGURE 14.   The CD of tropomyosin in phosphate and glycine buffers containing 0.08 M NaCl (ionic strength 0.1).   Concentrations: 0.82 mM protein (mean residue); 20 mM NaDodSO₄.

G.   Tropomyosin

Tropomyosin illustrates the diverse effects of surfactants on proteins. Lest we leave the impression that these denaturants either enhance the ordered structure or have no effect at all, tropomyosin consisting of two-stranded helical rods dissociates into two subunits and actually reduces its helicity in neutral NaDodSO₄ solution. Based on CD spectrum (Fig. 14), tropomyosin at neutral pH is more than 90% helical, addition of NaDodSO₄ immediately reduces the CD magnitude to about one half, but the process can virtually be reversed by lowering the pH to 4 or less (K. Ikeda, private communication). That protonation of Glu and Asp residues could stabilize the helical conformation is not too surprising. The sequence of one of the two chains shows many glutamyl and aspartyl residues in juxstapositions, for instance, –Glu–Glu–Glu–, –Glu–Asp–Glu–, –Asp–Glu–Glu–, –Glu–Glu–Ala–Glu–, –Glu–Asp–Ala–Asp–, –Glu–Ser–Asp–, and –Glu–Leu–Glu–Glu–Glu– (35).

In the same category of tropomyosin is the myosin rod,

which is also more than 90% helical. Again its helical content is decreased to about one half in neutral NaDodSO$_4$ solution. Similar reduction in helicity has also been reported
for myoglobin, which is 79% helical in the absence of surfactant (36).

 We are as yet unable to observe the β-turn in surfactant
solution. We have not quantitatively analyzed our results and
estimated the fractions of helix, β-form and, β-turn (31;
Chang, Wu and Yang, to be published) mainly because the reference spectra for these conformations in surfactant solution
may differ in magnitude from those without surfactant and the
end effects could be serious for short peptides. For those
compounds containing aromatic groups the CD contributions of
non-peptide chromophores may no longer be insignificant. So
far we also confine our discussion to peptides and proteins
with no cystine residues, the disulfide bonds could enhance or
restrict the ordered conformation. Whether our working hypothesis is applicable to globular proteins in general is a
subject for future investigation.

                          IV.   SUMMARY

 Based on circular dichroic spectra, poly(L-lysine) and its
homologs with side group R = -(CH$_2$)$_m$NH$_3^+$ (m = 3 to 7) at neutral pH become helical upon adding sodium dodecyl sulfate
(NaDodSO$_4$), except that (Lys)$_n$ adopts the β-form. Potentiometric titration of (Orn)$_n$ (m = 3) in neutral sodium decyl
sulfate solution indicates a cooperative binding isotherm.
Analysis by the Zimm-Bragg theory shows surfactant clusters on
the polypeptide chain. Several naturally-occurring oligo- and
polypeptides studied are mostly unordered in aqueous solution.
But surfactants may provide a proteinaceous environment in
which partial helix or β-form can be induced. Such ordered
structure or the lack of it depends on the amino acid sequence.
Thus, glucagon 19-29, Leu[15] gastrin I (at low pH), β-endorphin
and β-lipotropin are partially helical, insulin 23-29 adopts
the β-form, and sleep peptide remains unchanged in NaDodSO$_4$
solution. The conformation of renin substrate is equivocal,
as does its sequence-predictive structure. Excess surfactant
can dissociate the β-form. The high helicity of two-stranded
tropomyosin is reduced in NaDodSO$_4$ solution, but it can be
mostly restored by lowering the pH to minimize the coulombic
repulsion of the Glu⁻ and Asp⁻ residues and also between them
and surfactant anions.

ACKNOWLEDGMENTS

We thank Drs. G. C. Chen, K. Ikeda, S. Kubota, K. Shirahama and Y. W. Tseng for their valuable help and discussion.
Thanks are also due Ms. M. H. Stull and Mrs. S. Chao for their
technical assistance.
Figures 2 to 7 are reproduced with permission from Biopolymers through the courtesy of John Wiley & Sons and figure
13 is from the Proceedings of the National Academy of Sciences.

REFERENCES

1. Putnam, F. W., Adv. Protein Chem. 4:79 (1948).
2. Tanford, C., Adv. Protein Chem. 23:121 (1968).
3. Yang, J. T., and Foster, J. F., J. Amer. Chem. Soc. 75:
    5560 (1953).
4. Chaudhuri, S. R., and Yang, J. T., Biopolymers 7:1379
    (1968).
5. Sarkar, P. K., and Doty, P., Proc. Natl. Acad. Sci. USA
    55:981 (1966).
6. Grourke, M. J., and Gibbs, J. H., Biopolymers 5:586
    (1967).
7. Satake, I., and Yang, J. T., Biochem. Biophys. Res.
    Commun. 54:930 (1973).
8. Tseng, Y. W., and Yang, J. T., Biopolymers 16:921 (1977).
9. McCord, R. W., Blakeney, E. W., Jr., and Mattice, W. L.,
    Biopolymers 16:1319 (1977).
10. Satake, I., and Yang, J. T., Biopolymers 14:1841 (1975).
11. Green, N., and Fasman, G. D., Biochemistry 8:4108 (1969).
12. Feldshtein, M. M., Zezin, A. B., and Gragerova, J. J.,
    Biokhimia 37:305 (1972).
13. Zezin, A. B., Felsahtein, M. M., Merzlov, V. P., and
    Maletina, J. J., Molekul. Biol. 7:174 (1973).
14. Satake, I., and Yang, J. T., Biopolymers 15:2263 (1976).
15. Nagasawa, M., and Holtzer, A., J. Amer. Chem. Soc. 86:538
16. Grourke, M. J., and Gibbs, J. H., Biopolymers 10:795
    (1971).
17. Zimm, B. H., and Bragg, J. K., J. Chem. Phys. 31:526
    (1959).
18. Shirahama, K., Tsujii, K., and Takagi, T., J. Biochem.
    (Japan) 75:309 (1974).
19. Takagi, T., Shirahama, K., Tsujii, K., and Kubo, K.,
    Tampakushitsu, Kakusan, Koso [Proteins, Nucleic Acids,
    Enzymes] 21:811 (1976)
20. Lewis, K. E., and Robinson, C. P., J. Colloid Interfac.
    Sci. 32:539 (1970).

21. Arai, H., Murata, M., and Shinoda, K., J. Colloid Interfac.
     Sci. 37:223 (1971).
22. Shultz, G. E., Barry, C. D., Chou, P. Y., Finkelstein, A.
     V., Lim, V. I., Ptitsyn, O. B., Kabat, E. A., Wu, T. T.,
     Levitt, M., Robson, B., and Nagano, K., Nature 250:140
     (1974).
23. Chou, P. Y., and Fasman, G. D., Biochemistry 13:222
     (1974).
24. Fasman, G. D., Chou, P. Y., and Adler, A., Biophys. J.
     16:1201 (1976).
25. Blundell, T., Dodson, G., Hodgkin, D., and Mercola, D.,
     Adv. Protein Chem. 26:279 (1972).
26. Sasaki, K., Dockerill, S., Adamiak, D. A., Tickle, I. J.,
     and Blundell, T., Nature 257:751 (1975).
27. Srere, P. A., and Brooks, G. C., Arch. Biochem. Biophys.
     129:708 (1969).
28. Gratzer, W. B., and Beaven, G. H., J. Biol. Chem. 244:
     6657 (1969).
29. Beaven, G. H., Gratzer, W. B., and Davies, H. G., Eur. J.
     Biochem. 11:37 (1969).
30. Chou, P. Y., and Fasman, G. D., Biochemistry 14:2536
     (1975).
31. Chen, Y. H., Yang, J. T., and Chau, K. H., Biochemistry
     13:3350 (1974).
32. Li, C. H., and Chung, D., Proc. Natl. Acad. Sci. USA 73:
     1145 (1976).
33. Li, C. H., Barnafi, L., Chrétien, M., and Chung, D.,
     Nature 208:1093 (1965).
34. Yang, J. T., Bewley, T. A., Chen, G. C., and Li, C. H.,
     Proc. Natl. Acad. Sci. USA 74:3235 (1977)
35. Stone, D., Sodek, J., Johnson, P., and Smillie, L. B.,
     Proc. Fed. Eur. Biochem. Soc. 31:125 (1975).
36. Mettice, W. L., Riser, J. M., and Clark, D. S., Bio-
     chemistry 245:5161 (1976).

AN APPROXIMATE MODEL FOR PROTEIN FOLDING;
EXPERIMENTAL AND THEORETICAL ASPECTS[1]

Harold A. Scheraga

Department of Chemistry
Cornell University
Ithaca, New York

and

Biophysics Department
Weizmann Institute of Science
Rehovot , Israel

I. INTRODUCTION

Ever since Anfinsen showed that the amino acid sequence of
a protein determines its three-dimensional structure (1), protein chemists have endeavored to discover how proteins fold
into their native conformations (2-17). Both experimental and
theoretical methods have been used to gain information about
the folding pathway and about the interactions that stabilize
the native conformation. The experimental procedures have
been applied primarily to bovine pancreatic trypsin inhibitor
(BPTI) (4), bovine pancreatic ribonuclease (RNase) (6), and
hen egg-white lysozyme (11), whereas the theoretical methods

[1]This work was supported by research grants from the National
Science Foundation (PCM75-08691) and from the National
Institute of General Medical Sciences (GM-14312) and the
National Institute of Arthritis and Metabolic Diseases (AM-
13743) of the National Institutes of Health, U.S. Public
Health Service. Requests for reprints should be addressed to
the author at Cornell University.

(5,7,9,15,16) have usually focused on BPTI as a model because
of its small size. In the experimental approach, methods that
give structural information about limited portions of the
molecule in solution have been developed and, in the theoretical approach, recent trends in the use of empirical (rather
than quantum mechanical) methods to solve chemical problems in
large molecular systems have been exploited. The theoretical
methods, which usually involve minimization of an empirical
potential energy function, are based on the observation that,
at least for ribonuclease, there appear to be no insurmountable energy barriers that prevent the attainment of the global
minimum in conformational energy space (18). We shall consider here some of the experimental and theoretical aspects of
the protein folding problem, which lead to a model for the
process; for more extensive treatments of these subjects, reference should be made to two recent reviews (10,17). As
illustrations of these approaches, we shall consider some
experimental work on RNase and theoretical work on BPTI,
carried out in our laboratory.

II.   THEORETICAL APPROACH

The theoretical basis for the conformational analysis of
polypeptides, using empirical potential energy functions composed of terms which describe nonbonded, electrostatic,
hydrogen- and hydrophobic-bonding interactions and torsional
potentials (2,3,10,17,19-21), and a summary of the methodology
(17,22,23), have been discussed elsewhere. The major difficulty that remains to be overcome, in order to compute the most
stable conformation of a polypeptide or protein, is that which
arises from the existence of many minima in the multi-dimen-
sional conformational energy space (the multiple-minimum problem). In an earlier paper (22), we showed how this problem
has been surmounted for small open-chain oligopeptides, cyclic
peptides, and regular-repeating-sequence polypeptide analogs
of fibrous proteins (specifically, collagen). We shall con-
centrate here, therefore, on current efforts (17) to solve
this problem for globular proteins.
     Two types of computer programs are available for the cal-
culation of the conformational energy of a protein. These are
ECEPP (Empirical Conformational Energy Program for Peptides)
(24) and a united-atom version called UNICEPP (United-atom
Conformational Energy Program for Peptides) (25). A united-
residue program is also available to speed up the computation
of interaction energies between remote residues (26). If we
could direct the computation to the potential energy well of
the native protein, ECEPP or UNICEPP (and associated minimi-

zation procedures) could be used to reach the minimum.   Thus,
the multiple-minimum problem would have been circumvented.
The minimization procedures (within a given potential energy
well) consume large amounts of computer time when used with
ECEPP (and somewhat less so with UNICEPP) but, nevertheless,
the minimum can be attained (23).   Thus, the problem resolves
itself into being able to attain the correct potential energy
well, i.e. to avoid an unnecessary and unproductive search of
most of conformational space and, instead, direct the search
toward the region of space where the native structure lies.
Our approach has been, first, to try to use simplified approximate procedures to obtain an approximate structure (i.e. one
that would correspond to the potential energy well of the
native globular protein); then the approximate procedures
would be abandoned, -- and the "exact" one (based on ECEPP or
UNICEPP) would be introduced; i.e. complete energy minimization (taking into account the interactions between all atoms
in the molecule, and the effect of water thereon) can be
carried out to refine the approximate structure.

The approximate procedures, discussed in more detail elsewhere (17), are based on the observation that short-range
(intra-residue) interactions dominate in determining the conformational preferences of a protein. Thus, a variety of
(approximate) short-range interaction models, in which medium-
and long-range interactions are neglected, have been used to
predict the location of α-helical, extended, and β bend (hair-
pin turn) structures in protein molecules (17).   Very recently,
conformational energy calculations, with the effects of water
omitted (27) and included (28), have provided additional evidence that the tendency toward formation of β bends is dominated by short-range interactions.   In section III, we discuss,
as an illustrative example, an experimental approach for the
determination of the tendency of various segments of a protein
molecule to adopt the α-helical conformation.

In higher approximations, the structures deduced on the
basis of short-range interaction models are then modified by
the introduction of medium- and long-range interactions, using
simplified approximate representations of the polypeptide
chain (see section IV).   The use of these higher approximations (and ultimately an "exact" one) should compensate for
the inadequacies of the short-range interaction models.

III. EXPERIMENTAL DETERMINATION OF
HELIX-FORMING TENDENCY

Within the framework of the assumption that short-range
interactions dominate, the tendency toward helix formation can
be expressed in terms of the parameters s and σ of the Zimm-
Bragg theory for the helix-coil transition in homopolymers of
amino acids (29). These parameters can be evaluated from
experimental studies of helix-coil transitions in such homopolymer systems. However, for technical reasons discussed
elsewhere (30), homopolymers of most of the naturally occurring amino acids cannot be employed, and use is made, instead,
of random copolymers in which the desired amino acid, the
"guest", is incorporated at random into a nonionic water-
soluble homopolyamino acid, the "host" (host-guest technique).
The helix-coil stability constants (s and σ) of the guest
residue can then be determined from its influence on the melting behavior of the host homopolyamino acid, and a knowledge
of s and σ of the host homopolymer in water.

In this manner, the values of s and σ for many of the naturally occurring amino acids have been determined (30), and
experiments on the remaining ones are in progress. These
values constitute a quantitative scale of helix-forming tendency of the various types of amino acids (in water), and are
utilized, in a one-dimensional Ising model theory, to compute
the locations of α-helical segments in proteins, -- within the
framework of a short-range interaction model (30). For
several of the amino acids (viz., Gly, Ala and Val), the
values of s and σ were computed theoretically from potential
functions (31,32), with fairly good agreement with the experimental results (30). In addition, the various energetic factors that contribute to the conformational preferences of each
residue have been identified (33). Experiments are also in
progress to determine s and σ in non-aqueous solvents (34).

Conceivably, it may be possible to carry out similar
experiments with host-guest random copolymers to determine the
tendency of each of the naturally occurring amino acid residues to form extended structures. It may be possible to use
poly (L-tyrosine) as a host, since it appears to undergo a
transition between an extended structure and a disordered
state (35-37), and then examine the effect of various guest
residues on the order ↔ disorder transition of the poly(L-
tyrosine) host.

Similar data on the helix- and extended-structure-forming
tendencies (and also β bend-forming tendencies) of amino acids
can be obtained by observations of the frequencies of occurrence of the amino acid residues in these conformations in
proteins. Correlations between these tendencies, determined

from synthetic polypeptides on the one hand, and from protein
structures, on the other, provide a confirmation of the validity of the concept of the dominance of short-range interactions and, where this concept breaks down, provide information about the role of medium- and long-range interactions in
their influence on the conformation (30).

When considering the frequencies of occurrence of the
amino acid residues in various conformational states, it is
possible to include other states (discussed extensively in
ref. 17) besides those corresponding to α-helices, extended
structures, and β bends. For this purpose, several multi-
state models have been considered (5,38-40). Unfortunately,
the frequencies of occurrence of these additional states are
lower than those of the others, and these frequencies therefore have lower statistical significance. Further, it is
important to realize that there are two basic shortcomings in
short-range interaction models, even multi-state ones, because
of the omission of medium- and long-range interactions. First,
it is not possible to predict the conformational state with
100% accuracy. Second, even if 100% accuracy could be achieved, it is not possible to predict the precise conformation,
i.e. the actual values of the backbone dihedral angles, within
each conformational state with sufficient accuracy to obtain
a structure that resembles the native one (see, especially,
Fig. 11 of ref. 17 and the associated discussion). However,
as indicated in section II, the introduction of medium- and
long-range interactions should compensate for these two inadequacies of the short-range interaction models.

IV.   INCORPORATION OF LONG-RANGE INTERACTIONS

Using the results of short-range interaction models to
place an initial limitation on the conformational space to be
searched, the inadequacies of these models can be rectified by
subsequent inclusion of medium- and long-range interactions.
A variety of procedures (making use of simplified approximate
representations of the polypeptide chain) have been suggested
for this purpose (5,7,8,9,13-16,41). Essentially, they consist of energy minimization within limited portions of the
chain, Monte Carlo procedures, and molecular dynamics calculations. By starting with conformations of BPTI resulting from
a "perfect" short-range interaction model (solely for the purpose of testing the procedures for introducing medium- and
long-range interactions), it has been possible to alter these
conformations so that they avoid steric overlaps and acquire
the globularity of the native molecule (5,9); however, while
several features of the native structure are obtained, the

agreement is not yet perfect. If an actual, rather than the
"perfect", short-range interaction model is used with a Monte
Carlo procedure to introduce medium- and long-range interactions, the results are not as good (9). Some of the ideas
currently being tested to improve the accuracy of these approximate procedures are discussed in section V.

V.   POSSIBLE IMPROVEMENTS IN FOLDING ALGORITHMS

As a basis for discussion, let us consider an approximate
procedure in which a multi-state short-range interaction model
is used as a constraint in a Monte Carlo calculation (9).
This constraint places a limitation on the conformation space
to be searched, and the Monte Carlo procedure permits the
inclusion of medium- and long-range interactions. Further,
the Monte Carlo procedure enables the conformation of the molecule to pass over barriers between local minima (41). Several possible improvements can be introduced into this approximate procedure.
    First of all, it is possible to improve the short-range
interaction model, and hence its accuracy, by use of higher-
order probabilities, inclusion of medium-range electrostatic
interactions, and by modifying the treatment of the statistical weights that are the elements of the statistical-weight
matrices (42).
    Secondly, with the availability of more protein structural
data, it is possible to improve the values of the statistical
weights of the short-range interaction models, and also those
of the parameters (43) that represent the contact free
energies in the medium- and long-range interaction models
(44,45).
    Thirdly, various characteristic aspects of medium- and
long-range interactions can be included specifically in a
folding algorithm. Examples of these are the following.
    Nucleation sites for folding can be predicted on the basis
of a compensation of the entropy loss upon formation of a
pocket by the free energy of formation of the hydrophobic
bonds within the pocket (46). The amino acid sequence of the
protein is searched for pockets of nonpolar residues (47)
whose (negative) free energy of interaction compensates for
the increase in free energy that is required to bring them into contact to form hydrophobic pockets; the pocket of lowest
free energy is designated as the predicted initial nucleation
site (46). A constraint can be introduced to preserve this
nucleation site during the conformational changes of the Monte
Carlo procedure.

A similar constraint can be included, after helical and extended segments are introduced initially with the short-range interaction model, by incorporating a free energy of formation of a hydrogen bond in these ordered regions. This will reduce the likelihood of their melting once they are formed and make it more difficult to break a helical sequence in its interior than at its ends.

The inclusion of information as to the proper location of disulfide bridges will constrain them to form properly. Also, use can be made of experimental information on the distances between groups (other than disulfides) that are not covalently bonded, and of other conformational information (33,48); e.g. NMR data were used by Dygert et al. (49) in calculations on gramicidin S. Other constraints on folding can be introduced by using information (50-54) as to (a) which residues prefer to lie inside the molecule and which prefer to lie outside, or how far from the center of mass they prefer to be (54), and (b) which structures (i.e. helical, extended, or nonregular) prefer to lie inside the molecule and which prefer to lie out-side (53), and by using information about stabilities and locations of pleated-sheet structures (55-67) and separately-folding domains (68-71). Calculations on the interactions between regular and nonregular segments in simplified model structures provide information that may also serve to help direct the folding properly (72). Similarly, information on several proteins with homologous sequences (and, therefore, presumably similar three-dimensional structures) provides additional constraints (23,73,74). Finally, a recent applica-tion of differential geometry to polymer conformation includes information on short- and medium-range interactions in terms of the curvature and torsion at each residue (75); such a rep-resentation of the chain, which provides an objective method for comparing protein conformations (rather than the misleading reliance on the root-mean-square deviation), may prove to be of use in a folding algorithm.

In summary, this approach (with improvements such as those suggested here) can be characterized as follows. Use is made of a statistical mechanical short-range one-dimensional (Ising) model treatment to locate helical, extended, and other con-formations. The shortcomings of the one-dimensional model are then overcome by introducing medium- and long-range inter-actions (including hydration, and various constraints) by a Monte Carlo procedure (9). The following picture of folding has emerged thus far from tests of this procedure on BPTI: (a) ordered structures tend to form in particular regions of the chain because of short-range interactions; (b) medium-range interactions (including hydration) tend to bring nearby ordered and unordered regions together to form "contact" regions [nucleation sites also form as a result of steps (a)

and (b) (46)]; (c) long-range interactions (including hydration) bring the "contact regions together to form the globular structure of the native protein. This method is then a
Monte Carlo search that is constrained to "probable" regions
of conformation space by the statistical mechanical requirements to satisfy the Ising model, and by the other restric-
tions mentioned above. Experiments described in section VI,
in which the pathways of protein folding are studied, indicate
that this is a reasonable approximate model for protein
folding.

As an aside, it is worth noting that, on the basis of the
procedure just described, and the picture that has emerged
therefrom [i.e. points (a)-(c) above], it is possible to postulate pathways of protein folding (71) from a knowledge of
the triangular map of the protein structure (based on the
X-ray analysis), an example of which is shown in Fig. 2 of
ref. 71, even without doing any computations. The criterion
(that short-range interactions take precedence over medium-
range interactions, and medium-range interactions take precedence over long-range interactions) is satisfied by (postulated) pathways in which contacts near the diagonal of the
triangular (contact) map are formed first, and contact regions
are then formed in the order of increasing distance from the
diagonal. Thus, it has been possible, relatively easily, to
postulate folding pathways for rubredoxin, ferricytochrome c,
and lysozyme (71). Interestingly, in the case of lysozyme,
the postulated mechanism of protein folding leads naturally to
the formation of the two separate wings of the molecule, and
then to their subsequent association (71).

As indicated in section II, once the approximate procedure
(involving the Ising-Monte Carlo approach) leads to the correct potential well (i.e., to a structure that should resemble
the one to be anticipated for the native protein), then a complete energy minimization (taking into account the interactions between all atoms in the molecule, using ECEPP or
UNICEPP, and the effect of water thereon) can be carried out
to refine the approximate structure. (The criterion that "the
correct potential well has been reached" is that the free
energy can no longer be reduced by further trials in the Monte
Carlo procedure discussed above).

It cannot be emphasized too strongly that complete energy
minimization is intended for use only in the final stages of
a computation of the structure of a protein. It would be
very unproductive, and wasteful of computer time, to introduce
such a procedure at an earlier stage of the computations. The
earlier stages involve all the approximate empirical procedures described in the literature and carried out in numerous
laboratories (17); they also include the various constraints

described above. On the other hand, it must be recognized very
clearly that the approximations used in the early stages probably will not lead, by themselves, to the native structure.
These approximations constitute a presumably-necessary first
step that must be followed by complete energy minimization, --
in the manner used, for example, to reduce the energies of
structures obtained by X-ray diffraction analysis (23).

VI. EXPERIMENTAL STUDIES OF PROTEIN FOLDING

   In parallel with the computational approach described
above, experiments are being carried out to provide additional
evidence for this model of nucleation and folding in protein
molecules. A series of investigations on the folding of BPTI
has been reported by Creighton (4). For illustrative purpose
here, we describe some work from our own laboratory on the
pathway of folding of RNase since the folding of this molecule
has been studied in two types of experiments, *viz.* with the
disulfide bonds broken and intact, respectively. When the disulfide bonds are intact, the unfolding and refolding can be
accomplished reversibly by heating and cooling (76); the
process can be followed by a variety of optical methods, e.g.
ultraviolet-difference spectrophotometry or optical rotatory
dispersion (77). When the disulfide bonds are broken (i.e.
reduced), their re-oxidation accompanies the folding process
(1). In order to detect intermediates along the folding pathway, it is necessary to use methods that provide structural
information about limited portions of the molecule in solution. We shall consider first the thermally-induced conformational changes with the disulfide bonds intact.
   Using either chymotrypsin or trypsin as a proteolytic
probe to hydrolyze the peptide bonds of ribonuclease, as they
became accessible during the thermally-induced unfolding, it
was possible to determine which portions were unfolded at
various temperatures (78-81). As the temperature was raised,
more chain segments became unfolded and hence susceptible to
hydrolysis by these proteolytic enzymes. With this information, and with related nuclear magnetic resonance and ultraviolet difference spectral observations on the folding process,
it was possible to postulate a pathway for the thermally-
induced unfolding of RNase (6). This pathway is illustrated
in Fig. 3 of ref. 6. According to this hypothesis, there are
two initial nucleation sites for folding. These are joined in
subsequent stages by the N- and C- termini, and then the
remaining segments of the molecule fold against this compact
portion. Figure 4 of ref. 6 illustrates the role of nonpolar

residues in the C-terminal portion of the molecule in the
stabilization of the conformation of the C-terminus and of the
chain segments with which it interacts.

Burgess et al. (82) subsequently provided experimental
evidence for the early participation of the C-terminus in the
folding process by showing that the C-terminal valine residue
could not be liberated by carboxypeptidase A during the later
stages of folding, i.e. after the C-terminus had been involved
in stabilizing interactions with other chain segments.

Chen and Lord (83) studied the thermal transition in RNase
with laser Raman spectroscopy, and concluded that "the Raman
data are quantitatively consistent with the six-stage scheme
of unfolding of Burgess and Scheraga (6), except that no
change in the environment of the tyrosines is seen until 45°C".
The sequential character of the unfolding pathway of RNase (6)
is also supported by a flash-photochemical labeling technique
(84) and by electron spin resonance measurements on spin labels
attached to RNase (85).

Studies have also been carried out on the pathway of folding accompanying the re-oxidation of reduced disulfides (18,
86,87). Kinetic studies of the folding, using antibodies that
would bind to specific portions of the chain (when they attained the native conformation), indicated that the C-terminal
portion (i.e. an antigenic site in residues 80-124) folds
first (87). This, together with an interpretation of data on
the formation of disulfide bonds during the folding process
(18,86), has led to a postulated pathway for folding during
oxidation (86) that resembles the pathway proposed for the
folding when the disulfide bonds are intact (6). This suggests
that, even though disulfide bonds stabilize the native conformation, they do not influence the pathway of folding.

The various experiments that have been carried out on protein folding indicate that the pathways of folding are limited
and that the more stable native conformation is accessible
from a number of different regions of conformation space that
contain less stable species. These experiments support the
notion that the folding may involve nucleation at one or more
sites in the chain, accompanied initially by formation of contacts in certain regions, with a subsequent association of contact regions. This view, which is consistent with the three-
step mechanism that emerged from Monte Carlo calculations on
protein folding (9),(see section V), provides the basis for
current efforts to compute the folded (native) structure from
the unfolded one.

Finally, experimental data can provide information , not
only about the pathway of folding, but also about the structure of the native protein in solution (33,48). This information can be used, as indicated in section V, as constraints

in a theoretical folding algorithm. For example, by a series of chemical and physico-chemical experiments, it was possible to pair three of the six tyrosyl residues specifically with three of the eleven carboxyl groups (all aspartyl residues) in RNase (88). Similarly, it has been demonstrated that His-12 and His-119 are near each other (89-91) and close to Lys-41 (92,93), all three constituting part of the active site of RNase.

VII. CONCLUSIONS

A combined experimental and theoretical approach, each complementing the other, is being used to gain an understanding of the factors that determine how a polypeptide chain folds into the conformation of a native globular protein. Experiments carried out on the folding of RNase indicate that there are no insurmountable energy barriers in the attainment of the native structure (18) and lead to a hypothetical pathway for folding (6,86). Both experiment and theory suggest that folding begins by nucleation of a near-native conformation among nearby residues in the sequence. One theoretical method is based on the assumption that the initial nucleation site is a pocket formed by hydrophobic interactions (46). More than one nucleation site can form in different parts of the chain; in subsequent steps, these nucleation sites grow in size and then coalesce into the globular form of the native structure. The pathways from the unfolded to the folded form presumably traverse a limited hypervolume of conformation space. Monte Carlo calculations, with various constraints, simulate the conformational changes that the polypeptide chain undergoes in its passage from the unfolded to the native structure. It remains to be seen whether these constraints, and the details of the Monte Carlo procedure, can be combined properly to produce a computed conformation that resembles the native one, whose structure can then be refined by energy minimization.

REFERENCES

1. Anfinsen, C.B., in *"New Perspectives in Biology"* (M. Sela, ed.), p. 42. Elsevier, Amsterdam, 1964.
2. Ramachandran, G.N., and Sasisekharan, V., *Adv. Protein Chem. 23*:283 (1968).
3. Scheraga, H.A., *Adv. Phys. Org. Chem. 6*:103 (1968).
4. Creighton, T.E., *J. Mol. Biol. 87*:563, 579, 603 (1974); *95*:167 (1975); *96*:767, 777 (1975).
5. Burgess, A.W., and Scheraga, H.A., *Proc. Natl. Acad. Sci., U.S. 72*:1221 (1975).
6. Burgess, A.W., and Scheraga, H.A., *J. Theor. Biol. 53*:403 (1975).

7. Levitt, M., and Warshel, A., *Nature 253*:694 (1975).
8. Ptitsyn, O.B., and Rashin, A.A., *Biophys. Chem. 3*:1 (1975).
9. Tanaka, S., and Scheraga, H.A., *Proc. Natl. Acad. Sci., U.S. 72*:3802 (1975); *74*:1320 (1977).
10. Anfinsen, C.B., and Scheraga, H.A., *Adv. Protein Chem. 29*:205 (1975).
11. Anderson, W.L., and Wetlaufer, D.B., *J. Biol. Chem. 251*:3147 (1976).
12. Taketomi, H., Ueda, Y., and Gō, N., *Intntl. J. Peptide and Protein Res.* 7:445 (1975).
13. Gō, N., *Adv. Biophys. 9*:65 (1976).
14. Karplus, M., and Weaver, D.L., *Nature 260*:404 (1976).
15. Kuntz, I.D., Crippen, G.M., Kollman, P.A., and Kimelman, D., *J. Mol. Biol. 106*:983 (1976).
16. Levitt, M., *J. Mol. Biol. 104*:59 (1976).
17. Nemethy, G., and Scheraga, H.A., *Quart. Rev. Biophys. 10*:239 (1977).
18. Hantgan, R.R., Hammes, G.G., and Scheraga, H.A., *Biochemistry 13*:3421 (1974).
19. Scheraga, H.A., **Chem. Revs.** 71:195 (1971).
20. Flory, P.J., *Macromolecules 7*:381 (1974).
21. Gō, N., and Scheraga, H.A., *Macromolecules 9*:535 (1976).
22. Scheraga, H.A., in *"Proc. Fifth American Peptide Symposium"* (M. Goodman and J. Meienhofer, eds.), p. 246. Wiley, New York, 1977.
23. Swenson, M.K., Burgess, A.W., and Scheraga, H.A., in *"Frontiers in Physico-Chemical Biology"* (B. Pullman, ed.), in press, Academic Press, New York, 1978.
24. Momany, F.A., McGuire, R.F., Burgess, A.W., and Scheraga, H.A., *J. Phys. Chem. 79*:2361 (1975).
25. Dunfield, L.G., Burgess, A.W., and Scheraga, H.A., *J. Phys. Chem.*, submitted.
26. Pincus, M.R., and Scheraga, H.A., *J. Phys. Chem. 81*:1579 (1977).
27. Zimmerman, S.S., and Scheraga, H.A., *Biopolymers 16*:811 (1977); *17*:in press (1978).
28. Hodes, Z.I., Nemethy, G., and Scheraga, H.A., *Biopolymers*, submitted.
29. Zimm, B.H., and Bragg, J.K., *J. Chem. Phys. 31*:526 (1959).
30. Scheraga, H.A., *Pure and Applied Chem.*, in press (1978).
31. Gō, M., Gō, N., and Scheraga, H.A., *J. Chem. Phys. 52*:2060 (1970); *54*:4489 (1971).
32. Gō, M., Hesselink, F.T., Gō, N., and Scheraga, H.A., *Macromolecules 7*:459 (1974).
33. Scheraga, H.A., in *"Current Topics in Biochemistry, 1973"* (C.B. Anfinsen and A.N. Schechter, eds.), p.1. Academic Press, New York, 1974.
34. Sridhara, S., Ananthanarayanan, V.S., Taylor, G.T., and Scheraga, H.A., *Biopolymers 16*:2565 (1977).

35. Patton, E., and Auer, H.E., *Biopolymers* *14*:849 (1975).
36. Auer, H.E., and Patton, E., *Biophys. Chem.* *4*:15 (1976).
37. McKnight, R.P., and Auer, H.E., *Macromolecules* *9*:939 (1976).
38. Maxfield, F.R., and Scheraga, H.A., *Biochemistry* *15*:5138 (1976).
39. Robson, B., and Suzuki, E., *J. Mol. Biol.* *107*:327 (1976).
40. Tanaka, S., and Scheraga, H.A., *Macromolecules* *10*:9 (1977).
41. Gō, N., and Scheraga, H.A., *Macromolecules*, submitted.
42. Ueda, Y., and Scheraga, H.A., work in progress.
43. Tanaka, S., and Scheraga, H.A., *Macromolecules* *9*:945 *(1976)*.
44. Isogai, Y., Nemethy, G., and Scheraga, H.A., work in progress.
45. Anderson, J.S., and Scheraga, H.A., work in progress.
46. Matheson, R.R., Jr., and Scheraga, H.A., *Macromolecules*, submitted.
47. Poland, D.C., and Scheraga, H.A., *Biopolymers* *3*:315 (1965).
48. Scheraga, H.A., *"Protein Structure"*, p. 241, Academic Press, New York, 1961.
49. Dygert, M., Gō, N., and Scheraga, H.A., *Macromolecules* *8*:750 (1975).
50. Lee, B., and Richards, F.M., *J. Mol. Biol.* *55*:379 (1971).
51. Chothia, C., *J. Mol. Biol.* *105*:1 (1976).
52. Janin, J., *J. Mol. Biol.* *105*:13 (1976).
53. Wertz, D.H., and Scheraga, H.A., *Macromolecules* *11*:9 (1978).
54. Rackovsky, S., and Scheraga, H.A., *Proc. Natl. Acad. Sci.*, *U.S.* *74*:5248 (1977).
55. Chothia, C., *J. Mol. Biol.* *75*:295 (1973).
56. Rao, S.T., and Rossmann, M.G., *J. Mol. Biol.* *76*:241 (1973).
57. Rossmann, M.G., and Liljas, A., *J. Mol. Biol.* *85*:177 (1974).
58. Schulz, G.E., and Schirmer, R.H., *Nature* *250*:142 (1974).
59. Levitt, M., and Chothia, C., *Nature* *261*:552 (1976).
60. Nishikawa, K., and Scheraga, H.A., *Macromolecules* *9*:395 (1976).
61. Richardson, J.A., *Proc. Natl. Acad. Sci.*, *U.S.* *73*:2619 (1976); *Nature* *268*:495 (1977).
62. Rossmann, M.G., and Argos, P., *J. Mol. Biol.* *105*:75 (1976); *109*:99 (1977).
63. Sternberg, M.J.E., and Thornton, J.M., *J. Mol. Biol.* *105*:367 (1976); *110*:269, 285 (1977); *113*:401 (1977); *115*:1 (1977).
64. Levitt, M., and Greer, J., *J. Mol. Biol.* *114*:181 (1977).
65. Von Heijne, G., and Blomberg, C., *J. Mol. Biol.* *117*:821 (1977).
66. Chothia, C., Levitt, M., and Richardson, D., *Proc. Natl. Acad. Sci.*, *U.S.* *74*:4130 (1977).

67. Sternberg, M.J.E., and Thornton, J.M., *Nature* 271:15
    (1978).
68. Wetlaufer, D.B., *Proc. Natl. Acad. Sci.*, *U.S.* 70:697
    (1973).
69. Wetlaufer, D.B., Rose, G.D., and Taaffe, L., *Biochemistry*
    15:5154 (1976).
70. Hochman, J., Gavish, M., Inbar, D., and Givol, D.,
    Biochemistry 15:2706 (1976).
71. Tanaka, S., and Scheraga, H.A., *Macromolecules* 10:291
    (1977).
72. Rackovsky, S., and Scheraga, H.A., *Macromolecules* 11:1
    (1978).
73. Gabel, D., Rasse, D., and Scheraga, H.A., *Intntl. J.
    Peptide and Protein Res.* 8:237 (1976).
74. Tanaka, S., and Scheraga, H.A., *Macromolecules* 10:305
    (1977).
75. Rackovsky, S., and Scheraga, H.A., *Macromolecules*,
    submitted.
76. Harrington, W.F., and Schellman, J.A., *Compt. rend. trav.
    lab. Carlsberg, Ser. chim.* 30:21 (1956).
77. Hermans, J., Jr., and Scheraga, H.A., *J. Am. Chem. Soc.*
    83:3283 (1961).
78. Rupley, J.A., and Scheraga, H.A., *Biochemistry* 2:421 (1963).
79. Ooi, T., Rupley, J.A., and Scheraga, H.A., *Biochemistry*
    2:432 (1963).
80. Ooi, T., and Scheraga, H.A., *Biochemistry* 3:641, 648 (1964).
81. Klee, W.A., *Biochemistry* 6:3736 (1967).
82. Burgess, A.W., Weinstein, L.I., Gabel, D., and Scheraga,
    H.A., *Biochemistry* 14:197 (1975).
83. Chen, M.C., and Lord, R.C., *Biochemistry* 15:1889 (1976).
84. Matheson, R.R., Jr., Van Wart, H.E., Burgess, A.W.,
    Weinstein L.I., and Scheraga, H.A., *Biochemistry* 16:396
    (1977).
85. Matheson, R.R., Jr., Dugas, H., and Scheraga, H.A.,
    Biochem. Biophys. Res. Commun. 74:869 (1977).
86. Takahashi, S., and Ooi, T., *Bull, Inst. Chem. Res.,
    Kyoto Univ.* 54:141 (1976).
87. Chavez, L.G., Jr., and Scheraga, H.A., *Biochemistry*
    16:1849 (1977).
88. Scheraga, H.A., *Fed. Proc.* 26:1380 (1967).
89. Barnard, E.A., and Stein, W.D., *J. Mol. Biol.* 1:339, 350
    (1959).
90. Gundlach, H.G., Stein, W.H., and Moore, S., *J. Biol.
    Chem.* 234:1754, 1761 (1959).
91. Heinrikson, L., Stein, W.H., Crestfield, A.M., and Moore,
    S., *J. Biol. Chem.* 240:2921 (1965).
92. Heinrikson, L., *J. Biol. Chem.* 241:1393 (1966).
93. Hirs, C.H.W., Halmann, M., and Kycia, J.H., in *"Biological
    Structure and Function"* (T.W. Goodwin and O. Lindberg,
    eds.), Vol. 1, p. 41. Academic Press, New York, 1961.

EVOLUTIONARY ASPECTS OF THE STRUCTURE
AND
REGULATION OF PHOSPHORYLASE KINASE

E.H. Fischer[1]
J.O. Alaba[2]
D.L. Brautigan
W.G.L. Kerrick
D.A. Malencik
H.J. Moeschler[3]
C. Picton
Sitivad Pocinwong

Department of Biochemistry
University of Washington
Seattle, Washington

Among the multiple advantages that have been ascribed to
cascade systems (1-5) for the covalent control of enzymes, perhaps the most fundamental one is that they can be called upon
to link various metabolic pathways through the versatility of
the regulatory enzymes involved. This is certainly the case
in the control of glycogen metabolism. The process can be
triggered by a hormonal signal acting through the membrane-
bound adenylate cyclase (6). Cyclic AMP released will activate
the cAMP-dependent protein kinase which can act on a number of
enzymes including phosphorylase kinase and glycogen synthase
and the other enzymes of lipid metabolism such as the hormone
sensitive lipase or cholesterol esterase (for review see 7,8).

[1]Supported by grants from NIH, PHS (AM 07902, 17081 and
HL 13517), NSF (GB 3249) and the Muscular Dystrophy Association of America.
[2]Present address: Laboratory of Cell Biology, National
Cancer Institute, NIH, Bethesda, Maryland 20014
[3]Present address: Dept. de Biochimie, Case Postale 78
Jonction, 1211 Geneve 8, Switzerland

Phosphorylase kinase, the third member of this chain of events, is both under hormonal and neural control. On the one   hand, its activity can be modulated by phosphorylation-dephosphorylation (at least in higher vertebrates); on the other hand, it can be activated by the nerve impulse that triggers contraction because of an absolute requirement for calcium ions (see reviews 8,9). Whenever muscle contracts, glycogen must be degraded to provide for some of the energy needed to maintain contraction. Both processes are initiated by the release of calcium ions from the sarcoplasmic reticulum. This report will be restricted to a detailed study of the subunit structure and function of phosphorylase kinase which plays a crucial role in synchronizing these two physiological events.

Subunit Structure and Function of Phosphorylase Kinase

Since much of the earlier work on this enzyme had been carried out on mammalian muscle (9-12), we thought that it might be of interest to look at the structure of an enzyme isolated from a more primitive vertebrate. The Pacific dogfish (*Squalus acanthia*) was selected since it has separated from the main line of evolution leading to man over 400 million years ago, but is already endowed with a well developed endocrine system.

Some of the properties of dogfish and rabbit skeletal muscle phosphorylase kinase are listed in Table I. Both enzymes are heteropolymers made up of three types of subunits (α, β and γ) of molecular weight ca. 130,000, 118,000 and 45,000 respectively, with some differences between the two species. The holoenzyme $(\alpha\beta\gamma)_4$ is a tetramer of molecular weight ca. 1.3×10^6; some uncertainty exists in the ratio of the γ-subunit for reasons that will become apparent later.

While both enzymes demonstrate an absolute requirement for calcium ions and have approximately identical specific activities at alkaline pH, they display major differences in their enzymatic behavior under physiological conditions. The rabbit enzyme is essentially inactive even in its Ca-containing form; activity is increased approximately 40-fold by phosphorylation of the protein. By contrast, the Ca-containing dogfish enzyme displays approximately half maximum activity at neutral pH with no change observed whether it is acted upon by protein kinases or phosphatases. In this primitive vertebrate, therefore, activation and inhibition relies solely on the uptake and release of Ca-ions.

Four main questions can be asked. First, what is the role of the various subunits; more particularly, which subunit contains the catalytic site? Second, since the enzyme shows an absolute requirement for calcium, is the metal ion involved in catalysis or does it only serve a regulatory function? Third, it has been clearly established that in higher vertebrates,

TABLE I

Comparative Properties of Dogfish vs. Rabbit Skeletal Muscle Phosphorylase Kinase and its Chymotryptic Fragment

Properties	Phosphorylase Kinase		Rabbit Chymo-triptic Fragment
	Dogfish	Rabbit	
Concentration in Muscle (μM)[a]	4	10	
Molecular Weight	1.3×10^6	1.28×10^6	33,000
Subunit MW: α	130,000	130,000	
β	118,000	118,000	
γ	45,000	37,000	
Specific Activity (pH 8.2) minus Ca^{2+}	0	0	280
plus Ca^{2+}	30	37	280
pH 6.8/8.2 activity ratio			
dephospho-form	0.5	0.01	0.8
phospho -form		0.40	0.9
K_m Rabbit phosphorylase $\underline{b}$ (μM)			
dephospho-form 6.8/8.2, resp.	167/84	140/40	
phospho -form 6.8/8.2, resp.		37/17	
K_m ATP (mM) pH 6.8 or 8.2	1.9	4.3	
Substrate Sepcificity	% Relative Rates		
Phosphorylase $\underline{b}$	100	100	100
TN-I	0	10	10
TN-T	0	2	10
ATP	100	100	100
dATP	55	96	80
GTP	0	4	3

[a] Calculated on MW of 1.3×10^6. Intracellular water taken as 75% tissue wet weight.

phosphorylation involves the α and β subunits; does the same
situation prevail in early vertebrates? Lastly, what could be
the role of the smallest, 45,000 MW γ-subunit?

Most regulatory enzymes exist as heteropolymers in which
a regulatory subunit affects the activity of a catalytic component. In fact, in the early days of biochemical evolution,
it must have been a beneficial event whenever the catalytic
activity of an enzyme could be abolished by chance interaction
with another protein. Whenever this occurred, the basic requirements for regulation were established: any condition
that would lead to a dissociation-or losening-of this complex
would generate activity. This is the situation one finds,
for instance, in aspartyl transcarbamoylase (13,14) or the
cAMP-dependent protein kinase (7). Both enzymes are activated
or desensitized by dissociation of the complex. In fact,
there is good evidence that certain single chain enzymes must
have originally been constructed as heteropolymers but the
two separate components have become covalently linked in the
course of time. This appears to be the case for the cGMP-
dependent protein kinase which exists as a monomer but of a
size approximately equal to the sum of the regulatory plus
catalytic subunits of the cyclic AMP-dependent enzyme (15).
The two domains can be recognized in terms of their specific
function by proteolytic cleavage which generates a fully active
catalytic subunit (16). The same situation probably applies
to phosphorylase, one of the two enzymes directly involved in
glycogen metabolism. Most regulatory functions (the phosphoryl group introduced during the phosphorylase b to a conversion, sites for the allosteric effectors, AMP, G1P or G6P)
are located on a 30,000 MW amino terminal portion of the molecule whereas the catalytic site is buried in the remaining
70,000 MW fragment (17,18). These two domains are losely
linked by a segment readily accessible to proteolytic cleavage
(19). It is therefore conceiveable that at one time, the
molecule existed as two separate subunits, regulatory and
catalytic, that became covalently linked in the course of
time. This would account for the unusally large size of the
enzyme monomer. On this basis, then, it was safe to assume
that one of the subunits of phosphorylase kinase would contain
the catalytic site while one if not both of the other subunits
would serve certain regulatory functions.

The Catalytic Subunit

Limited proteolysis of rabbit phosphorylase kinase results in an increase in catalytic activity. Since this activation is accompanied by a rapid destruction of the α and β
subunits while γ is resistant, it was originally hypothesized
that γ might represent the catalytic subunit of the enzyme.

However, this view appeared inconsistent with observations made
on the subunit composition of dogfish phosphorylase kinase.
When different purified preparations of this enzyme were examined by SDS gels, it was clear that the proportion of γ-subunit could vary considerably while the ratio of α to β remained
constant at 1:1. Yet, the specific activities of these preparations were essentially the same (20). This seemed to preclude the possibility that the γ-subunits, at least by themselves, could be catalytically active. Furthermore, these
data indicated that under certain conditions, the γ-subunit
can bind to the enzyme or undergo aggregation.

Better arguments suggesting that the catalytic site did
not reside in the γ-subunit but, actually, might originate from
the β-subunit came from a limited proteolysis study of dogfish
phosphorylase kinase as illustrated in Figure 1. Within minutes, one observes a rapid activation of the enzyme comcomi-
tant with the disappearance of the α-subunit which happens to
be very susceptible to tryptic attack. This suggests that α
possesses a regulatory function, maintaining the enzyme in its
inactive form. Then, as β is degraded, activity disappears
while the γ-subunit remains essentially untouched. In rare
instances, when partially degraded solutions of phosphorylase
kinase were passed through Bio gel columns, an active fraction
containing a main component corresponding in size to that of
the β-subunit could be isolated. While these data would appear
convincing at first glance, it should be emphasized that they
are indirect and could be misleading: enzymatic activity is
measured under nondenaturing conditions while gels as illustrated in Figure 1 were obtained in the presence of SDS. Obviously, a few nicks of the peptide chain that would alter considerably the gel patterns might hardly be seen in terms of
the quarternary structure of the enzyme under assay conditions.

Origin of Catalytic Site and Regulatory Role of Calcium

Most convincing evidence that the catalytic site could
not originate from the γ-subunit was obtained from a recent
study of <u>rabbit</u> phosphorylase kinase. By limited chymotryptic
attack, a small, <u>fully</u> active fragment of the enzyme was
generated and isolated. Its properties are summarized in the
third column of Table I. The material has a molecular weight
of approximately 33,000; its specific activity is <u>ca</u>. 10 times
that of the native enzyme which is what one would expect if
90% of the remainder of the molecule had been eliminated.
Most importantly, a) the fragment no longer displays an requirement for Ca^{2+} for activity; naturally, it had also lost
all susceptibility to EDTA inhibition. This clearly established
that calcium ions play a regulatory role only and are not in-

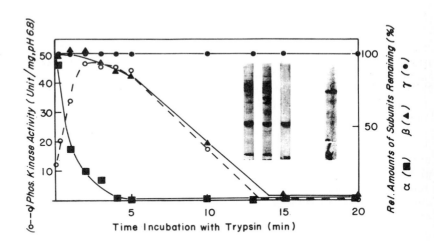

FIGURE 1. Changes in activity of dogfish phosphorylase
kinase by limited proteolysis. Trypsin (3.8 μg/ml) was added
to phosphorylase kinase (1.9 mg/ml); at various times, samples
were removed, treated with soybean trypsin inhibitor (50 μg/ml)
and examined for kinase activity and subunit distribution.
Three gels at various time intervals of trypsin treatment are
shown: (left to right) 5 sec; 3 min; and 15 min. The gel
on the far right is the isolated β-subunit.

volved in catalysis. b) The fragment obtained undergoes auto-
phosphorylation as does the native enzyme. At first glance,
this would seem to have excluded the possibility that the
fragment had originated from the γ-subunit since no phosphorylation of that component has ever been observed. On the other
hand, it was shown that denatured proteins were more susceptible
to phosphorylation by the cAMP-dependent protein kinase than
their native counterparts (21). Of course, a partially denatured γ-subunit could have been isolated.

To exclude this last possibility, the fragment was
isolated from native phosphorylase kinase pre-phosphorylated in
its α and β subunits. Since all the labelled ATP had been
eliminated prior to the chymotryptic attack, any radioactivity
found in the active fragment could have only originated from
the α or β subunit. As shown in Figure 2, a ^{32}P-labelled
active fragment was isolated from the native enzyme. Partial
dephosphorylation with E. coli alkaline phosphatase had no

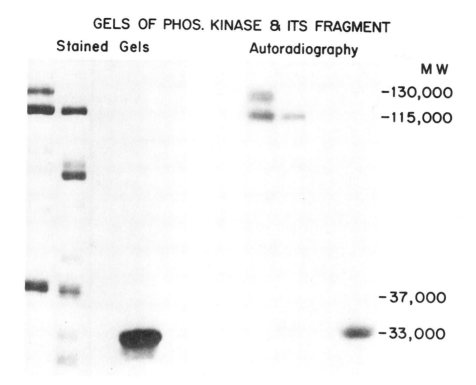

GELS OF PHOS. KINASE & ITS FRAGMENT

Stained Gels   Autoradiography

M W
—130,000
—115,000

—37,000
—33,000

FIGURE 2.   Isolation of active fragment from a P^{32} labelled phosphorylase kinase. The three gels on the left are stained with Coomassie Blue, to their right is an autoradiograph. They show: a) original phosphorylase kinase; b) digest after chymotrypsin; and c) isolated active fragment.

effect on its enzymatic activity. To prove beyond doubt that the catalytic site belongs to the β rather than to the α subunit, one should compare the electrophoretic characteristics (peptide maps) of the labelled phosphopeptides obtained from the various species or better yet, determine their amino acid sequences. This work is presently being carried out.

The α- *and* β-*Subunit*

As shown above, there is an increase in dogfish phosphorylase kinase activity as the α-subunit is degraded. Since no
phosphorylation occurs in this organism, one can assume that
the α-subunit was first introduced into the architecture of
the enzyme as a regulatory component, and this might still be
its main function. Binding of Ca^{2+} must reduce the interaction
between the catalytic and regulatory components thereby allow-
ing enzymatic activity to express itself. In the mammalian
enzymes, phosphorylation of the α and β subunit could further
relax the system resulting in additional increases in activity.
At which point of vertebrate evolution did the phosphorylation
reaction appear and what was its original purpose? Obviously,
it places the enzyme under hormonal control, which adds
flexibility in the modulation of enzyme activity. Phosphorylation of α could assist in the dephosphorylation of the β-
subunit along the lines proposed by the "second site phosphorylation" hypothesis of Cohen et al. (22). Be as it may,
one wonders why the enzyme found it necessary to incorporate
into its structure a component as large as the α-subunit.
Obviously, it must contain a great amount of information that
is still unrecognized at the present time.

A purified preparation of phosphorylase phosphatase (23)
showed only 5% activity on intact phosphorylase kinase as compared to phosphorylase a. It was totally inactive on the
active chymotryptic fragment. Dephosphorylation of phosphorylase kinase would thus seem to require another specific phosphatase - it might constitute another point of hormonal control.

The γ-*Subunit*

As indicated earlier, the stoichiometric ratio of γ-subunit was often found to vary from preparation to preparation
of purified dogfish phosphorylase kinase, as if an excess of
this subunit could bind to the enzyme or undergo aggregation.
Electron microscopy performed by Breck Byers showed long
filaments rather similar to those displayed by F-actin. Both
the γ-subunit and actin have molecular weights ca. 45,000
and they aggregate as a single band on dodecyl sulfate gel
electrophoresis. Amino acid analysis of the two proteins gave
identical compositions with the only difference that whereas
dogfish actin contains one residue of 3-methyl histidine as
expected, the γ-subunit always contains less than one. Finally, it can be "decorated" by heavy meromyosin just as well
as actin.

Several questions must be asked. First, one could wonder whether the γ-subunit belongs to the enzyme at all, which boils down to asking which criteria must be fulfilled so that a subunit can be considered part of an enzyme. Surely, the best answer would be provided if one could show that the subunit in question affects - one way or the other - the activity or localization of the enzyme. One would have to dissociate the enzyme under non-denaturing conditions, then reassemble it with or without this particular subunit. At this time, a pure αβ complex that can be reconstituted with the γ-subunit has not been obtained, so that the role of this third component in the structure or activity of the enzyme can not be really assessed. On the other hand, two other criteria are fulfilled: neither the rabbit nor the dogfish enzyme has ever been isolated without this third component. Furthermore, the stoichiometry of the α:β:γ complex has never been less than 1:1:1.

A second possibility is that actin is present merely as a contaminant; that is, that the enzyme would have a strong affinity for actin and would often be isolated as a mixture with this protein (αβγ + actin). Alternatively, one could have a form of the enzyme in which actin had actually displaced the γ-subunit (αβ-actin), which could be the case if the two molecules had certain properties in common. It is well known that actin is a "sticky" protein that can bind to a number of structural components or enzymes (24,25); if it were present in excess, its property would naturally overshadow those of the γ-subunit. However, if this were the case, one would expect that heavy meromyosin (26) or DNase I (27) would interact with the contaminating actin. But as shown in Figure 3, neither reaction occurs. In both instances, the holoenzyme does not affect the enzymatic activities of heavy meromyosin or DNase I whereas the isolated γ-subunit does, almost as much as purified actin. Therefore, actin can not be present as a mere contaminant. Furthermore, if it had replaced the γ-subunit and were incorporated in the core of the enzyme (i.e., in the form of αβ-actin), it would have to be so tightly bound that it could not be displaced by either heavy meromyosin, one of its natural partners, or DNase I. In any event, one would still have to explain why its 3-methyl-histidine content would be so low.

If the enzyme could co-polymerize with actin, perhaps by interacting with the γ-subunit, or exchange its γ-subunit for actin, then one could be dealing with five types of oliomeric species as depicted in Figure 4.

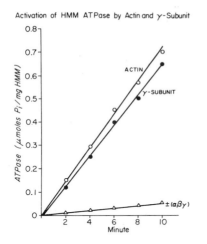

Activation of HMM ATPase by Actin and γ-Subunit

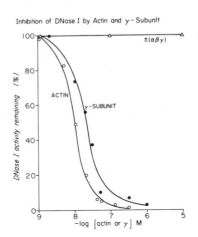

Inhibition of DNase I by Actin and γ-Subunit

FIGURE 3:   LEFT: Activation of heavy meromyosin (HMM)
ATPase by dogfish actin or the γ-subunit of phosphorylase
kinase.   To 0.14 mg HMM in 14 mM imidazole, pH 7.0, 30°C, was
added either 0.2 mg actin, 0.2 mg of γ-subunit or 1.28 mg of
intact phosphorylase kinase.   The molar ratio of HMM to actin or
γ-subunit was 1:10.   The intact enzyme was without effect.
RIGHT: Inhibition of DNase I by actin and the γ-subunit.
DNase I (5.4 x 10^{-9} M) in 10 mM Tris, pH 7.4 was treated with
increasing amounts of dogfish actin or phosphorylase kinase
γ-subunit.   The intact enzyme gave no inhibition.

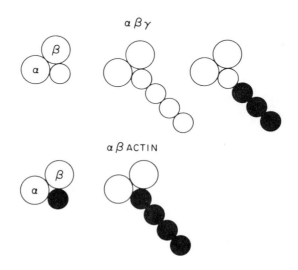

Figure 4.   Five types of oliomeric species.

In models I, II and III, actin could react with the γ-subunit, or with a γ-subunit side chain; in models IV and V, it could actually displace the γ-subunit from the core of the enzyme. If pure γ-subunit were free of 3-methyl histidine, the proportion of this residue in the enzyme would vary from 0 to 1 per "γ-subunit" whether one would be dealing with a type I or type IV model, respectively. In analyzing several preparations for their 3-methyl histidine content, values averaging around 0.3 were obtained. Attempts are being made to obtain these various species in order to characterize their enzymatic and regulatory properties. Some comparative properties of the γ-subunit and actin are summarized in Table II.

Table II

Comparative Properties of Actin and γ-Subunit

	Dogfish Actin	Dogfish γ-Subunit
M.W. (SDS gel)	45,000	45,000
A.A. Composition	Same	Same
3-Methyl Histidine	1	0-0.7 (avg. 0.3)
Electron Microscopy	filaments	± filaments
Decoration with HMM	+	+
HMM ATPase activation	+	+
DNase Inhibition	+	+
Nucleotide binding	1	1

The similarities between the γ-subunit and actin are not as evident in rabbit phosphorylase kinase: whereas actin has been remarkably conserved through the ages and has changed very little in the several hundred million years separating the dogfish from the rabbit, the γ-subunit has rapidly evolved. The rabbit γ-subunit has a lower molecular weight of approximately 37,000; its amino acid composition, though different, is clearly homologous. No 3-methyl histidine residue has been found. What might have been the evolutionary pressure that led to the divergence of the two molecules? Perhaps, there was a need to prevent the interaction of this soluble, or cytoplasmic

form of the enzyme with structural elements of the cells. It
is conceivable that another form of the enzyme exists in
association with the cell membrane that would display similar
structural features as the dogfish kinase. Phosphorylase
kinase has been shown to be, at least in part, associated with
the cell membrane (28).
      What would be the function of the γ-subunit and why should
it be related to actin? One possibility is that the enzyme
simply borrowed this structural protein as a convenient building block - just as creatine phosphokinase was utilized as one
of the components of the M-line proteins (29). Here, a con-
verse process might have taken place, i.e., a contractile
protein was used for the assembly of an enzyme.
      Far more probable is that actin was utilized to confer
Ca^{2+}-sensitivity to this enzyme. Actin is known to bind one
equivalent of calcium very strongly (30). It would seem
natural for the very enzyme that sits at the crossroads between glycogenolysis and contraction to introduce actin into
its structure and utilize this Ca^{2+}-binding component as a
regulatory subunit.
      Alternatively, the γ-subunit might have been introduced,
not to affect the activity of the enzyme, but its localization
within the cell. It could, for instance, serve as a recognition subunit that would allow the enzyme to interact with
muscle thin filaments. Against this hypothesis is the fact
that no phosphorylation of dogfish troponin I or troponin T
could be demonstrated with dogfish phosphorylase kinase (20).
Conversely, components of the thin filament did not affect the
activity of phosphorylase kinase.
      A further possibility is that this subunit might have
allowed (or still allows) phosphorylase kinase to interact
with cytoplasmic actin - that is, with the assembly of the
microtubule-microfilament system ubiquitously associated with
eukaryotic cell membrane (31,32). As of now, these questions
cannot be answered.
      In the isolation of phosphorylase kinase, it has long been
noticed that the enzyme can be obtained - both as a "clear"
and "turbid" fraction. Examination of the turbid fraction
by SDS gels shows that it always contains a high proportion of
actin and a 96,000 MW component. The latter has been tentatively identified as α-actinin: among various properties,
it causes the G to F actin conversion under conditions where
G-actin alone would not polymerize. Soluble enzyme from the
"clear fraction" does not bring about this reaction.

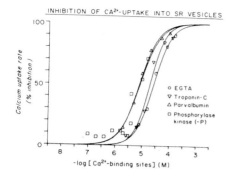

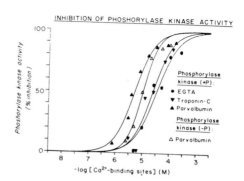

FIGURE 5: LEFT (A): Inhibition of calcium uptake by purified fragments of sarcoplasmic reticulum in the presence of increasing concentration of Ca^{2+}-binding compounds. The reaction was carried out in 30 mM imidazole, pH 7.0, containing 50 mM KCl, 5 mM $MgCl_2$, 5 mM potassium oxalate, 5 mM ATP and 10 μM $^{45}Ca^{2+}$. After 15 sec., the suspension was filtered through a Millipore filter and $^{45}Ca^{2+}$ was determined in the filtrate. Symbols are: (o) EGTA: (∇) troponin C; (Δ) mouse muscle parvalbumin; (◘) dephospho form of rabbit muscle phosphorylase kinase. RIGHT (B): Inhibition of phosphorylase kinase by increasing concentrations of Ca^{2+}-chelators, including (●) EGTA; (▼) troponin C; and (▲) parvalbumin. The same data were obtained with the dephospho form of phosphorylase kinase (not illustrated) except when parvalbumin was present (Δ). Incubation was carried out at 30° in 13.5 mM imidazole buffer, pH 7.0, 22.5 mM KCl; 14.25 mM $MgCl_2$; 7.25 potassium oxalate; 3.6 mM ATP, 10 μM $CaCl_2$, 8 mg/ml phosphorylase b and the Ca-binding compound. After one minute incubation, the reaction was started by addition of phosphorylase kinase (∼10^{-9} M) and the formation of phosphorylase a measured.

Concerted Regulation of Glycogenolysis and Muscle Contraction

For a synchronous regulation of these two processes, Ca^{2+} must bind simultaneously to TN-C and phosphorylase kinase as its concentration increases in the myoplasm. In both the dog-fish and the rabbit, this requirement is clearly satisfied as shown by the use of skinned muscle fibers which lack a functionally intact sarcolemma. In such preparations, the ionic environment surrounding the contractile proteins can be precisely controlled while isometric tension is monitored. In all instances, tension occurs almost simultaneously with the uptake of this metal ion by phosphorylase kinase. Computer fitting of the data indicates a single class of Ca^{2+} binding sites for the enzyme with a K_D of ca. 6 μM (35).

While this synchronous regulation is reasonably well understood, many questions remain. For instance, how can epinephrine trigger glycogenolysis in the relaxed muscle, when the intracellular concentration of Ca is extremely low? What could be the role of the low molecular weight, Ca-binding parvalbumins (36) which are also present in higher vertebrates (37). For instance, could parvalbumin serve as a calcium sink to sop up low levels of calcium that might enter the sarcoplasm, and otherwise cause uncontrolled contraction? This hypothesis would seem all the more attractive if phos-phorylase kinase, upon phosphorylation, could remove calcium from this pool and become activated in the absence of con-traction. Finally, do any of the muscle Ca-binding proteins interact with one another or with the sarcoplasmic reticulum system?

To investigate the calcium binding characteristics of the various proteins, a kinetic approach utilizing the sarcoplasmic reticulum (SR) was selected in the hope that it would reveal certain specific protein-protein interactions. The rate of Ca^{2+} uptake by the SR increases with Ca^{2+} con-centration. As one adds competing Ca-binding proteins, this rate decreases. What one measures is the extent of inhibition of calcium uptake. Thus, in expressing the data as a function of Ca-binding sites for each protein, an estimation of their relative affinities for Ca^{2+} can be made from the curve mid-points.

As can be seen from Figure 5A, relative Ca^{2+}-affinities decreased in the order: phosphorylase kinase ≃ parvalbumin $(2.7 - 3.4 \times 10^6 M)$ > EGTA $(1 \times 10^6 M)$ > Troponin ≃ Troponin C $(0.3 - 0.6 \times 10^6 M)$. The values obtained from troponin C are within the range observed for the two high affinity Ca-binding sites (38); likewise, data for parvalbumin were consistent with those obtained by equilibrium procedures (39). Essentially no difference in Ca-affinity was found between the phosphory-lated forms of the enzyme.

These data were confirmed in a second approach, where the inhibitory effects of parvalbumin and troponin C on the activity of phosphorylase kinase were determined. The curves obtained for TN-C and EGTA were roughly the same whether experiments were carried out with the phosphorylated or non-phosphorylated form of the enzyme. On the other hand, there was a small but consistent difference when parvalbumin was added: it inhibited more readily the phosphorylated than the non-phosphorylated form of phosphorylase kinase, shown in open symbols in Figure 5B; this result cannot be explained solely on the basis of the competitive binding of Ca^{2+} by the two proteins. This means that one cannot envision a scheme in which, for instance, the enzyme could withdraw calcium from a Ca^{2+}-parvalbumin pool once it had become phosphorylated under the influence of epinephrine. On the contrary, if anything, parvalbumin seems to withdraw Ca^{2+} more readily from the phosphorylated form of the enzyme.

One of the first events that occurs during cell function, transformation or development (e.g., membrane activation by depolarization or deformation by lectins or surface antigens (40)), is an increase in the intracellular concentration of cGMP and calcium (41,42). In fact, Ca^{2+} might emerge as the prime effector by acting through the microfilament-microtubule system. Because phosphorylase kinase has an absolute requirement for calcium ions and, perhaps, structural features that might enable it to interact with membrane or cytoplasmic components, it is tempting to suggest that it might be associated with some of the early events that occur after cell activation. Calcium would then be the messenger of cell membrane stimulation, just like cAMP is the messenger by which certain hormonal signals are transmitted. Phosphorylase kinase, or an analgous calcium-requiring protein kinase (43), would then serve as a vehicle by which calcium would express its physiological function, just as protein kinase expresses the function of cAMP. Such a hypothesis would imply that phosphorylase kinase would have a far wider substrate specificity extending beyond its role in the control of carbohydrate metabolism. We propose to investigate certain aspects of this hypothesis.

ACKNOWLEDGMENTS

The assistance of Dr. B. Byers and Dorr Tippens is gratefully acknowledged. We also thank Richard Olsgaard for the amino acid analysis and Cheryl May for the typing of the manuscript. The work was supported by grants from the National Institutes of Health, PHS (AM 07902, 17091 and HL 13517), National Science Foundation (GB 3249) and the Muscular Dystrophy Association, Inc.

REFERENCES

1. Fischer, E.H., Heilmeyer, L.G., Haschke, R.H., Curr. Top.
    Cell. Regul. 4:211 (1971).
2. Helmreich, E., and Cori, C.F., Adv. Enzyme Reg. 3:91
    (1966).
3. Stadtman, E.R. and Chock, P.B., Proc. Natl. Acad. Sci. USA
    74:2761 (1977).
4. Chock, P.B. and Stadtman, E.R., Proc. Natl. Acad. Sci. USA
    74:2766 (1977).
5. Cohen, P. in "Control of Enzyme Activity" Chapman and
    Hall, London, 1976.
6. Helmreich, E.J., Zenner, H.P., Pfeuffer, T. and Cori, C.F.,
    Curr. Top. Cell. Regul. 10:41 (1976).
7. Krebs, E.G., Curr. Top. Cell. Regul. 5:99 (1972).
8. Fischer, E.H., Becker, J.-U., Blum, H.E., Kerrick, W.G.L.,
    Lehky, P., Malencik, D.A. and Pocinwong, S., in "Molecular
    Basis of Motility" (L. Heilmeyer, J.C. Rüegg, and T.
    Wieland, eds.), p. 136, Springer-Verlag, Berlin-Heidel-
    berg, 1976.
9. Cohen, P., Biochem. Soc. Symp. 39:51 (1974).
10. Fischer, E.H. and Krebs, E.G., J. Biol. Chem. 216:113
    (1955).
11. Hayakawa, T., Perkins, J.P., Krebs, E.G., Biochemistry
    12:574 (1973a).
12. Hayakawa, T., Perkins, J.P., Walsh, D.A., Krebs, E.G.,
    Biochemistry 12:567 (1973b).
13. Gerhart, J.C. and Schachman, H.K., Biochemistry 4:1054
    (1965).
14. Gerhart, J.C., Curr. Top. Cell. Regul. 2:275 (1970).
15. Lincoln, T.M., Dills, W.L. and Corbin, J.C., J. Biol.
    Chem. 252:4269 (1977).
16. Inoue, M., Kishimoto, A., Takai, Y. and Nishizuka, Y.,
    J. Biol. Chem. 251:4476 (1976).
17. Titani, K., Koide, A., Hermann, J., Ericsson, L.H.,
    Kumar, S., Wade, R.D., Walsh, K.A., Neurath, H. and
    Fischer, E.H., Proc. Natl. Acad. Sci., USA 74:4762
    (1977).
18. Sygusch, J., Madsen, N.B., Kasvinsky, P.J., and Fletterick,
    R.J., Proc. Natl. Acad. Sci. USA 74:4757 (1977).
19. Raiband, O. and Goldberg, M.E., Biochemistry 12:5154
    (1973).
20. Pocinwong, S., Ph.D. Thesis, University of Washington,
    1975.
21. Bylund, D.B., and Krebs, E.G., J. Biol. Chem. 250:6355
    (1975).
22. Cohen, P., Antoniw, J.F., Davison, M., and Taylor, C.,
    in "Proc. 3rd Intl. Conference on Metabolic Intercon-

versions of Enzymes (E.H. Fischer, E.G. Krebs, H. Neurath, E.R. Stadtman, eds.), p. 221, Springer-Verlag, Berlin-Heidelberg-New York, 1973.

23. Gratecos, D., Detwiler, T.C., Hurd, S. and Fischer, E.H. Biochemistry 16:4812 (1977).
24. Arnold, H. and Pette, D., Eur. J. Biochem. 6:163 (1968).
25. Bray, D., Nature 256:616 (1975).
26. Goldman, R.D., J. Histochem. Cytochem. 23:539 (1975).
27. Lazarides, E., Lindberg, U., Proc. Natl. Acad. Sci. USA 71:4742 (1974)
28. Heilmeyer, L.M.G., in "Molecular Basis of Motility" (L. Heilmeyer, J.C. Rüegg, T. Wieland, eds.), p. 154, Springer-Verlag, Berlin-Heidelberg, 1976.
29. Turner, D.C., Wallinmann, T., and Eppenberger, H.M., Proc. Natl. Acad. Sci., USA 70:702 (1973).
30. Kuehl, W.M., and Gergely, J., J. Biol. Chem. 244:4720 (1969).
31. Forer, A., in "Cell Cycle Controls" (G.M. Podilla, I.L. Cameron, A.M. Zimmerman, eds.), p. 319, Academic Press, New York, 1974.
32. Goldman, R., Pollard, T. and Rosenbaum, J. (eds.) "Cell Motility" Volumes I, II, III, Cold Spring Harbor, Mass., 1976.
33. Kerrick, G.W. and Krasner, B., J. Applied Physiol. 39:1055, (1975).
34. Kerrick, G.W. and Donaldson, S.K., Biochim. Biophys. Acta 275:117 (1972).
35. Kerrick, W.G.L, Hoar, P., Malencik, D.A., Pocinwong, S. and Fischer, E.H., Third Joint US-USSR Symposium on Myocardial Metabolism, 1977.
36. Pechere, J.F., Capony, J.P., and Demaille, J., Syst. Zool. 22:533 (1973).
37. Lehky, P., Blum, H.E., Stein, E.A. and Fischer, E.H., J. Biol. Chem. 249:4332 (1974).
38. Potter, J.D., and Gergely, J., J. Biol. Chem. 250:4628 (1975).
39. Benzonana, G., Capony, J.P., Pechere, J.F., Biochim. Biophys. Acta. 278:110 (1972).
40. Edelman, G.M., Science 192:218 (1976).
41. Nathanson, J.A., Physiol. Rev. 57:157 (1977).
42. Rasmussen, H. and Goodman, D.B.P., Physiol. Rev. 57:421 (1977).
43. Dabrowska, R., Sherry, J., Aromatorio, D. and Hartshorne, D., Biochemistry 17:253 (1978).

POLY(ADP-RIBOSE) AND ADP-RIBOSYLATION OF PROTEINS[1]

O. Hayaishi, K. Ueda, H. Okayama, M. Kawaichi,
N. Ogata, J. Oka, K. Ikai, S. Ito and Y. Shizuta

Department of Medical Chemistry
Kyoto University Faculty of Medicine
Kyoto, Japan

Introduction

Poly(ADP-ribose) and the ADP-ribosylation of proteins constitute a novel type of covalent modification of proteins. During the last few years, these reactions have attracted the attention of a number of molecular biologists, because they are ubiquitously distributed in nature and are implicated in the regulation of cell proliferation, protein synthesis and DNA as well as RNA metabolism (1). In this presentation, I shall attempt to summarize recent developments in this field of research and to present some of our current ideas and experimental results. In order to facilitate comprehension and discussion, I should like to first review briefly, mono ADP-ribosylation of proteins, in which only a single ADP-ribosyl moiety is transferred to a protein acceptor. Then, I would like to cover a more complex reaction, in which the ADP-ribosyl units are polymerized in the chromosome of nuclei. The structure, the biosynthesis and degradation, and the possible function of poly(ADP-ribose) will be discussed and brief concluding remarks made. In all the reactions I shall speak about, the ADP-ribosyl moiety is derived from the coenzyme, nicotinamide adenine dinucleotide.

Nicotinamide adenine dinucleotide (NAD), formerly termed

---

[1]Supported in part by grants-in-aid for Cancer Research from the Ministry of Education, Science and Culture, Japan.

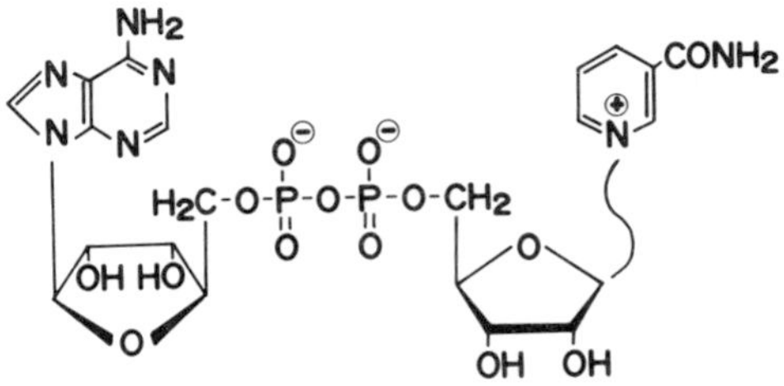

NH₂ ... (Figure 1 structure)

FIGURE 1.   Structure of NAD.

DPN, is the most abundant of the respiratory coenzymes. The major biochemical function of NAD has always been known to be an electron carrier in various biological oxidation-reduction systems. The structure of NAD can be envisaged as the ADP-ribosyl moiety, attached covalently to a vitamin, nicotinamide, through a β-N-glycosidic linkage. This linkage is a so-called high energy bond, since its free energy of hydrolysis is reported to be approximately -8.2 kcal per mole at pH 7 and 25° (2). The energy of this bond supplies the driving force for the various ADP-ribosylation reactions under discussion in this paper. However, the biological significance of this bond energy was not fully appreciated until 10 years ago when we demonstrated the enzymic transfer of the ADP-ribosyl moiety of NAD to a protein acceptor.

Mono ADP-Ribosylation

Diphtheria toxin is an exotoxin produced by <u>Corynebacterium diphtheriae</u>. This toxin is the first example of an ADP-ribosyl transferase, which catalyzes mono ADP-ribosylation of proteins. Elongation Factor 2, abbreviated as EF2, is an enzyme, which is involved in the protein synthesis in eukaryotic cells. It catalyzes the translocation of peptidyl t-RNA on the ribosome. In 1968, we reported that diphtheria toxin catalyzed the transfer of the ADP-ribosyl moiety of NAD to Elongation Factor 2 with the concomitant release of nicotinamide and a proton (3). The ADP-ribosylated EF2 thus formed was catalytically inactive. Diphtheria toxin, therefore, prevents protein synthesis and kills the host cells. In contrast to the nuclear poly(ADP-ribose) synthetase system, which will be discussed later, only one ADP-

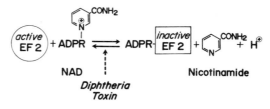

FIGURE 2.   ADP-ribosylation of Elongation Factor 2
by diphtheria toxin.

TABLE I.   Mono ADP-Ribosylation Reactions.

Enzyme	Acceptor
Diphtheria toxin	Elongation factor 2
Pseudomonas toxin	Elongation factor 2
T4 phage { Viral	*E. coli* proteins
{ Induced	RNA polymerase (α-subunits)
N4 phage	*E. coli* proteins
Cholera toxin	Membrane proteins (?)

ribosyl unit is transferred specifically onto each molecule of Elongation Factor 2 and the reaction is reversible.

Table I summarizes the enzymes and acceptors presently reported to be involved in mono ADP-ribosylation reactions. Subsequent to our finding of diphtheria toxin, Pseudomonas toxin was shown to catalyze exactly the same reaction (4). Furthermore, the virions of E. coli phage T_4 (5) and N_4 (6) were shown to contain NAD: protein ADP-ribosyl transferases with a broad acceptor specificity. Infection of E. coli with T_4 phage induces another enzyme that catalyzes a more specific ADP-ribosylation of arginyl residues of α-subunits of RNA polymerase (7, 8). More recently, cholera toxin was reported to catalyze mono ADP-ribosylation of membrane proteins (9). So far all the enzymes that have been reported to catalyze mono ADP-ribosylation reactions are found in prokaryotes rather than eukaryotes.

Poly ADP-Ribosylation

So much for the mono ADP-ribosylation reaction of proteins. Now I would like to move on to the main theme of this lecture, the poly(ADP-ribose) system in nuclei and discuss first the structre of this unique macromolecule and its acceptor.

Poly(ADP-ribose) is a unique homopolymer composed of a linear sequence of repeating ADP-ribose units linked together by ribose (1" → 2') ribose glycosidic bonds and attached to an acceptor protein (Fig. 3). The chain length of this polymer, which is synthesized

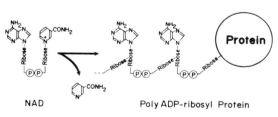

NAD   Poly ADP-ribosyl Protein

FIGURE 3.   Poly ADP-ribosylation of nuclear proteins.

both in vitro and in vivo,
ranges from 1 up to about 100
monomer units depending on
the conditions.

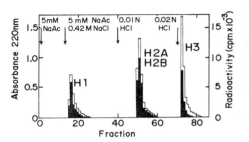

I should like now to
discuss in some details the
nature of the acceptor protein and the linkage through
which the terminal ribose of
poly(ADP-ribose) is attached
to the acceptor molecule.

FIGURE 4. Association of
ADP-ribose with histones (CM-
cellulose column).

As early as 1968, we re-
ported that ADP-ribose might
be covalently linked to histones (10). A typical exper-
imental result is presented in Fig. 4. When ^{14}C-labeled NAD
was incubated with rat liver nuclei and the product of the reaction was extracted with dilute HCl and analyzed by CM-cellulose column chromatography, the radioactivity which is represented by black columns appeared to coincide exactly with the
location of carrier histones, H1, H2A, H2B and H3 represented
by the white columns. These results have been confirmed and
extended by a number of investigators, but the acceptor molecules of poly(ADP-ribose) are by no means limited to histones.
Several other proteins and peptides in chromatin have also
been reported to be ADP-ribosylated. However, definitive characterization of these acceptor proteins has not been successful because of the heterogeneity of acceptor proteins and the
polymer size, their tendency to aggregate with one another and
the instability of the ADP-ribosyl protein linkages.

Recently, we made use of affinity chromatography on a borate polyacrylamide gel column and was able to make more definitive identification of acceptor proteins (11). Borate is

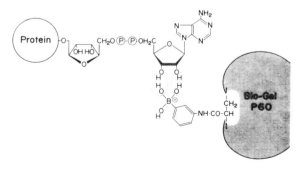

FIGURE 5. Affinity chromatography of ADP-ribosyl protein
on borate-polyacrylamide gel.

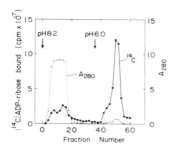

FIGURE 6. Borate gel column chromatography of HCl-extracted proteins.

known to interact specifically with cis-diol-containing compounds (Fig. 5). This procedure works well in the presence of a strong dissociating agent such as guanidine-HCl. Rat liver nuclei were incubated with ^{14}C-NAD and extracted with dilute HCl. When the extract was subjected to the chromatography in the presence of 6 M guanidine, the modified proteins are separated from the bulk protein as shown in Fig. 6. When pH was maintained at 8.2 where borate was negatively charged, more than 95% of the protein passed through the column but ADP-ribosylated proteins were retained on the gel. At pH 6.0 these compounds were eluted because there was no interaction when boric acid was uncharged. The ADP-ribosylated proteins thus isolated were further purified by CM-cellulose column chromatography in the presence of 7 M urea, and separated into 4 fractions A, B, C and D.

The fractions were then analyzed by SDS-gel electrophoresis (Fig. 7). NC stands for total rat liver nuclear proteins and STD stands for standard proteins of known mol. weights for reference. Under the solubilization conditions ADP-ribose is completely removed from the proteins. Fractions A and B appear to contain several non-histone proteins. Fraction C contained H2A, H2B, H3 or H4 plus small amounts of non-histone proteins. Fraction D appears to contain H1 almost exclusively.

The proteins in Fraction C were further analyzed by acid urea polyacrylamide gel electrophoresis together with authentic samples of various histone preparations. As shown in Fig. 8, the proteins in Fraction C, which is in the center gel, had the same mobility as the authentic histone H2B.

The major protein in Fraction C, which was purified by this procedure, had an amino acid composition almost identical to that of calf thymus histone H2B and possessed the same NH$_2$-terminal amino

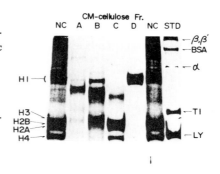

FIGURE 7. SDS-gel electrophoresis of ADP-ribosylated proteins.

acid, proline (Table II).

The protein in Fraction D was composed of almost the same amino acids as the authentic whole histone H1 of calf thymus. No NH$_2$-terminal amino acid was detected by the dansyl technique in agreement with the fact that the NH$_2$-terminal of H1 is blocked by acetylation.

These results indicate that the major acceptor proteins in Fractions C and D were histones H2B and H1, respectively. Quantitative analyses revealed that about 35 and 15% of the acid-insoluble ADP-ribose recovered from the CM-cellulose column were bound to histones H2B and H1, respectively, and approximately 50% was bound to non-histone proteins.

The use of a borate gel column thus provided a simple and convenient procedure for the isolation of ADP-ribosylated proteins and enabled us to make more definitive characterization of the acceptor proteins. Furthermore, the ADP-ribosylated histone preparations thus obtained became a useful tool for the study of enzymatic synthesis and degradation of this polymer as will be discussed shortly.

The site of attachment of poly(ADP-ribose) to histones has also been studied by a number of investigators. Based on the instability to neutral hydroxylamine treatment and the alkali lability, we proposed in 1969 an ester linkage involving the carboxyl group of either a glutamate or aspartate residue of histones (12)(Fig. 9). Dixon reported a similar observation in trout testis (13). Koide in collaboration with Moore also confirmed our conclusion (14). On the other hand,

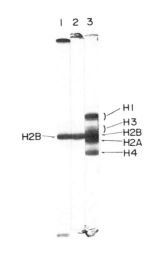

FIGURE 8. Acid urea gel electrophoresis of Fraction C.

TABLE II. Amino acid composition and NH$_2$-terminal of the major proteins in Fractions C and D.

Amino acid	Mole percent			
	Fr. C	Calf thymus H2B[+]	Fr. D	Calf thymus H1[*]
Lys	16.0	16.0	28.9	27.9
His	2.6	2.4	trace	0.0
Arg	7.7	6.4	2.5	1.8
Asp	4.9	4.8	2.0	2.0
Thr	5.8	6.4	5.3	6.0
Ser	10.5	11.2	6.6	6.2
Glu	8.2	8.0	3.6	3.5
Pro	4.4	4.8	9.7	9.2
Gly	6.2	5.6	6.5	6.6
Ala	10.7	10.4	24.4	25.1
Half-cys	0.0	0.0	0.0	0.0
Val	6.9	7.2	4.4	4.0
Met	1.3	1.6	trace	0.0
Ileu	4.8	4.8	1.5	1.0
Leu	5.3	4.8	4.0	4.8
Tyr	4.0	4.0	0.0	0.5
Phe	1.7	1.6	0.5	0.5
NH$_2$-terminal	Pro	Pro	N.D.	N-acetyl

[+] Calculated from the known sequence (Iwai *et al.* 1972)

[*] From Rasmussen *et al.* 1962

Smith and Stocken of Oxford isolated serine phosphate associated with poly(ADP-ribose) from a proteolytic digest of ADP-ribosylated histone Hl (15). Such a linkage is reported to be hydroxylamine resistant and alkali labile. More recently Ogata in my laboratory examined the ADP-ribosylated histone Hl fraction. Preliminary evidence indicates that glutamate residues No. 2, 14 and 116 are ADP-ribosylated.

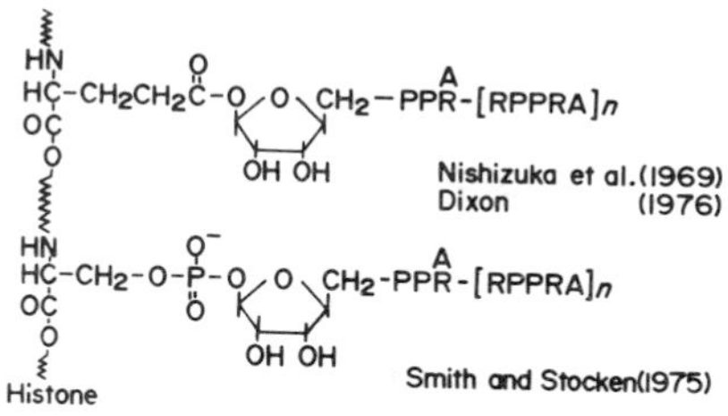

FIGURE 9. Two proposed structures for poly ADP-ribosyl histone.

So much for the structure of poly(ADP-ribose) and the acceptor proteins; now I would like to turn to the enzymic synthesis and degradation of poly(ADP-ribose). Among the enzymes that will be discussed in this presentation, poly(ADP-ribose) synthetase has been most extensively purified and investigated. The highest purification of the synthetase was recently achieved from rat liver in my laboratory (16). As shown in Table III, an approximately 7,000-fold purification was achieved with an overall yield of about 15%. A similar preparation was also obtained from calf thymus and was shown to be essentially homogeneous. The unique feature of this enzyme is its requirement for DNA for activity, and its further stimulation by histone, as shown in Table IV.

When the most highly purified enzyme was incubated with ^{14}C-NAD in the presence of dithiothreitol, $MgCl_2$, and buffer, poly(ADP-ribose) synthesis did not take place unless DNA was also included in the reaction mixture. The addition of histone alone was ineffective, but when histone was included together with DNA, the activity was further stimulated severalfold. Curiously enough the addition of histone did not increase the number of chains but increased the average chain length of the

TABLE III. Purification of poly(ADP-ribose) synthetase.

Step	Total Activity	Specific Activity
	nmol/hr	nmol/hr/mg protein
Rat Liver Nuclei	1150	1
Chromatin	3100	4
0.6 M KCl Extract	1780	7
Hydroxylapatite	730	37
Amm. Sulfate	300	35
Sephadex G-150	320	660
P-cellulose	170	6900

product. If histone were an acceptor of poly(ADP-ribose), one would expect to see an increase in the chain number rather than the chain length. These results therefore provided the first clue indicating that histones may not be ADP-ribosylated by the highly purified synthetase. This interpretation was further supported by the results of analyses of the product, which are presented in Fig. 10.

Table IV. Effect of DNA and histone on poly(ADP-ribose) syn-thetase.

Addition	Poly(ADPR) synthesized	Chain number	Chain length
	pmol	pmol	units
None	O		
Histone (10μg)	O		
DNA (10μg)	531	293	1.6
Histone a DNA	1,343	297	3.7

10μM [Ade-^{14}C] NAD, 1mM DTT, 10mM MgCl$_2$,
50mM Tris·HCl(pH8.0), Synthetase (0.2μg) in 0.5ml
(Incubation 37°, 5min)

A highly purified synthetase preparation was incubated with ^{14}C-NAD under three different sets of experimental condi-tion: (1) in the presence of DNA alone, (2) DNA plus histones H1 and H2B or (3) H3, H2A, and H4. The product of the reaction was dissolved in SDS and analyzed by SDS gel electrophoresis. The gel was then cut in half axially and one piece was stained with Coomassie blue. This is shown in the upper part of this figure. The other half was sliced and the radioacitivity de-termined. As can be seen in the lower part of this figure, the radioactivity was found to be localized at Rf 0.58 with respect to the position of the indicator dye, bromophenol blue. (BPB). The addition of histones increased the radioactivity present in this peak. However, no other radioactive peaks were found which corresponded to the positions of the histones, again indicating that none of these histones were ADP-ribosylated by the highly purified synthetase. As I mentioned before, histones H1 and H2B were clearly shown to be ADP-ribosylated in the rat liver nuclei. These results, therefore, may be interpreted to

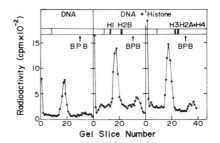

FIGURE 10. SDS-gel electrophoresis of ADP-ribosylated acceptor.

mean that the synthetase is not
capable of initiating the chain
formation but only capable of
catalyzing the elongation of
the preexisting poly(ADP-ribose) attached to histones.
In support of this hypothesis,
when ADP-ribosylated histones
prepared in nuclei and isolated by the borate gel chromatography were employed in
such experiments, the radioactivity was found to be incorporated into the histone
fractions.

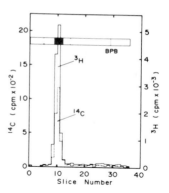

FIGURE 11. SDS-gel electrophoresis of reaction products
with ADP-ribosyl-histone H1.

^{14}C-labeled ADP-ribosyl
histone was prepared using
rat liver nuclei and purified
by the borate gel column procedure. It was then incubated with
tritium-labeled NAD and the purified synthetase. The incorpor-
poration of tritium-labeled ADP-ribose into the ^{14}C-ADP-ribosyl
histone is clearly seen from the result presented in Fig. 11.
In order to make sure that the preexisisting ADP-ribose is
elongated by the synthetase, the following experiment was carried out.

Again ^{14}C-labeled ADP-ribosyl histone was prepared, isolated by the borate gel chromatography and further purified by
CM-cellulose column chromatography. Such a modified histone
was then incubated with the highly purified synthetase in the
presence of DNA and cold NAD. After the reaction was over, the
product of the reaction was subjected to the action of snake
venom phosphodiesterase that specifically cleaves the pyrophosphate linkage, as shown by the arrows in Fig. 12.

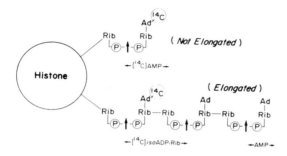

FIGURE 12.   Phosphodiesterase digestion of ADP-ribosyl
histone.

When the chain was not elongated, the phosphodiesterase digestion should yield ^{14}C-labeled AMP from the terminus. On the other hand, when the chain was elongated with cold ADP-ribose from NAD by the action of the synthetase, this terminal AMP was transposed into the inside of the polymer and would be recovered as a part of _iso_ADP-ribose; the new terminus was no longer radioactive.

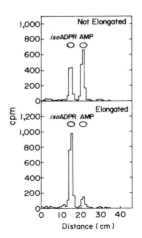

Typical experimental results are shown in Fig. 13. The original ADP-ribosyl histone yielded ^{14}C-labeled _iso_-ADP-ribose as well as AMP upon phosphodiesterase digestion, whereas the same histone which was subjected to the action of the synthetase in the presence

FIGURE 13. Evidence for elongation of ADP-ribosyl histone H1.

of cold NAD yielded mainly _iso_ADP-ribose but little AMP. Thus the synthetase catalyzed the elongation of the pre-exisisting ADP-ribose chains rather than the initiation of the new chains. These results are interpreted to mean that the initiation reaction may be catalyzed by an as yet unidentified enzyme or may require some other cofactors or conditions. So much for the biosynthesis of this polymer.

Figure 14 summarizes the enzymic reactions involved in the degradation of poly(ADP-ribose) in the nuclear system. Until recently two different types of enzymes had been reported.

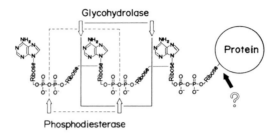

FIGURE 14. Modes of degradation of poly(ADP-ribose).

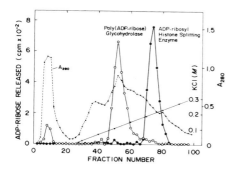

FIGURE 15. DEAE-cellulose column chromatography of ADP-ribosyl histone splitting enzyme.

Poly(ADP-ribose) glycohydrolase cleaves the ribose-ribose linkage yielding ADP-ribose as the reaction product (17, 18). This enzyme appears to be responsible for the degradation of poly(ADP-ribose) in vivo and has been partially purified from calf thymus, rat liver and testis. The other type of enzyme is snake venom phosphodiesterase, which cleaves the pyrophosphate linkage yielding isoADP-ribose. A similar enzyme was isolated from rat liver by Futai and Mizuno (19), but its physiological function is at present unknown. The enzyme which catalyzes the cleavage of the ADP-ribosyl histone linkage has been looked for by a number of investigators without success. We recently found an enzyme which liberated ADP-ribose from ADP-ribosyl histones and tentatively termed it "ADP-ribosyl histone splitting enzyme" (20). Purification of this enzyme is illustrated in Fig. 15.

The 105,000 x g supernatant of rat liver homogenate was precipitated by 45% saturated ammonium sulfate and then chromatographed on a DEAE-cellulose column. The dotted line represents the amount of protein. The ADP-ribosyl histone splitting enzyme activity was eluted in a single peak at the NaCl concentration of about 0.2 M and was clearly separated from the well-known poly(ADP-ribose) glycohydrolase which came out at 0.13 M NaCl.

In contrast to the latter enzyme, this enzyme catalyzed the splitting of a bond between ADP-ribose and a protein portion in mono ADP-ribosylated histone H1 or H2B but little, if any, of the glycosidic bonds within poly(ADP-ribose). Analyses of the reaction product by paper chromatography and Dowex 1 column chromatography indicated that the split product contained the entire ADP-ribose moiety but was not exactly identical with ADP-ribose. Preliminary evidence suggested that it was either an altered ADP-ribose molecule produced by

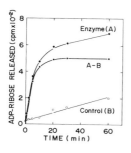

FIGURE 16. Time course of splitting of ADP-ribosyl histone.

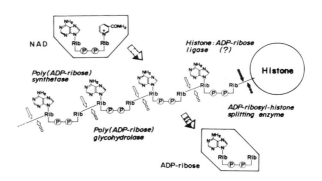

FIGURE 17.   Synthesis and degradation of poly ADP-ribosyl histone.

structural rearrangement, or it may be ADP-ribose bound to an unidentified small molecule.   The enzyme had an optimal pH of about 6.0 and was strongly inhibited by 5 mM ADP-ribose.

The time course of this splitting reaction is illustrated in Fig. 16.   Even at pH 6.0 and 37°, the radioactive ADP-ribose was gradually released by non-enzymatic cleavage indicating that the ADP-ribosyl histone linkage is rather unstable.   The addition of the enzyme accelerated the rate of the reaction at least 50-fold.   The non-enzymatic cleavage was much accelerated under alkaline conditions, and the product of the non-enzymatic reaction appeared to be identical to that of the enzymatic reaction.   Further studies on this new enzyme are now in progress and may provide a clue concerning the nature of the ADP-ribosyl protein linkages.

I should like now to summarize the talk so far presented and briefly speculate its possible biological functions.   Poly(ADP-ribose) synthetase was extensively purified and characterized.   It catalyzes the synthesis and elongation of poly(ADP-ribose) from the ADP-ribosyl moiety of NAD.   As presented here, our current results suggest the presence of another enzyme, a kind of ligase, or a factor which attaches the initial ADP-ribose residue to the acceptor molecule.

Table V summarizes the enzymes and acceptors reportedly involved in poly ADP-ribosylation reactions in eukaryotes. The nuclear enzymes catalyze the ADP-ribosylation of histones as well as several non-histone proteins.   A Mg^{++}, Ca^{++}-dependent endonuclease and RNA polymerase were also reported to serve as acceptor proteins (21).   In addition, a recent report by Kun indicated the presence of a mitochondrial enzyme which ADP-ribosylated a mitochondrial protein (22).   Smulson and co-workers described a cytoplasmic, probably ribosomal, enzyme,

TABLE V.   Poly ADP-ribosylation reactions.

Enzyme	Acceptor	ADP-ribose Units
Nuclear	Chromosomal proteins	
	Histones	Polymer
	Non-histone proteins	Polymer
	Mg^{++}, Ca^{++}-endonuclease	Oligomer
	RNA polymerase	Polymer (?)
Mitochondrial	Mitochondrial proteins	Oligomer
Cytoplasmic	Histones	Polymer

which ADP-ribosylated histones (23).

In the chromatin of eukaryotic cells, histones Hl and H2B are demonstrated to be the major target of poly ADP-ribosylation. Histone Hl is associated with a stretch of 40-60 base pairs of the DNA between adjacent nucleosomes (Fig. 18). It has been suggested to play a role in the condensation and dispersion of chromatin during cell division. Two molecules each of H2A, H2B, H3 and H4 are found in nucleosome cores which resemble beads on a string of DNA as shown in Fig. 18. Their specific functions are, however, not clearly understood at present. So far poly(ADP-ribose) has been considered to be involved mainly in the regulation of DNA metabolism, but there have been reports indicating a role of poly(ADP-ribose) in the regulation of RNA metabolism, cell proliferation, differentiation or the NAD level in the cell. Further experiments are required before definite conclusions can be drawn as to the true function of poly(ADP-ribose).

The increasing number of poly and mono ADP-ribosylation reactions in both eukaryotes and prokaryotes reported during the last few years, indicates that these reactions constitute a new and a rather general way by which proteins are modified covalently and such reactions may play a major role in the regulation of various activities of the cell, particularly the chromatin function and the metabolism of proteins and nucleic acids.

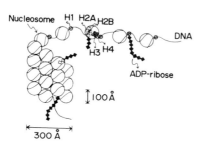

FIGURE 18.   Chromatin structure and poly ADP-ribosylation.

REFERENCES

1. Hayaishi, O., and Ueda, K., Ann. Rev. Biochem. 46:95-116.
    (1977)
2. Zatman, L. J., Kaplan, N. O., and Colowick, S. P., J. Biol.
    Chem. 200:197-212 (1953).
3. Honjo, T., Nishizuka, Y., Hayaishi, O., and Kato, I., J.
    Biol. Chem. 243:3553-3555 (1968).
4. Iglewski, B. H., and Kabat, D., Proc. Natl. Acad. Sci.
    U.S.A. 72:2284-2288 (1975).
5. Zillig, W., Mailhammer, R., Storko, R., and Rohrer, H.,
    Curr. Top. Cell. Regul. 12:263-271 (1977).
6. Pesce, A., Casoli, C., and Schito, G. G., Nature 262:412-
    414 (1976).
7. Goff, C. G., J. Biol. Chem. 249:6181-6190 (1974).
8. Rohrer, H., Zillig, W., and Mailhammer, R., Eur. J. Bio-
    chem. 60:227-238 (1975).
9. Moss, J., and Vaughan, M., J. Biol. Chem. 252:2455-2457
    (1977).
10. Nishizuka, Y., Ueda, K., Honjo, T., and Hayaishi, O., J.
    Biol. Chem. 243:3765-3767 (1968).
11. Okayama, H., Ueda, K., and Hayaishi, O., Proc. Natl. Acad.
    Sci. U.S.A. In press (1978).
12. Nishizuka, Y., Ueda, K., Yoshihara, K., Yamamura, H., Ta-
    keda, M., and Hayaishi, O., Cold Spring Harbor Symp. Quant.
    Biol. 34:781-786 (1969).
13. Dixon, G. H., Wong, N., and Poirier, G. G., Fed. Proc. 35:
    1623 (1976).
14. Requelme, P., Burzio, L., and Koide, S. S., Fed. Proc. 36:
    785 (1977).
15. Smith, J. A., and Stocken, L. A., Biochem. Biophys. Res.
    Commun. 54:297-300 (1973).
16. Okayama, H., Edson, C. M., Fukushima, M., Ueda, K., and
    Hayaishi, O., J. Biol. Chem. 252:7000-7005 (1977).
17. Ueda, K., Oka, J., Narumiya, S., Miyakawa, N., and Hayai-
    shi, O., Biochem. Biophys. Res. Commun. 46:516-523 (1972).
18. Miwa, M., Tanaka, M., Matsushima, T., and Sugimura, T.,
    J. Biol. Chem. 249:3475-3482 (1974).
19. Futai, M., and Mizuno, D., J. Biol. Chem. 242:5301-5307
    (1967).
20. Okayama, H., Honda, M., and Hayaishi, O., Proc. Natl. Acad.
    Sci. U.S.A. In press (1978).

21. Yoshihara, K., Tanigawa, Y., Burzio, L., and Koide, S. S., Proc. Natl. Acad. Sci. U.S.A. 72:289-293 (1975).

22. Kun, E., Zimber, P. A., Chang, A. C. Y., Puschendorf, B., and Grunicke, H., Proc. Natl. Acad. Sci. U.S.A. 72:1436-1440 (1975).

23. Roberts, J. H., Stark, P., Giri, C. P., and Smulson, M., Arch. Biochem. Biophys. 171:305-315 (1975).

STRUCTURE AND FUNCTION OF COLICIN E3

Kazutomo Imahori, Shigeo Ohno, Yoshiko Ohno-Iwashita and Koichi Suzuki

Department of Biochemistry, Faculty of Medicine
The University of Tokyo
Tokyo Japan

I. INTRODUCTION

Colicin E3 is a protein produced by a certain strains of coliform bacteria that carry a specific episome (colicinogenic cell). It kills sensitive strain of E.coli by inactivating ribosomes through the specific cleavage of their 16S RNA (1,2). Recently it was found that colicin E3 molecules inactivated the ribosomes in the in vitro system, just in the same way as in the in vivo system (3,4). These findings may suggest that colicin E3 is a kind of endonuclease and inactivates intracellular ribosomes by penetrating through cell membrane. However this idea is too simple, since it cannot explain the following contradicting experimental observations. (i) The amount of colicin E3 necessary for ribosome inactivation in the in vitro system was about 1000 times as much as that required in the in vivo system. (ii) Colicin E3 inactivates the ribosomes of other cells but not ribosomes of the host cell. If the ribosomes of colicin-producing cells were inactivated by colicin E3 itself, the cell could not synthesize colicin E3. (iii) Colicin E3 cleaves 16S rRNA at a specific position. However such cleavage takes place only when the RNA is integrated into 70S ribosomal particles. Neither isolated 16S RNA nor RNA in 30S subunits is susceptible to the fragmentation of colicin E3.

We thought that the answer to solve these
contradictions must be installed in the structure
of colicin E3.   It was expected that colicin E3
molecule should have some mechanism to facilliate
its penetration into cell interior.   Through our
recent works we succeeded in explaining these
contradictions and requirement by the structure of
colicin E3 molecule.   These results would furnish
an example how protein molecule would change its
property and function by complex formation or
fragmentation.

Results and Discussion

(A)   Colicin E3 is a complex of two proteins.
     For a long time colicin E3 was thought to be of
a single polypeptide.   In 1974 Jakes and Zinder
revealed that it is a complex of two components (5).
However, since they used SDS for dissociation of
colicin E3 and they failed to remove SDS from the
products they could not characterize the products
in detail.   In accordance with their findings, we
purified each component by several methods (6,7).
The most successful one was the gel filtration in
the presence of urea (7).   As shown in Fig 1, two
protein fractions, A and B appeared.   Each fraction
was pooled separately and dialyzed against phosphate
buffer.   As judged by SDS-gel electrophoresis,
each component was homogeneous.

     Next we characterized each component.   The
molecular weights of proteins A and B were estimated
as 50,000 and 10,000 respectively.   The amino acid
composition of each component is shown in Table I.

     When assayed by in vitro ribosome-inactivating
activity, protein A was more than 1000 times as
active as native E3, as judged by 50% inhibition of
ribosome activity (Fig 2).   This result suggested
that protein A is an active component of colicin E3
and protein B is an inhibitor of protein A.   This
idea was supported by the results of Fig 3, where
the activity of protein A was titrated with protein
B.   The results clearly indicated that the activity
of protein A was neutralized by an equal molar
amount of protein B.   The results also suggest
that one to one mixture of two proteins led to a
tight complex.   Truly when such a mixture was

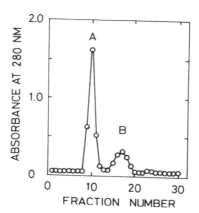

FIGURE 1.   Gel filtration of native colicin E3 in the
presence of 6 M urea.   Native colicin E3 was freshly prepared
in buffer (6 mg/1.5 ml) and was applied to a Sephadex G-100
column (1×50 cm) equilibrated with the same buffer.   The
column was developed with the same buffer at room temperature
and each fraction of 2 ml was monitored in terms of absorbance
at 280 nm.

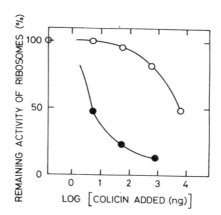

FIGURE 2.   In vitro ribosome inactivation by native
colicin E3 and protein A.   Ribosomes (1.0 A_{260} unit) were
incubated with various amounts of colicin samples in buffer
(30 μl) for 15 min at 37°C.   The remaining activity of the
incubated ribosomes was measured using a poly (U)-directed
polyphenylalanine synthesizing system.   The activity of
ribosomes incubated without colicin was taken as 100% activity.
O, Native colicin E3; ●, protein A.

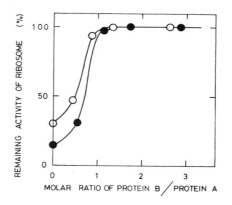

FIGURE 3.   Titration of protein A with protein B.   Protein
A was mixed with various amounts of protein B at the molar
ratios shown on the abscissa in buffer C at 0°C; aliquots
containing 15 ng (○) or 140 ng (●) of protein A were withdrawn and their ribosome-inactivating activities were measured
as described in the legend to Fig.2.

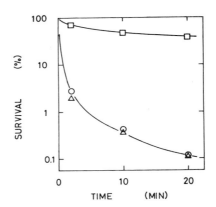

FIGURE 4.   Survival time course of cells treated with
protein A, native colicin E3, and reconstituted A-B complex.
Each colicin sample (the amount was adjusted with respect to
protein A) was added to 1 ml of log-phase culture (ca. 2×10^8
cells/ml) and the mixture was incubated at 37°C.   At the
times shown on the abscissa aliquots were withdrawn and diluted
with 0.85% NaCl solution to stop further adsorption of colicin.
Numbers of surviving colony-forming cells were measured on agar
plates.   □ , 13.6 ng of protein A; ○, 16.4 ng of native
colicin E3; △, reconstituted A-B complex (13.6 ng of protein A
+ 4.5 ng of protein B).

applied to disc-electrophoresis they migrated in one band with the same mobility as native colicin E3. The interaction of two proteins should be very strong. If we assume that the activity of native E3 in the in vitro assay is due to its dissociated product (protein A) the dissociation constant of colicin E3 can be estimated as $3 \times 10^{-12} M^{-1}$.

These results will explain why Nature has provided colicin E3, which is a complex, rather than protein A alone. If proteins A and B were synthesized simultaneously in the cell they would form a complex spontaneously and the inactivation of ribosomes in the same cell would be excluded. The results also explain the discrepancy of colicin E3 activities in the in vivo and in vitro systems. If we assume that protein B would be removed at the cell surface of the infected cell and only protein A could penetrate cell interior, it is reasonable that the activity in the in vivo system is 1000 times higher than in vitro system.

The complex formation has another advantage. In Fig 4, the bacteriocidal activities of protein A and colicin E3 were compared by survival time course. Evidently protein A is far less efficient than colicin E3 in killing sensitive cells. In other words, protein B stimulates the bacteriocidal activity of protein A. In Fig 4, the reconstituted complex of proteins A and B was assayed for its survival time course. Apparently it cannot be distinguished from the one of native colicin E3.

In the experiments above, we could purify two components of colicin E3 in native state. By using such native proteins, we have revealed the property and function of each protein. We found out that these two components can be reconstituted to colicin E3 spontaneously. These findings could give explanations for following dilemma or discrepancy which has been mistrerious so far. Colicin E3 inactivates ribosomes of invading cells but not those of colicin producing cells. Colicin E3 inactivates the ribosomes in the cells but rarely those in the in vitro system. Colicin E3 may be regarded like trypsin-trypsin inhibitor complex. However it is not strictly so. This complex has been designed to perform delicate regulation in accordance with needs. The complex is even

designed to promote the killing activity.   Recently
the importance of oligomeric proteins has been
recognized.   The present system might provide
another example of oligomeric proteins, which might
play an important role in the biological system.

(B)   Architecture of colicin E3.

    Even before the subunit structure of colicin E3
had been revealed we had an idea that some kind of
activation of colicin E3 should take place in the
cell membrane at the stage of infection.   One
candidate was the one analogous to the conversion
of chymotrypsin from chymotrypsinogen.   Thus we
examined the limited proteolysis of colicin E3 (8).
As expectedly the in vitro activity of colicin E3
increased by such a treatment.   Recnetly Lau and
Richards conducted a similar experiment and found
that trypsin cleaved colicin E3 into two fragments
(9).   However since these fragments have not been
characterized in detail, we carried out further
experiments.

    Native colicin E3 was treated with trypsin and
the resulting sample was chromatographed on Sephadex
G-100 column.   As shown in Fig 5 two major peaks,
T1 and T2 appeared in accordance with the results of
Lau and Richards.   However by analyzing each peak
fraction on SDS-gel electrophoresis, we found that
T2 fraction was composed of two conponents, T2A and
T2B.   T2B migrated faster.   Using the same methods
applied for separation of proteins A and B, T2A and
T2B were separated from each other.   Each one of
T1, T2A and T2B gave a single band on SDS-gel
electrophoresis.

    Next we characterized each of these fragments.
T2B co-migrated with protein B in SDS-gel electrophoresis and the former could neutralize the
activity of protein A to the same extent as the
latter.   We could not distinguish T2B from protein
B in amino acid composition and N-terminal amino
acid (Gly).   Thus T2B was assigned as protein B
itself.

    T2A is a basic protein as judged by the amino
acid composition shown in Table I.   Because of its
basic property the molecular weight of T2A is hard

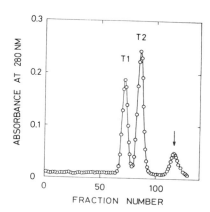

FIGURE 5.   Gel filtration of colicin E3 digested with trypsin.   Digestion products of colicin E3 with trypsin (6 mg/1 ml of buffer B) were applied to a Sephadex G-100 column (1.8×130 cm) equilibrated with buffer and developed with the same buffer.   Each fraction of 2.5 ml was monitored in terms of absorbance at 280 nm.   The arrow indicates the bed volume.

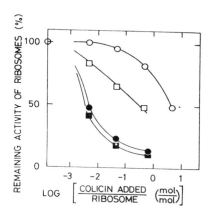

FIGURE 6.   In vitro ribosome inactivation by native colicin E3 (O), T2 (□), protein A (●), and T2A (■).   Colicin samples were incubated with ribosomes at the molar ratios shown on the abscissa in buffer for 15 min at 37°C.   The remaining activity of the incubated ribosomes was measured as described in the legend to Fig.2.   Molar ratios of ribosomes/ colicin samples were calculated on the assumption that 1.0 A_{260} unit of ribosomes corresponds to 25 pmol.

TABLE I.   Amino acid compositions of components
and fragments of colicin E3.   Values were calculated
from the results of one set of experiments on the
basis of the molecular weights shown on the bottom
row.

Amino acid	T1	T2A	A	B
Lys	11	16	35	5
His	6	4	10	1
Arg	11	5	22	1
Asp	48	12	70	18
Thr	19	4	21	2
Ser	39	4	38	6
Glu	21	8	42	8
Pro	27	8	30	3
Gly	57	16	65	7
Ala	35	2	51	1
1/2Cys	—	—	0	1
Val	36	0	32	6
Met	5	0	7	1
Ile	13	4	16	3
Leu	18	6	25	6
Tyr	3	5	9	4
Phe	9	2	12	8
Trp	—	2	5	3
Total	358	98	490	84
Molecular weight×10^{-3}	36	11	50	10

to determine. However we estimated it as 11,000 from its co-migration with cytochrome c and from amino acid analysis.

It is interesting that such a basic fragment as T2A could be recovered after trypsinolysis. When protein A was subjected to trypsinolysis only T1 fragment was obtained. When T2A was digested by trypsin, no trace of large peptide was observed. These results suggest that T2A could survive from trypsinolysis because it was protected by protein B, which was also resistant to trypsin. Truly SDS-gel electrophoresis of T2A-B complex gave the same pattern before and after trypsin treatment.

T2A had no bacteriocidal activity. However it had a strong activity for the inactivation of ribosomes in the in vitro system. As can be seen in Fig 6, the activity of T2A is comparable to or somewhat higher than that of protein A. Thus T2A is an active fragment of protein A. Like protein A, the activity of T2A was neutralized by protein B, as shown in Fig 7. However, it could not be neutralized completely unless large excess amount of protein B was added. This indicates that the binding of T2A with B is weaker than that of protein A with B. Truly, as shown in Fig 6, the in vitro activity of T2 fraction was higher than native colicin E3 but lower than T2A or protein A. This can be explained by that under this condition, T2 was partly dissociated into B and T2A and the activity could be accounted by the latter.

The molecular weight of T1 was estimates as 35,000 from SDS-gel electrophoresis. As shown in Table I, this protein is quite rich in hydrophobic amino acid residues. This hydrophobic property may prevent this protein from trypsinolysis. It is also concievable that this fragment is originated from a hydrophobic and protease resistant region (T1 region) of protein A, since trypsinolysis of native colicin and protein A yielded the same T1 fragment. T1 fragment indicated no activity both in vivo and in vitro assay system.

By summarizing these results, we can figure out a model for the architecture of colicin E3 molecule. Colicin E3 is a complex of proteins A and B.

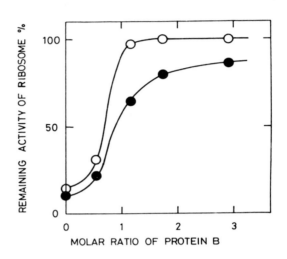

FIGURE 7.   Titrations of proteins A and T2A with protein B.
Protein A or T2A was mixed with various amounts of protein B
in buffer at the molar ratios shown on the abscissa.   Aliquots
containing 140 ng of protein A (○) or 50 ng of T2A (●) were
withdrawn and their ribosome-inactivating activities were
measured as described in the legend to Fig.2.   When excess
amounts of protein B (more than 10-fold molar excess) were
added, 100% inhibition of the activity of T2A was also
observed.

The protein A molecule consists of two regions, T1
and T2A.   The T1 region is folded into a compact
conformation because of its high hydrophobicity
and is resistant to tryptic digestion.   Although
T1 fragment had no apparent biological activity
T1 region of E3 molecule may play an important role
at the stage of attachment to or penetration into
the cell membrane, since T2 fragment had neither
bacteriocidal activity nor activity to inhibit
native colicin E3 in vivo.   The machinery necessary
for ribosome inactivation is all located in the T2A
region.   Protein B binds to T2A region firmly and
specifically and directly neutralizes the activity
of protein A or its active fragment, T2A fragment.
The T2A region is folded into a compact structure
and is resistant to trypsin provided protein B is
bound to it.   This idea can be supported by the
fact that N-terminal sequence of T2A fragment is
Lys.Gly.Phe.Lys...   However, when protein A is free
from protein B, the T2A region is unfolded and can
be completely digested by trypsin.   This suggestion
is reasonable because the T2A region is very rich
in lysine.   This basicity of the T2A region may
play an important role in the binding of protein A
or its active fragment with protein B, and also with
ribosomal RNA.

Probably a peptide segment between T1 and T2A
regions exists in native colicin E3, which can be
cleaved very easily by several proteases.   This
peptide segment would be specifically designed to
produce T2 fragmetn.   Although the physiological
significance of the fragmentation has not been
cleared out, we assume that fragmentation takes
place at the stage of penetration into cell interior.
Since the dissociation of protein B easier from T2
than from native colicin E3 fragmentation would
faciliate the production of T2A free from protein B,
in the cell membrane.   Once T2A fragment would be
liberated in the cell membrane, it would be easily
transferred to cytoplasm because of hydrophilic
nature.

(C)   Action mechanism of colicin E3.

Colicin E3 inactivates 70S ribosomes by
cleaving its 16S rRNA at a specific position and
releasing so-called E3 fragment of 50nucleotides.

This fact may suggest that colicin E3 is a kind of
endonuclease or ribonuclease.   However this conclusion needs reservation since it has been reported
that neither 30S subunits nor 16S rRNA were
susceptible to the fragmentation reaction of E3 (10,
11).   There exists a possibility that some of
ribosomal proteins may have a latent nuclease activity and colicin E3 may induce this activity.

However, since previous works were carried out
by using weakly active colicin E3 itself, we re-
examined the possibility of fragmentation of 16S
rRNA in the isolated 30S subunit system.   In the
present experiments specific cleavage of 16S rRNA
was measured directly by gel electrophoresis (12).
The results are shown in Fig 8.   Fig 8-c indicates
the appearance of E3 fragments when enough protein
A was interacted with isolated 30S subunits.   That
the peaks indicated in Fig 8-b,c and e are not due
to artefacts can be proved by the following facts.
The position of these peaks coincided with the
peak obtained when 70S ribosomes were treated with
protein A (Fig 8-f,g).   This peak disappeared when
protein B was added with protein A (Fig 8-d).

However, isolated 30S subunits were far less
sensitive to protein A compared with 70S ribosomes
as shown in Fig 9, where fragmentation percentage
is plotted against the amount of protein A added.
Percent of fragmentation means the molar ratio of
E3 fragments to input 30S subunits.   In the case
of 30S subunits almost equal molar amount of protein
A was required to obtain saturation level of E3
fragments.   On the contrary, in the case of 70S
ribosomes the addition of 1/40 molar amount of
protein A is enough to saturate the amount of E3
fragments.

These results may suggest that the reaction of
protein A with 30S subunits is stoichiometric and
not catalytic.   However, when 1/40 molar amount
of protein A was incubated with isolated 30S subunits for a prolonged time, the amount of E3
fragment approached to the saturation level.   Thus
the difference observed for the fragmentation
efficiency should be attributed to the difference of
reaction velocities.

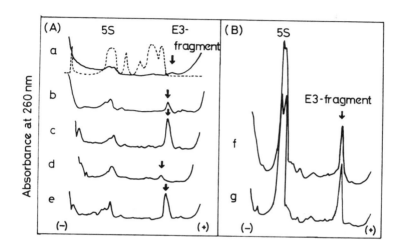

FIGURE 8.   Analysis of E3 fragments by polyacrylamide gel
electrophoresis.   (A) Samples (30 µl) of 30S subunits (1.7
A_{260} units) (a-d) or 30S subunits plus 0.254 A_{260} units of
70S ribosomes (e) were incubated for 30 min at 37°C in buffer
with various amounts of protein A.   (B) Samples of 30S
subunits (1.7 A_{260} units) plus 50S subunits (3.3 A_{260} units)
were incubated in the same conditions as (A).   The amounts
of protein A were zero (a), 0.125 µg (b,f), 5.0 µg (c,e,g),
and a preincubated mixture of 5.0 µg of protein A and 2.16 µg
of protein B (d).   The samples were analysed by UV scan
after electrophoresis.   The dotted line shows the pattern of
proteins detected by coomassie brilliant blue staining.

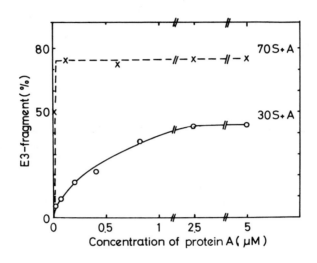

FIGURE 9.   Effect of protein A concentration on the amount
of E3 fragments.   Various amounts of protein A were incubated
for 30 min at 37°C with 30S subunits (1.7 A_{260} units) (**O**),
or with 70S ribosomes (5.1 A_{260} units) (**X**).   E3 fragments
were detected as described in Fig.1.   In the latter case (**X**),
the amount of E3 fragments was expressed by the molar ratio
to 5S RNA, assuming that extinction coefficients per nucleotide
of 5S RNA and E3 fragments were equal.   In the former case
(**O**), the amount of E3 fragments was calibrated by the amount
of E3 fragments released from 70S ribosomes.

A small amount of 5S RNA can be observed in Fig 8-a to 8-e. This indicates that the present preparation of 30S subunits is contaminated with 50S subunits. This fact may suggest that E3 fragments obtained from 30S preparation are derived from contaminating 70S ribosomes. This possibility, however, can be ruled out because the molar amount of E3 fragments in Fig 8-c is far in excess of that of contaminating 70S ribosomes. In addition, native 30S subunits, which cannot bind to 50S subunits due to the presence of dissociation factor, were fragmented to the same extent as other 30S preparations.

These results indicated that colicin E3 attacks directly 30S subunits and 50S subunits merely stimulate the reaction. However detailed mechanism on 30S subunits has not been elucidated. Although protein A cleaves 16S rRNA at a specific position neither unusual bases nor a specific base sequences were found around the position of cleavage.

Thus the specificity of cleavage cannot be explained by a primary structure of RNA. Alternatively some specific conformation around the point of cleavage may be responsible for the specificity of fragmnetation. It is also probable that protein A or its active fragment would bind specifically to a ribosomal protein which exists close to the breakage point of RNA. Another possibility is that protein A or its active fragment may induce a latent nuclease activity of some ribosomal protein.

Further works are in progress to elucidate the structure and function of colicin E3, its components and its fragments.

REFERENCES

(1) Senior, B.W.,and Holland, I.B. (1971). Proc. Natl
      Acad. Sci. U.S. 68:959-963.
(2) Bowman, C.M., Dahlberg, S.E., Ikemura, T.,
      Konisky, J., and Nomura, M. (1971). Proc.
      Natl. Acad. Sci. U.S. 68:946-968.
(3) Boon, T. (1971). Proc. Natl. Acad. Sci. U.S.
      68:2421-2425.
(4) Bowman, C.M., Sidikaro, J.,and Nomura, M. (1971).
      Nature New Biol. 234:133-137.
(5) Jakes, K.S., and Zinder, N.D. (1974).Proc. Natl.
      Acad. Sci. U.S. 71:3380-3384
(6) Hirose, A., Kumagai, J., and Imahori, K. (1976).
      J. Biochem. 79:305-311.
(7) Ohno, S., Ohno-Iwashita, Y., Suzuki, K., and
      Imahori, K. (1977). J. Biochem. 82:1045-1053.
(8) Ohsumi, Y., and Imahori, K. (1974). Proc. Natl.
      Acad. Sci. U.S. 71:4062-4066.
(9) Lau, C., and Richards, F.M. (1976). Biochemistry
      15:3856-3863.
(10) Boon, T. (1972). Proc. Natl. Acad. Sci. U.S.
      69:549-552.
(11) Bowman, C.M. (1972). FEBS Lett. 22:73-75.
(12) Ohno-Iwashita, Y., and Imahori, K. (1977).
      J. Biochem. 82:919-922

A CHROMATIN-BOUND NEUTRAL PROTEASE AND ITS INHIBITOR IN RAT PERITONEAL MACROPHAGES[1]

Takashi Murachi
Yukio Suzuki

Department of Clinical Science
Kyoto University Faculty of Medicine
Kyoto, Japan

INTRODUCTION

Biological regulation as mediated by proteolytic enzymes attracts increasing interests of
bioscientists from various fields. The choice
of the kind of cells, tissues or systems to be
studied often seems to be a subject of chance.
Our choice of rat peritoneal macrophages was not
made with any specific aim at first, except some
vague idea that the study in proteases of macrophage
might be interesting in view of its function in
immunity. Since macrophages are typical phagocytes,
the occurrence of powerful intracellular protease
is reasonable and, in fact, such has been reported
to be the case (1). However, these proteases were
only poorly characterized when we began performing
a series of experiments several years ago (2).
The discovery and application of a group of
specific protease inhibitors of microbial origin
by Umezawa and his associates have immensely
facilitated the characterization of proteases in
complex biological materials (3,4). Thus we were

[1] Supported in part by a grant from the Ministry
of Education, Science and Culture of Japan.

183

able to demonstrate the occurrence of a chymotrypsin-
like enzyme in the freeze-thaw lysate of rat
peritoneal macrophages, without knowing then its
intracellular localization (2). The subsequent
study revelaed the occurrence in the same lysate
of a factor or factors inhibitory to the neutral
protease (5). Furthermore, two lines of unexpected
results have followed: that both of the neutral
protease and its inhibitor are associated with
chromatin of the cells and that the inhibitor is
not a protein but of nucleic acid-like nature.
Experimental evidence available so far indicates
that the inhibitor is a compound closely related
to, if not identical with, poly(ADP-ribose).
    This article reviews the experimental approaches
we have been making for these several years.

EXPERIMENTAL SYSTEM

    Peritoneal macrophages were obtained from male
adult rats of the Wistar strain weighing 300-350 g
given 15 ml of 0.1-0.5% glycogen in saline by
intraperitoneal injection. Animals were sacrificed
four days after the injection, and the exudate
cells were collected in Hanks' solution. More than
80 per cent of the cells collected were found to be
large mononuclear cells. Macrophages could be
separated from other cells using a glass-adherence
technique (2), but the experimental results concerning proteases were practically unchanged before
and after such separation. Neutral and acid protease activities were determined using 2% alkali
and heat-treated casein and acid-denatured hemoglobin, respectively, as substrates. The trichloroacetic acid-soluble products released were measured
spectrophotometrically either directly at 280 nm or
at 750 nm after the addition of Folin-Ciocalteau
reagent. Protease inhibitors of microbial origin
were gifts from Drs. T. Aoyagi and H. Umezawa.
The details of these and other experimental procedures have been published previously (2,5).

OCCURRENCE OF A NEUTRAL PROTEASE

The lysate of macrophage cells was prepared by freezing and thawing treatments six times and then centrifugation at 3000 rpm for 10 min. Figure 1 shows that the protease activity in the lysate is maximal at pH 4.0 and also detected in the neutral pH range (2). Inhibition experiments revealed that the majority of the hemoglobin-hydrolyzing activity at pH 4.0 was inhibited by pepstatin while approximately 70 per cent of the caseinolytic activity at pH 7.0 was sensitive to chymostatin (2).

When the lysate was centrifuged at 105,000 g for one hour, the acid protease activity was quantitatively recovered in the supernatant. The neutral protease activity was found to be associated with

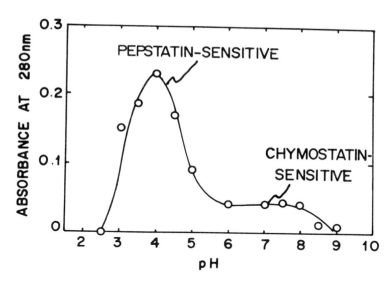

FIGURE 1. Dependence on pH of the hemoglobin-hydrolyzing activity of rat peritoneal macrophages. An incubation mixture made of 1 ml of 1% denatured hemoglobin solution adjusted to desired pH, 0.3 ml of buffer at various pH values, and 0.2 ml of the enzyme solution was allowed to stand at 38° for 2 hours. The reaction was terminated by the addition of 3 ml of 5% trichloroacetic acid and the absorbance at 280 nm of the filtrate was measured. The ordinate is for the increment in absorbance at 280 nm after 2 hour incubation (2).

TABLE I. Distribution of proteases in the rat peritoneal macrophages[a]

Fraction	Protein (mg)	Neutral protease			Cathepsin D		
		Total Activity (units)	Specific activity (units/mg protein)	Yield (%)	Total activity (units)	Specific activity (units/mg protein)	Yield (%)
Lysate	16.0	321	20.1	100	12.4	0.78	100
Water extract	5.6	0	0	0	13.1	2.34	106
KCl extract	6.0	154	25.7	48	0.3	0.05	2
Final precipitate	4.7	107	22.8	33	0	0	0

[a] The pellet of cells was suspended in distilled water and subjected to freezing and thawing six times. The lysate was centrifuged at 105,000 x $\underline{g}$ for 1 hour. The supernatant was removed and named the "water extract" fraction. The precipitate was suspended in 0.05 M Tris-HCl buffer, pH 8.5, containing 1 M KCl, and then centrifuged at 105,000 x $\underline{g}$ for 1 hour. The supernatant fluid was concentrated by dialysis against polyethylene glycol 6,000, and further dialysis against distilled water overnight. The product was named the "KCl extract" fraction. The precipitate that remained after extraction with 1 M KCl was suspended in distilled water, and this suspension was named the "final precipitate" fraction.

the pellet fraction from which it was extracted
with 1 M KCl (Table I). The latter enzyme has an
optimum at pH 8.5 and is strongly inhibited by
diisopropylphosphorofluoridate and by chymostatin,

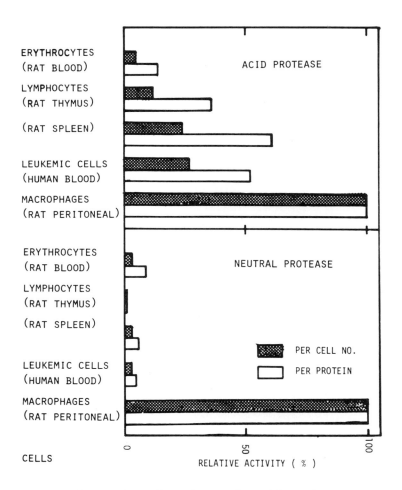

FIGURE 2. Distribution of acid and neutral
proteases among several different kinds of blood
cells and phagocytes. Cathepsin D (acid protease)
activity was measured at pH 3.5 with acid-denatured
hemoglobin as a substrate, and neutral protease at
pH 7.2 with casein. Shaded bars are for relative
specific activity per cell; open bars are for that
per mg protein (7).

but not by p-chloromercuribenzoate, ethylene-
diaminetetraacetic acid, leupeptin, antipain, or
pepstatin (5).

    Figure 2 illustrates the distribution of acid
and neutral proteases among several different kinds
of blood cells and phagocytes. It is apparent that
rat peritoneal macrophages show an outstandingly
high level of the neutral protease activity (6).
The rationale of such high activity of the neutral
protease in macrophage remains to be elucidated.

OCCURRENCE OF AN INHIBITOR ON THE NEUTRAL PROTEASE

    The first hint on the occurrence of protease in-
hibitor was given by an experiment on the effect of
salts. As shown in Figure 3 (open circles), the
neutral protease activity was found to be markedly

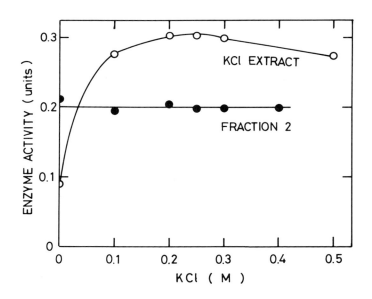

FIGURE 3.   The effects of KCl concentration on
the activity of the neutral protease in macrophages.
o, KCl extract (46 µg protein per tube); ●, fraction
2 (20 µg protein per tube) obtained by gel filtration
of the KCl extract through Sephadex G-75 equilibrated with 0.05 M Tris-HCl, pH 8.5, containing 1 M KCl (5).

enhanced when assayed in the presence of KCl, attaining the maximum at a final concentration of
0.25 M.   Enhancement to comparable degrees was
also noted in the presence of other salts such as
KBr, KI, NaCl, MgCl$_2$, etc. (5).   The finding could
have been explained either by conformational change
of the protease molecule so as to induce activation
or by dissociation of an inhibitor from the enzyme
in the presence of a high concentration of salts.
Subsequent experiments demonstrated that the
latter possibility was true.
    When the KCl extract (Table I) was subjected to
gel filtration through a Sephadex G-75 column in
the presence of KCl at a lower (0.1 M) concentration, the neutral protease activity appeared almost
close to the void volume of the column (Figure 4A).

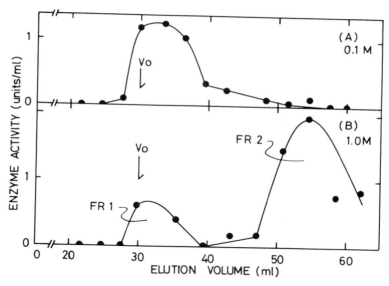

FIGURE 4.   The elution patterns of the neutral
protease of macrophages in the KCl extract fraction
from a Sephadex G-75 column.   The Sephadex G-75
column (2.5x25.0 cm) was equilibrated and eluted
with 0.05 M Tris-HCl, pH 8.5, containing 0.1 M KCl
in (A) and 1.0 M KCl in (B).   Flow rate, 15 ml per
hour; fraction volume, 3 ml.   The KCl extract fraction (2 ml containing 0.52 mg protein), with the
KCl concentration made up to 0.1 M in (A) and 1.0 M
in (B), was applied to the column.   Vo donotes
the void volume of the column (5).

On the contrary, the bulk of the activity appeared in lower molecular weight fractions when the gel filtration was performed in the presence of a higher (1.0 M) concentration of KCl (Figure 4B). Fraction 1 in Figure 4B had a much higher absorbance value at 260 nm as compared to fraction 2. Both fractions 1 and 2 were separately pooled, concentrated, and dialyzed against water.

It was now revealed that fraction 2 contained the neutral protease, which was no longer activated by the addition of KCl to the assay medium (solid circles in Figure 3). Fraction 1, on the other hand, was found to exhibit a distinct and progressive inhibition of the caseinolytic activity of fraction 2, as shown in Figure 5. Trypsin and

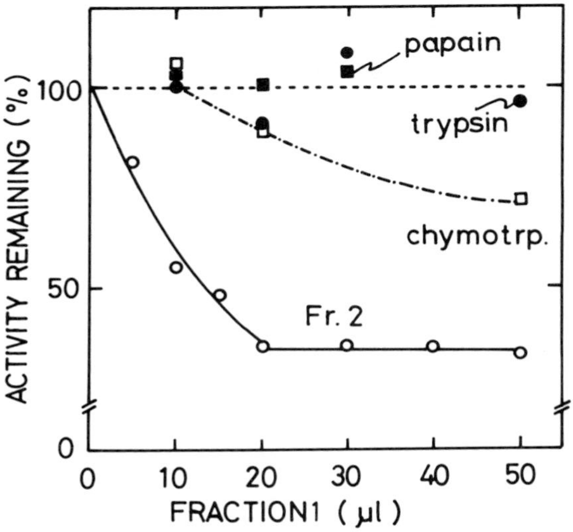

FIGURE 5. The effect of fraction 1 on the neutral protease activities. Fractions 1 and 2 were separately pooled (see Fig. 4), concentrated to approximately the same volume as that of the original extract, then dialyzed against distilled water overnight at 4°. Increasing amounts of fraction 1 were incubated at pH 8.5 and at 37° in the reaction mixture which contained 20 µg of fraction 2 (o); 5µg of trypsin (●); 10µG of α-chymotrypsin (□); 20µg of papain activated with 5 mM cysteine (■) in total volume of 0.2 ml (5).

papain were not inhibited by fraction 1, while
chymotrypsin was only partially inhibited.

CHROMATIN-BOUND PROTEASE AND INHIBITOR

All of the preceding experiments were carried
out using the freeze-thaw lysate of macrophages.
For studying intracellular localization of the
neutral protease as well as of its inhibitor, the
disruption of the cellular membrane had to be performed by much milder procedures. Table II summarizes the lines of evidence for the occurrence of
a neutral protease and its inhibitor as bound to
chromatin thus isolated. The purity of the nuclear
fraction was also confirmed by electron-microscopy
of the specimen. The enhancement of the caseinolytic activity by the addition of 0.25 M NaCl to the
assay medium is indicative of the simultaneous
presence of the enzyme and the inhibitor which can
be dissociated at a high ionic strength.

The occurrence of a histone-degrading enzyme
in rat liver chromatin was reported by Chong et al.
(7). However, Heinrich et al. (8) later revealed
that the enzyme was actually localized in inner
membrane of mitochondria which happened to con-
taminate the chromatin preparation. The possibility
of such contamination is unlikely in the present
experiment as shown by the pattern of distribution
of marker enzymes (Table II).

The chromatin preparation from the nuclei was
further fractionated on a hydroxyapatite column in
a medium containing 2 M NaCl and 5 M urea as shown
in Figure 6. The eluted fractions were numbered
H1 to H4 (9), separately pooled, concentrated, and
dialyzed. The results of analyses on these fractions
are shown on Table III, which indicated that fraction H1 only contains the protease. The caseinolytic activity in fraction H1 was no longer sensitive to KCl or NaCl in the medium, but in its absence it was progressively inhibited by the addition of increasing amount of fraction H3 as shown
in Figure 7 (open circles). No such inhibitory
effect was found with fractions 2 and 4.

It should be noted that complete inhibition
took place even when a large amount of fraction H3
was added, a fact in entire accord with what was
observed earlier with fractions 1 and 2 shown in
Figure 5. This may imply identity of the two

TABLE II. Intracellular distribution of the protease and its inhibitor[a]

Fraction	Protein (%)	DNA (%)	Cytochrome c oxidase (%)	Cathepsin D (%)	Glucose-6-phosphatase (%)	Neutral protease (%)	Salt activation, fold
Homogenate	100	100	100	100	100	100	3.37
Post-nuclear	39	3	64	85	92	10	
Nuclear	39	90	8	7	14	95	3.26
Chromatin						85	2.73

[a] Rat peritoneal exudate cell suspension (1.2×10^7 cells per ml) in 0.5% Triton X-100 containing 0.25 M sucrose and 3.3 mM $CaCl_2$ was homogenized and fractionated by centrifugation at 900 g for 5 min. Chromatin was isolated from the nuclear fraction by the method of Huang and Huang (13). Salt activation means the ratio of caseinolytic activity in the presence of 0.25 M NaCl to that in its absence.

pairs of protease and inhibitor concerned, though
they were prepared by procedures different from
one another.

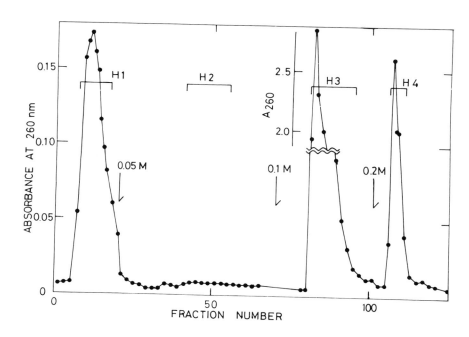

FIGURE 6.   Elution profile of 260 nm-absorbing
materials in the chromatin extract from a hydroxyapatite column.   Chromatin extract (11.8 A_{260} in
8.5 ml) was applied to a hydroxyapatite column
(1.5 cm x 17 cm) equilibrated with a medium of
2 M NaCl-5 M urea containing 1 mM sodium phosphate
buffer, pH 7.0, and 2 mM Tris.   After the column
was washed with 50 ml of the same medium, the
absorbed materials were eluted stepwise by increasing concentration of sodium phosphate buffer, pH
7.0, to 0.05 M, 0.1 M, and 0.2 M, at the positions
shown with arrows in the figure.   Three-ml fractions were collected at a flow rate of approximately
10 ml per hour.   The fractions marked by brackets
in the figure were pooled, and named fractions H1,
H2, H3 and H4, according the description made by
Rickwood and MacGillivray (9).

PROPERTIES OF THE NEUTRAL PROTEASE

Fraction Hl from hydroxyapatite column (Figure 6) was subjected to gel filtration through a Sephadex G-75 column, yielding a protease fraction apparently homogeneous on polyacrylamide gel electrophoresis. The protease has a molecular weight of approximately 26,000 daltons as determined by gel filtration and by gel electrophoresis in the presence of sodium dodecylsulfate. It also hydrolyzes histones and synthetic substrates such as benzoyl-L-tyrosine ethyl ester and benzoyl-L tyrosine p-nitroanilide.

In Table IV are summarized the effects of various inhibitors on the caseinolytic activity of the protease, indicating that this enzyme is a chymotrypsin-like enzyme as it is the case for a number of organella-bound proteases (10).

TABLE III. Fractionation of rat peritoneal macrophage chromatin on a hydroxyapatite column[a]

Fraction	Protein, mg	A_{280}/A_{260}	Neutral protease		
			Total units	%	Specific activity
Chromatin extract	2.90	1.36	378	100	130
H1	0.88	1.13	160	43	184
H2	0.15	0.73	7	2	45
H3	0.62	0.63	2	1	3
H4	0.23	0.62	1	0	3

[a] Fractions Hl, H2, H3, and H4 were eluted from the column (Fig. 6) and separately concentrated and dialyzed against water overnight.

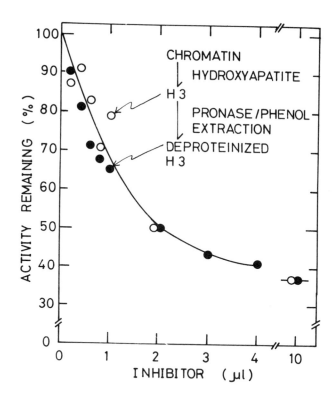

FIGURE 7. Effects of fractions H3 and deproteinized H3 on the neutral protease activity.
Deproteinized H3 (●) (1.2 A_{260} per ml) was obtained
from fraction H3 (o) (1.5 A_{260} per ml) by digestion
with Pronase followed by extractions with phenol.
Increasing amount of these fractions were incubated
with 2.6 µg of fraction H1 at pH 8.5 and at 37° in
the reaction mixture of a total volume of 0.2 ml.
The assay was carried out in the absence of NaCl.

CHARACTERIZATION OF THE INHIBITOR

Fraction H3 from hydroxyapatite column (Figure
6) did not loose its inhibitory potency either by
heat treatment at 100° for 5 min. or by trypsin
digestion at 37° for one hour. These facts
together with the elution position of this
material from hydroxyapatite gel (Figure 6) and the

pattern of its ultraviolet absorption (Table III)
led us to speculate if the inhibitory principle
were of nucleic acid-like nature, although no pre-
vous reports as far as we were aware of had ever
been published to demonstrate that a nucleic acid
could be a protease inhibitor.

Attempts were then made to eliminate any pro-
teinous contaminants in fraction H3 by incubating
it with Pronase followed by extractions with phenol.
The deproteinized fraction H3 thus obtained was
found to be as inhibitory as is the parent fraction
H3 (solid circles in Figure 7). When deproteinized
H3 was further treated with DNase 1, nuclease Pl,
or snake venom phosphodiesterase, it did not loose
inhibitory potency as shown in Table V, suggesting
non-identity of the inhibitory principle in question

TABLE IV. Effects of various compounds on the
caseinolytic activity of neutral protease[a]

Compound	Final concentration	per cent activity
None		100
Diisopropyl phosphoro-fluoridate	1 mM 10 mM	12 0
Ethylenediamine-tetraacetic acid	1 mM	95
p-Chloromercuri-benzoate	0.5 mM	87
Chymostatin	10 μg/ml 100 μg/ml	9 2
Pepstatin	30 μg/ml	100
Leupeptin	60 μg/ml	97
Antipain	60 μg/ml	98

[a] Assays were performed in 0.05 M Tris-HCl buffer,
pH 8.5, containing 1 M KCl and various inhibitors;
incubation for 1 hour at 37°.

with a DNA or RNA species ordinarily encountered.
After a number of trials we came across with poly-
(ADP-ribose) glycohydrolase which Sugimura and his
associates recently isolated from calf thymus and
reported to be specific on poly(ADP-ribose) (11).
As shown in the bottom row of Table V, this glycohydrolase was markedly effective in inactivating deproteinized H3. The inactivation was found to be
dependent on the dose of glycohydrolase added to
the incubation medium, while heat-inactivated
glycohydrolase was without effect.

TABLE V. Effect of digestion of deproteinized
fraction H3 by various nucleases and glycohydrolase
on the inhibitory potency of fraction H3 [a]

Nuclease and glycohydrolase	pH	Volume required for 20% inhibition, μl	Relative inhibitory potency
None (control)	6.5		1.0
DNase 1 from bovine pancreas	7.0	6.5	1.0
Nuclease P1 from *Penicillium citrinum*	5.7	7.0	0.9
Venom phospho-diesterase	8.0	6.1	1.1
Poly(ADP-ribose) glycohydrolase from calf thymus	7.2	16.0	0.4

[a] The deproteinized inhibitor (100 μl, 0.14 A_{280})
was digested with each 1μg of enzyme (100 μl)
at 37° for 1 hour at pH values indicated above. The
inhibitory effect of each digest on the caseinolytic
activity of fraction H1 was plotted in the same way
as that in Fig. 7, and from the straight portion of
each plot the volume of the digest required for 20%
inhibition was calculated. The relative inhibitory
potency was calculated by taking the value for the
control as unity.

Table VI summarizes the effects of various
nucleotides, naturally occurring or synthetic, on
the caseinolytic activity of the neutral protease
from macrophage chromatin. The results indicate
that poly(ADP-ribose) is the only nucleotide among
those tested which actually inhibits the protease.
The specimen of poly(ADP-ribose) used in this
experiment was a gift from Dr. T. Sugimura and said
to have a mean chain length corresponding to approximately 30 repetitions of ADP-ribose units
(Figure 8). It is interesting to note that a monomeric ADP-ribose has no effect. On the other hand,

TABLE VI. Effects of various nucleotides and
related compounds on the caseinolytic activity
of fraction Hl [a]

Effector	Remaining activity, %	
	at 2µg of effector/tube	at 5µg of effector/tube
Calf thymus DNA, native	95	106
Calf thymus DNA, denatured	94	104
Yease RNA	97	106
Polyadenylate	104	106
Polyuridylate	102	104
Polycytidylate	100	89
Poly(ADP-ribose)	64	38
ADP-ribose	93	99

[a] The protease activity of fraction Hl was
measured with and without the effector in the absence of NaCl. To prepare the denatured DNA, the
native DNA (100 µg per ml) was heated at 100° for
10 min and then rapidly cooled.

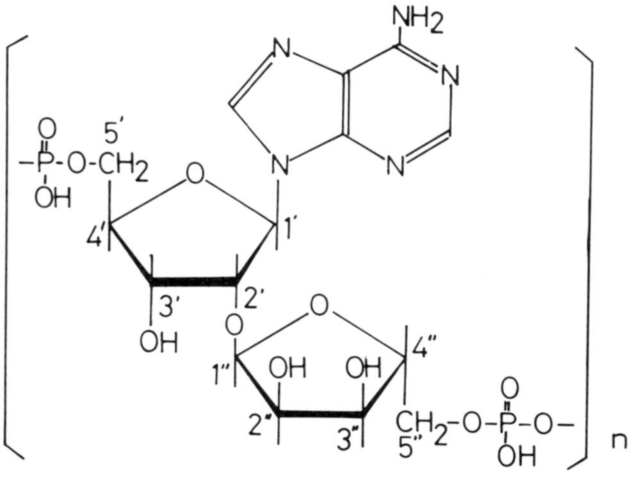

FIGURE 8. A repeating unit of poly(ADP-ribose).

the poly(ADP-ribose) preparation used did not exhibit any effect on trypsin, chymotrypsin and
papain. All of these data lend support to believe
that the protease inhibitor present in fraction H3
is a compound very closely related to poly(ADP-
ribose), except for the finding that the inhibitor
was not inactivated by phosphodiesterase. An
ordinary poly(ADP-ribose) preparation should have
been very readily cleaved by phosphodiesterase.
Therefore, further investigation will be necessary
to elucidate whether the inhibitor would be a somehow modified derivative of poly(ADP-ribose) or an
entirely different nucleotide which happened to be
susceptible to poly(ADP-ribose) glycohydrolase.

REGULATORY ASPECTS

On the basis of the experimental findings so far
described, one could construct a hypothetical
mechanism for the regulation involving the neutral
protease in question. Figure 9 diagrammatically
illustrates a unit of nucleosome representing an
octameric associate of histones with a coil of
double-stranded DNA. In close association with
the nucleosome are the neutral protease, the poly-
(ADP-ribose)-like inhibitor, and poly(ADP-ribose)

glycohydrolase. As shown in the present study, the
hydrolysis of histones by the neutral protease is
suppressed by the inhibitor, while the cleavage of
the inhibitor by glycohydrolase is known to be
suppressed by histones (11). More than a decade
has elapsed since poly(ADP-ribose) was discovered
in nuclei and a number of biological roles were
described or proposed (12). Our present finding
may suggest a novel role of poly(ADP-ribose) in the
mechanism of gene expression through functioning as
an inhibitor of a chromatin-bound, histone-
degrading protease.

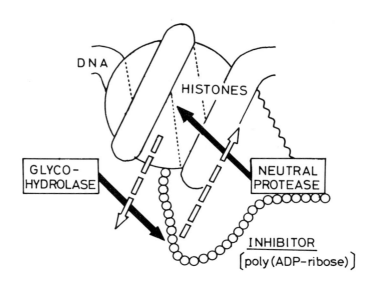

FIGURE 9. Hypothetical mechanism for the regulation involving chromatin-bound neutral protease,
its substrate (histones in nucleosome) and its
inhibitor (poly(ADP-ribose)-like compound), and
poly(ADP-ribose) glycohydrolase. Solid arrows,
hydrolytic cleavages; broken arrows, inhibitions.

SUMMARY

The glycogen-induced macrophages in rat peritoneal exudate contain several different proteolytic enzymes. The acidic protease(s) is present in the lysosomes, while a neutral protease is localized in the nuclei. The latter activity in the macrophage lysate was found to be enhanced approximately three-fold in the presence of 0.25 M KCl or NaCl. The mechanism of such enhancement was studied, and the results indicated the occurrence of a dissociable inhibitor of the neutral protease in the lysate. Subcellular fractionation revealed the simultaneous presence of the protease and the inhibitor in chromatin. The extract of chromatin in 2 M NaCl with 5 M urea was further fractionated on a hydroxyapatite gel in the same medium, yielding separation of the protease from its specific inhibitor. The partially purified enzyme has a molecular weight of approximately 26,000 daltons and a pH optimum of 8.5 toward casein. It digests histones. It is no longer sensitive to KCl, but is completely inhibited by DFP or chymostatin. The inhibitor is resistant to heat inactivation or digestion by proteolytic enzymes and nucleases. Available lines of evidence have indicated that the inhibitor is a compound closely related to, if not identical with, poly(ADP-ribose).

ACKNOWLEDGMENTS

We wish to thank Drs. T. Sugimura and M. Miwa, National Cancer Center Research Institute, Tokyo, for their generosity in supplying us with poly(ADP-ribose) and its glycohydrolase preparations.

REFERENCES

1. Cohn, Z., and Wiener, E., J. Exp. Med., 118:991 (1963).
2. Kato, T., Kojima, K., and Murachi, T., Biochim. Biophys. Acta, 289:187 (1972).
3. Aoyagi, T., Takeuchi, T., Matsuzaki, A., Kawamura, S., Kondo, S., Hamada, M., Maeda, K., and Umezawa, H., J. Antibiot., 22:283 (1969)

4.   Aoyagi, T., and Umezawa, H., in "Proteases and
     Biological Control" (E. Reich, D. Rifkin, and
     E. Shaw, eds.), p. 429. Cold Spring Harbor
     Laboratory, 1975.
5.   Suzuki, Y., and Murachi, T., J. Biochem.
     (Tokyo), 82:215 (1977).
6.   Suzuki, Y., and Murachi, T., Abstracts, 16th
     International Congress of Hematology, Kyoto,
     p. 190, (1976).
7.   Chong, M. T., Garrad, W. T., and Bonner, J.,
     Biochemistry, 13:5128 (1974).
8.   Heinrich, P. C., Raydt, G., Puschendorf, B.,
     and Jusic, M., Eur. J. Biochem., 62:37 (1976).
9.   Rickwood, D., and MacGillivray, A. J., Eur.
     J. Biochem., 51:593 (1975).
10. Katsunuma, N., Kominami, E., Kobayashi, K.,
     Bonno, Y., Suzuki, K., Chichibu, K.,
     Hamaguchi, Y., and Katsunuma, T., Eur. J.
     Biochem., 52:37 (1975).
11. Miwa, M., Tanaka, M., Matsushima, T., and
     Sugimura, T., J. Biol. Chem., 249:3475 (1974).
12. Harris, M. (ed.), "Poly(ADP-ribose)." Fogarty
     International Center Proceedings No. 26, 1974.
13. Huang, R.C.C., and Juang, P. C., J. Mol. Biol.
     39:365 (1969).

LOCAL CONFORMATION AND
THE FUNCTIONAL PROPERTIES OF ENZYMES[1]

Bert L. Vallee
James F. Riordan

Biophysics Research Laboratory
Department of Biological Chemistry
Harvard Medical School
and the
Division of Medical Biology
Peter Bent Brigham Hospital
Boston, Massachusetts

I. INTRODUCTION

Catalysis by enzymes is the result of a series of molecular, structural events that starts with the formation of three-dimensional structure during protein synthesis and ends with a conformational change accompanying enzyme-product dissociation. Numerous ingenious experimental approaches have been devised to explore and elucidate the individual aspects of this structure-function relationship. However, it has been especially difficult to examine those subtle conformational changes that occur within the time span of the actual catalysis process itself. Such investigations require both knowledge of the substrate binding and catalytic groups and some means to render them appropriate probes of the catalytic reaction. If these groups can signal changes in their relative orientations at rates at least as fast as catalysis, it is possible to observe directly the accompanying dynamic steps. This information is clearly essential to the comprehension and verification of enzyme mechanisms.

Our present knowledge of enzyme-substrate interactions

---

[1]Supported by NIH grant GM-15003.

203

derives largely from x-ray structure analysis of what are
thought to be appropriate enzyme-substrate and/or -inhibitor
complexes. Such studies have even led to deductions regarding the dynamics of enzyme action, much as the time averaging
nature of this analytical technique would obviously preclude
the direct observation of transient intermediates. These deductions are based on the hypothesis that the structure of an
enzyme in the crystalline state is the same as that in solution. However, numerous investigations have indicated that
enzymes in solution have multiple, rapidly interconverting conformations (Harrison et al., 1975; Weber, 1975; Blake, 1976)
as predicted some time ago by Linderstrøm-Lang (1959). In as
much as crystallization itself generates interactions, e.g.,
crystal packing forces, with magnitudes comparable to those
necessary to stabilize particular conformations, the three-
dimensional structures of enzymes in crystals and solutions
need not always be identical. Indeed, different crystal forms
could well be made up of multiple and/or different populations
of enzyme structures, as has been shown for hexokinase whose
different crystal forms exhibit variable catalytic properties;
in fact, one of them is completely inactive (Anderson et al.,
1974). Ideally, any conclusions about the structural basis
for enzyme activity should come from structure and activity
determinations performed on the same entity, particularly if
these measurements are to serve to define a mechanism of action. Catalytic activity of enzymes is generally determined
with enzyme solutions, but structural analyses are not yet
feasible in that physical state. Hence, determination of the
activity of enzyme crystals would seem to be a reasonable alternative at this time. This approach would indicate minimal-
ly whether or not activity changes during the process of crystallization. Detailed kinetic studies on carboxypeptidase A
(Quiocho and Richards, 1966; Spilburg et al., 1974, 1977),
carboxypeptidase B (Alter et al., 1977) and glycogen phosphorylase (Kasvinsky and Madsen, 1976) e.g., have revealed that
the predominant effect of crystallization is a large decrease
in k_{cat}. However, such reductions in k_{cat} are not universal.
The activity of pig heart lactate dehydrogenase for example is
not affected by changes in physical state (Bayne and Ottesen,
1976) and similar observations have been reported for other
systems (Rossi and Bernhard, 1970; Sawyer, 1972). Thus, it is
not possible to generalize regarding the effects of crystallization on enzyme function.
        The integration of functional data obtained in solution
with the three-dimensional structure derived from crystals is
still one of the most important means to discern the mode of
action of enzymes. Consequently, it is clear that mechanistic
interpretations derived from crystal structures must be based
firmly on a detailed examination of the activity associated

with the structure in the particular crysal form examined. Fortunately, such kinetic analysis of catalysis by crystals is readily feasible and, in fact, can provide a valuable guide to the choice of crystals suitable for x-ray structure analysis.

II. CHROMOPHORIC METAL ATOMS AS ACTIVE SITE PROBES

Activity measurements of enzyme crystals in comparison with those obtained in solution is only one method for studying the relationship of conformation to function. We have also searched for means that could simultaneously measure activity and the structural dynamics of the active center in solution under a wide range of conditions. The presence of a chromophoric metal atom at the active site of a metalloenzyme provides one such approach to the detection and study of local conformational changes that might occur during the course of catalysis. If the active site metal has a d^{10} structure, as in the case of zinc, a common component of metalloenzymes, it will not absorb radiation in the visible spectral region. However, a chromophoric metal, particularly cobalt(II), can generally be substituted for zinc to result in catalytically active derivatives (Vallee and Latt, 1970; Lindskog and Nyman, 1964; Simpson and Vallee, 1968; Sytkowski and Vallee, 1976). Analyses of the absorption, circular dichroic, magnetocircular dichroic and EPR spectra of the resultant cobalt metalloenzyme define features of metal coordination. The spectra are typically perturbed by environmental changes induced by inhibitor or substrate binding and can therefore signal the formation of intermediates during catalysis.

A. Cobalt Carboxypeptidase

Cobalt carboxypeptidase A, e.g., is active both as a peptidase and an esterase. Its visible absorption spectrum at room temperature (Figure 1) has a shoulder near 500 nm and maxima at 555 and 572 nm, both with molar absorptivities of about 150, and its infrared spectrum has bands at 940 and 1510 nm ($\varepsilon \sim 20$). Lowering the temperature to 4°K increases resolution of the visible bands but does not reduce absorptivity. Overall, the spectrum is indicative of irregular coordination geometry and tight bonding (Latt and Vallee, 1971). It is entirely dependent on the three-dimensional structure of the protein, since it is abolished by the addition of denaturing agents. It has been suggested that the metal and its ligands might reflect an entatic environment in which the components of the active site are poised to act in catalysis (Vallee and Williams, 1968).

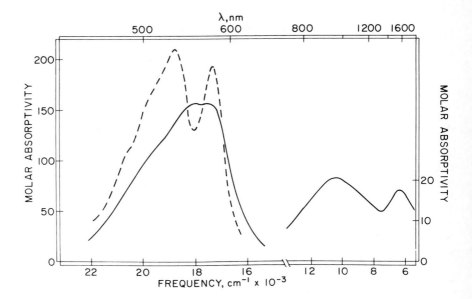

FIGURE 1. Absorption spectra of cobalt carboxypeptidase.
For the visible region the enzyme (1mM) was dissolved in 1 M
NaCl, 0.05 M Tris-Cl, pH 7.1, 20° (——). Another enzyme
solution about 3 mM was diluted with glycerol to 45% v/v and
cooled to 4.2°K (---) for spectral measurements.

For the near infrared region apocarboxypeptidase (1.5 mM)
was dissolved in 1 M NaCl, 0.005 M [D]Tris-Cl in D_2O, pD 7.2.
The sample cuvette contained 1.5 mM enzyme plus $CoSO_4$ in D_2O
buffer to yield a final total cobalt concentration of 2.0 mM,
and hence a 0.5 mM excess of free Co(II) ions: the reference
cell contained 1.5 mM apoenzyme brought to volume with buffer.

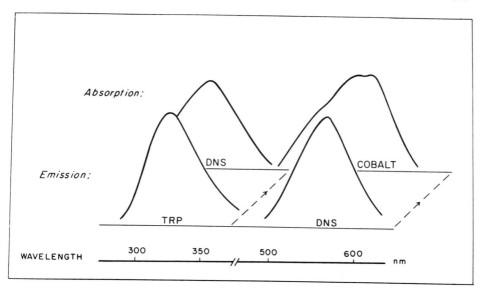

FIGURE 2. Schematic representation of the overlap relationships between enzyme tryptophan, substrate dansyl, and cobalt absorption and emission bands that constitute the energy
donor-acceptor relay system critical to observation of the
E·S complexes.

The cobalt enzyme has a circular dichroic spectrum with a negative ellipticity band at 538 nm and a shoulder at about 500 nm. In a magnetic field only the absorption band at 572 nm becomes optically active. The circular dichroic spectrum is perturbed by many inhibitors and by pseudo-substrates such as Gly-L-Tyr. The changes suggest that significant rearrangements of the electron distribution about the cobalt atom occur on interaction with substrates in accord with a direct role for the metal in peptide hydrolysis. The metal is thought to coordinate with the carbonyl oxygen atom of, and thereby destabilize, the peptide bond that is destined for cleavage (Vallee et al., 1963; Lipscomb et al., 1969). The spectral data obtained at equilibrium imply formation of an inner-sphere complex without a change in the overall metal coordination number or geometry.

The inner-sphere nature of the enzyme-substrate complex has been recognized by oxidation studies in which the substitution-labile Co(II) is converted in situ to substitution-inert Co(III). Van Wart and Vallee (1977) found that m-chloroperbenzoic acid, a competitive inhibitor analog, can selectively convert the Co(II) to the Co(III) enzyme with concomitant abolition of both esterase and peptidase activities. Formation of the Co(III) enzyme was established by absorption and EPR spectroscopy, by gel filtration, and by studying rates of metal dissociation. Loss of esterase activity was shown to be due to loss of substrate binding, whereas peptides could still bind to the Co(III) enzyme but were not hydrolyzed. Moreover, peptide substrates do not undergo a single turnover, thus ruling out the possibility of an outer-sphere hydrolytic mechanism or a so-called metal-hydroxide mechanism as suggested by others (Lipscomb et al., 1968; Wells and Bruice, 1977).

B. Metal to Metal Resonance Energy Transfer

Topography beyond the immediate environment of the active site can also be probed by cobalt substitution. Thermolysin, the neutral protease from B. thermoproteolyticus, contains four calcium atoms in addition to a single, active site zinc atom. Replacement of zinc by cobalt yields a product with twice the native activity (Holmquist and Vallee, 1974). Also, three of the four calcium sites can be occupied by lanthanide ions thereby excluding all calcium. X-ray structure analysis indicates that the most readily substituted calcium site is only 1.37 nm from the active site zinc (or cobalt) atom. Certain lanthanides, e.g., Tb(III) and Eu(III), exhibit markedly enhanced fluorescence when bound to proteins, and Tb(III) fluorescence overlaps the absorption of the cobalt chromophore.

Hence, Co(II)/Tb(III) substituted thermolysin provides a
favorable system for inter-ion resonance energy transfer which
can be used to measure distances (Horrocks et al., 1975). Addition of Tb(III) to zinc or zinc-free thermolysin can be monitored by exciting at 280 nm and observing the Tb(III) fluorescence emitted at 545 nm. Addition of Co(II) to the zinc-
free Tb(III) enzyme decreases the terbium fluorescence intensity by 89.5%. Treatment with N-bromosuccinimide quenches
enzyme tryptophan and Tb(III) fluorescence to a similar extent
and suggests that a tryptophan → Tb(III) → Co(II) resonance
energy relay system operates in the enzyme. The distance
between the Tb(III) and Co(II) atoms measured by this means
coincides exactly with that for the corresponding Ca(II) and
Zn(II) sites of the native enzyme, as gauged from x-ray structure analysis (Horrocks et al., 1975). This type of radia-
tionless energy transfer between two different metal atom
sites of a protein provides a new means to probe intramolecular distances of enzymes in solution and can be employed to
monitor conformational changes that occur during catalysis.

C. Syncatalytic Measurement of Distances

Transfer of electronic excitation energy by fluorescence
can be detected rapidly enough to allow examination of events
that are synchronous with catalysis. In particular, the
fluorescence of the dansyl group (5-N-dimethylaminonaphthyl-
1-sulfonyl) which can serve as an N- terminal blocking group
for peptide substrates of carboxypeptidase (or other endo- or
exopeptidases), overlaps cobalt absorption. Such N-dansylated
oligopeptides and ester substrates can therefore be used to
measure distances between the cobalt atom and the substrate
dansyl group within the enzyme active center while simultaneously signaling other aspects of active center topography as
the enzymatic reaction is in progress (Latt et al., 1970).
The relay system, analogous to the metal-metal energy transfer
just described, involves enzyme tryptophanyl residues, the
dansyl group of bound--but not of free--substrate, and the
cobalt atom. The overlap of the fluorescence emission and
absorption spectra are shown in Figure 2. The dansyl group
plays a dual role in this system: it accepts energy transferred from tryptophan and then transfers it to the cobalt
atom. Dissociation of the product after cleavage of the
susceptible peptide bond disrupts the energy relay system
(Latt et al., 1970, 1972). The degree of quenching of the
dansyl emission is a sensitive function of the dansyl-cobalt
distance. When the dansyl group is relatively close to the
cobalt atom, as for the dipeptide Dns-Gly-Phe, its emission is
quenched almost totally. But with longer peptides, e.g., Dns-

Gly_2-Phe or Dns-Gly_3-Phe, the dansyl group is progressively
farther away from the cobalt atom. The degree of dansyl
quenching depends on the inverse sixth power of the distance
between them, in accord with Förster's formulation of resonance energy transfer (Förster, 1948; 1965).

Energy transfer, T, from the dansyl group to the cobalt
atom is calculated from the relative fluorescence efficency,
F'_{Co}/F'_{Zn} of the dansyl moiety of the substrate when bound to
either the cobalt or zinc enzyme.

$$1 - T = F'_{Co} / F'_{Zn}$$

The ratio of donor-acceptor separation, R, to the critical
distance for 50% efficient energy transfer, R_o, calculated
from Förster's equation is

$$1 - T = 1/[1 + (R_o/R)^6]$$

and

$$R_o^6 = 8.78 \times 10^{-25} \kappa^2 Q \, J/n^4$$

where the donor quantum yield in the absence of transfer, Q,
and the overlap integral, J, are quantities determined experimentally; the index of refraction of the solvent, n, is 1.33;
and a value of 2/3 is employed for the random dipole orientation factor κ^2 (Latt et al., 1970).

Assignment of the probable value of κ^2 has consistently
posed problems in the determination of R. κ^2 can range from
0, when all vectors are mutually at right angles, to 4, when
all vectors are parallel. The value of random donor-acceptor
orientation is 2/3. The choice of cobalt as an energy accept-
or has the advantage that the nearly triply degenerate visible
absorption transitions of cobalt limit the possible range of
κ^2 values from 1/3 to 4/3. The average remains 2/3, but the
probability that this is valid is greatly enhanced for this
case (Latt et al., 1970, 1972).

The distances between the cobalt atom, acting at the
susceptible peptide bond, the dansyl group of bound substrates,
determined experimentally by means of energy transfer (Latt et
al., 1970, 1972) are within the limits of those measured on
CPK models of such peptides, assumed to be in an extended conformation (Table I). Further, the increases in distance as a
function of peptide chain length are consistent for both the
Phe and Trp sets of substrates and for corresponding members
of the two sets as well.

The dansyl-cobalt energy transfer measurements determine the radii of arcs about the cobalt atom on which the dansyl group might lie. The intersection of these arcs with the enzyme surface define the regions surveyed by the dansyl peptide substrates, but substrate orientation within such contours remains to be determined.

D. Detection of Intermediates

The enzyme substrate complex can be observed directly by resonance energy transfer between fluorescent enzyme tryptophanyl donors and the substrate dansyl group acceptor even in the zinc enzyme. Quantitatively, the degree of energy transfer depends on the distance between and orientation of the donor-acceptor pair and on the environment of the acceptor. Differences in tryptophan to dansyl transfer efficiencies and/or dansyl quantum yields can be used to characterize the resultant E·S species. Thus, if a set of reversible enzyme·substrate complexes, E·S, and/or a covalent intermediate, EA, is formed in the course of an enzyme catalyzed reaction, as in the following scheme:

$$E + S \rightleftarrows (ES)_1 \rightleftarrows (ES)_2 \rightleftarrows (ES)_n \rightleftarrows EA + P_1 \rightleftarrows E + P_2$$

it is possible to determine the minimal number of significantly populated states, the rates of interconversion of molecules among them, and the equilibrium constants determining the relative proportion of the populations.

Application of this technique to carboxypeptidase A and Dns-(Gly)$_3$-L-OPhe reveals a rapid equilibration to form an E·S complex (Figure 3). Previous kinetic data obtained with blocked oligopeptide substrates also indicate that this enzyme behaves entirely in accord with the classical Michaelis-Menten kinetic scheme:

$$E + S \underset{\longleftarrow}{\overset{K_S}{\longrightarrow}} E·S \overset{k_2}{\longrightarrow} P + E$$

Of course, it is possible that at 25°, the temperature employed for these studies, the initial signal reflects a distribution of E·S complexes that have reached equilibrium within the mixing time of the instrument. Such rapidly equilibrating species might only be detected by lowering the temperature and employing rapid mixing techniques.

This approach has also been applied successfully to delineate mechanisms of inhibition of carboxypeptidase A (Auld et al., 1972), the enzymatic consequences of metal substitution, and differences between ester and peptide hydrolysis (Auld and

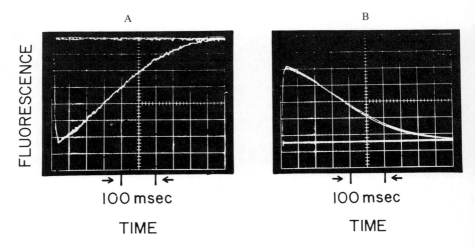

FIGURE 3. Enzyme tryptophan (A) and substrate dansyl (B) fluorescence during the time course of zinc carboxypeptidase, 2.5 x 10^{-6}M, catalyzed hydrolysis of Dns-(Gly)$_3$-L-OPhe, 1 x 10^{-4}M, in NaCl-0.03M Tris, pH 7.5, 25°. The fluorescence of either tryptophan (A) or dansyl (B) was measured as a function of time under stopped-flow conditions. Oscilloscope traces of duplicate reactions are shown in each case. Excitation was at 285 nm. Enzyme tryptophan fluorescence was measured by means of band-pass filter peaking at 360 nm and dansyl emission was measured by a 430 nm cut-off filter. Scale sensitivities for (A) and (B) are 50 mV/div and 500 mV/div. The existence of the E·S complex is signaled either by (A) the suppression of enzyme tryptophan fluorescence (quenching by the dansyl group) or (B) enhancement of the substrate dansyl group fluorescence (energy transfer from enzyme tryptophan).

Table I.   Dansyl-Cobalt Distances in Carboxypeptidase-
           Substrate Complexes[a]

Substrate	$R, (\overset{\circ}{A})$[b]	$R, (\overset{\circ}{A})$ C-P-K Models
Dns-Gly-L-Phe	>8	7
Dns-Gly-L-Trp	>8	
Dns-Gly-Gly-L-Phe	11.1-11.3	10
Dns-Gly-Gly-L-Trp	10.8-12.3	
Dns-Gly-Gly-Gly-L-Phe	11.7-12.7	13
Dns-Gly-Gly-Gly-L-Trp	12.9-14.4	
Dns-Gly-Gly-Gly-Gly-L-Phe	14.1-14.7	16

[a] 1 M NaCl-0.02 M Tris, pH 7.5, 25° (Latt et al, 1972).
[b] The range of experimental values for the distance, R,
reflects that of possible bound substrate quantum yields.
[c] Measured on Corey-Pauling-Koltun molecular models from
the center of the dansyl group to the cobalt atom assuming the peptide to be in an extended conformation and the
metal to bind the oxygen of the C-terminal peptide bond.

Table II.   Carboxylic Acid Inhibitors of the
            Carboxypeptidase A Catalyzed Hydrolysis
            Of Esters and Peptides[a]

Substrate	Inhibitor	$K_I \times 10^4$, M	Type of Inhibition
Dns-(Gly)$_3$-L-Phe	Phenylacetate	3.3	Non-competitive
Dns-(Gly)$_3$-L-OPhe	Phenylacetate	3.2	Competitive
Bz-(Gly)$_2$-L-OPhe	β-Phenylpropionate	1.2	Non-competitive
Bz-(Gly)$_2$-L-Phe	β-Phenylpropionate	1.2	Competitive
Cbz-(Gly)$_2$-L-Phe	Indole-3-acetate	1.7	Non-competitive
Bz-(Gly)$_2$-L-OLeu	Indole-3-acetate	1.6	Competitive

[a] Assays performed at 25°, pH 7.5, 1.0 M NaCl, 0.05 M Tris
except for the phenylacetate study where the conditions
were pH 6.5, 1.0 M NaCl, 0.03 M Mes (Auld and Holmquist,
1974).

Table III. Hydrolysis of Matched Ester and Peptide Pairs
            By [(CPD)Zn] and [(CPD)Cd][a]

	[(CPD)Zn]		[(CPD)Cd]	
Substrate	k_{cat} (min^{-1})	$10^4 K_m$ (M)	k_{cat} (min^{-1})	$10^4 K_m$ (M)
Bz-(Gly)$_2$-L-OPhe	30,000	3.3	34,000	79.0
Bz-(Gly)$_3$-L-OPhe	31,000	3.4	45,000	29.0
Dns-(Gly)$_3$-L-OPhe	11,000	0.25	28,000	2.2
Bz-(Gly)$_2$-L-Phe	1,200	10.0	41	8.0
Bz-(Gly)$_3$-L-Phe	2,600	37.0	86	41.0
Dns-(Gly)$_3$-L-Phe	4,200	8.0	400	8.1

[a]Assays performed at 25°, pH 7.5, 1.0 M NaCl, and a buffer
concentration of 0.05 M Tris for peptide hydrolysis and
10^{-4}M Tris for ester hydrolysis. The Anson enzyme was
used in all cases.

Holmquist, 1974). One of the earliest pieces of information demonstrating a difference in the esterase and peptidase activities of the enzyme was obtained by substituting Cd for Zn (Coleman and Vallee, 1962). Cadmium carboxypeptidase is active toward Bz-Gly-L-OPhe, but is virtually inactive toward Cbz-Gly-L-Phe. However, the latter is an inhibitor of esterase activity, indicating that the cadmium enzyme can still bind peptides.

Stopped-flow fluorescence studies of E·S complexes provide a direct comparison of the peptide binding affinities of the zinc and cadmium enzymes and, simultaneously, an explanation for the different roles of metals in peptide and ester hydrolysis. Cadmium carboxypeptidase binds the peptide Dns-(Gly)$_3$-L-Phe as readily as does zinc carboxypeptidase, but catalyzes its hydrolysis at a considerably reduced rate. In fact, the catalytic rate constants of the Cd enzyme are markedly decreased for all peptides examined, but the association constants (K_M^{-1} values) of the Cd enzyme are identical to those of the Zn enzyme. In marked contrast, the catalytic rate constants of the Cd enzyme for esters are nearly the same as those of the Zn enzyme while the association constants of the Cd enzyme are greatly decreased (Table II).

Substitution of cobalt or manganese for zinc also affects peptide and ester hydrolysis (Table III). The k_{cat} values for Bz-(Gly)$_2$-L-Phe hydrolysis follow the order Co > Zn > Mn > Cd, cobalt carboxypeptidase being 150 times more active than cadmium carboxypeptidase. However, the corresponding association constants, K_M^{-1}, are essentially identical. On the other hand, metal substitution markedly alters the binding affinity of the exact ester analog, Bz-(Gly)$_2$-L-OPhe. The K_M^{-1} values decrease in the same order as do the k_{cat} values of peptide hydrolysis, i.e., Co > Zn > Mn > Cd, the affinity of the ester for cobalt carboxypeptidase being 30 times greater than for cadmium carboxypeptidase. Metal substitution, however, has no significant effect on the catalytic rate constants for ester hydrolysis.

Stopped-flow fluorescence studies show that the apoenzyme binds peptides as tightly as does the zinc enzyme; their initial E·S complex concentrations are nearly equal. However, the apoenzyme·peptide complex is stable and there is no formation of products during the time needed for complete hydrolysis of the peptide by the zinc enzyme. Thus, although the apoenzyme cannot catalyze the hydrolysis of peptide substrates, it binds them to the same degree as does the zinc enzyme. On the other hand, removal of zinc from the native enzyme decreases binding of the exact ester analog by at least an order of magnitude. Similarly, as discussed above, the cobalt(III) enzyme, which is inactive toward both peptide and ester substrates (Van Wart and Vallee, 1977), binds peptides, but not esters. All of these studies indicate that the binding of peptides to metallo-carboxypeptidases must differ from that of esters.

A positively charged residue in the active center has
long been thought to be a major determinant of the specificity
of carboxypeptidase for which a free C-terminal carboxyl group
of the substrate is mandatory (Waldschmidt-Leitz, 1931; Smith,
1949; Vallee et al., 1963). Since acylation experiments have
excluded lysines (Riordan and Vallee, 1963), it seemed possible that an arginyl residue might function as the binding
locus for the carboxylate group (Vallee and Riordan, 1968).
Based on x-ray crystallographic studies of the Gly-L-Tyr
complex of the crystalline enzyme, it has since been concluded
that the carboxyl group of this pseudosubstrate is bound to
Arg-145. It has been inferred further that this arginine is
the determinant of enzyme specificity for interaction with the
C-terminal free carboxyl group of all substrates and is thus
indispensable to catalysis by carboxypeptidase (Lipscomb et
al., 1968). Chemical modification of arginine in carboxypeptidase diminishes the rate of Cbz-Gly-L-Phe hydrolysis, consistent with the hypothesis that peptide substrates can bind to an
arginyl residue (Riordan, 1973). However, this modification
does not diminish esterase activity toward Bz-Gly-L-OPhe suggesting that a positively charged group other than arginine
serves as the recognition site and binding locus for the free
carboxyl group of esters.
     Direct binding of substrates to the metal through their
carboxyl group was first suggested by Lumry and Smith (1955).
In view of the above results, the active site metal may indeed
serve as a primary binding locus for esters, but such binding
appears unlikely for peptides. Direct binding studies and
metal ion exchange data led Coleman and Vallee (1962, 1964) to
propose that binding of β-phenylpropionate involves an interaction of the carboxyl group with the metal. Later studies
with p-iodo-β-phenylpropionate (Navon et al., 1970) have confirmed this postulate. Presumably, the carboxyl groups of
both esters and carboxylic acid inhibitors compete for the
metal whereas peptides seem to bind to the guanido group of an
arginyl residue (Riordan, 1973).
     X-ray diffraction studies of the binding of ester sub-
strates or ester pseudosubstrates to the crytalline enzyme have
not been successful. However, a coordinated series of conformational changes is thought to occur when the peptide pseudosubstrate, Gly-L-Tyr, binds to the crystalline enzyme (Lipscomb et al., 1968). The x-ray structure analysis of the non-
productive enzyme·Gly-Tyr complex led to the suggestion that
an initial interaction of the carboxyl group of Gly-Tyr with
Arg-145 brings about a substrate-induced conformational change
of 1.2 nm in Tyr-248, moving the phenolic hydroxyl group into
the vicinity of the susceptible bond of the substrate (Lipscomb et al., 1968). Two facts are at variance with this in-

terpretation. First, as indicated above, Tyr-248 can be coordinated to the active-site metal in the absence of substrate.
A reexamination of initial x-ray data (Lipscomb et al., 1968)
subsequently uncovered and confirmed the existence of such a
Tyr-248·Zn interaction in the native crystals (Lipscomb, 1973).
Second, a conformational change of Tyr-248, induced by interaction of the substrate carboxyl group with Arg-145, and proposed solely on the basis of x-ray analysis of the enzyme Gly-
L-Tyr complex, is inconsistent with the findings with esters.
The evidence presented by Auld and Holmquist (1974) indicates
that esters bind to the metal, not to arginine. Hence, the
postulated conformational change of Tyr-248 cannot be triggered by an ester-arginine interaction. In fact, it may be concluded that esters and peptides are hydrolyzed by different
catalytic pathways and that neither binds productively to induce the inward movement of Tyr-248.

III. CHEMICAL MODIFICATIONS TO PROBE
THE ACTIVE SITES OF ENZYMES

A. Monoarsanilazo Tyr-248 Carboxypeptidase

Chemical modification of active site residues is another
means to study local conformational changes and we have found
azotyrosyl and nitrotyrosyl residues to be especially helpful
in this regard. The location of Tyr-248 of carboxypeptidase A
relative to the active-site zinc atom has been thought to be
an important structural feature of the enzyme pertinent to its
catalytic mechanism (Lipscomb et al, 1968; Quiocho et al.,
1972). Tyr-248 reacts specifically with diazotized arsanilic
acid to generate a visible absorption band with a maximum at
510 nm that reflects an intramolar coordination complex between
the azotyrosyl residue and the zinc atom (Johansen and Vallee,
1971, 1973). This complex dissociates on crystallization of
the enzyme, on removal of the metal, or on addition of substrates or inhibitors (Johansen and Vallee, 1975). Absorption
and circular dichroism-pH titrations of the modified zinc
enzyme demonstrate two pK_{app} values, 7.7 and 9.5, character-
izing the formation and dissociation of the complex, respectively. Titrations of the apoenzyme, which completely lacks
the 510 nm absorption band at all pH values, reveal a single
pK_{app}, 9.4, due to the ionization of the azophenol (λmax, 485
nm). Substitution of other metals for zinc results in analogous complexes with absorption maxima and circular dichroism
extrema characteristic of the particular metal. Studies with
nitrocarboxypeptidase (Riordan and Muszynska, 1976) also provide evidence consistent with the interaction of tyrosine-248
and the zinc atom.

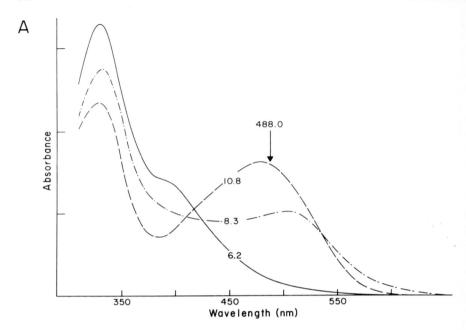

FIGURE 4.   (A) Absorption spectra of arsanilazotyrosyl-248 carboxypeptidase A at the pH values indicated.   (B) Resonance Raman spectra of the three species of arsanilazotyrosyl-248 carboxypeptidase A at the pH values indicated.   At intermediate pH values, the spectra constitute a superposition of these three basic spectra.   (From Scheule et al., 1977).

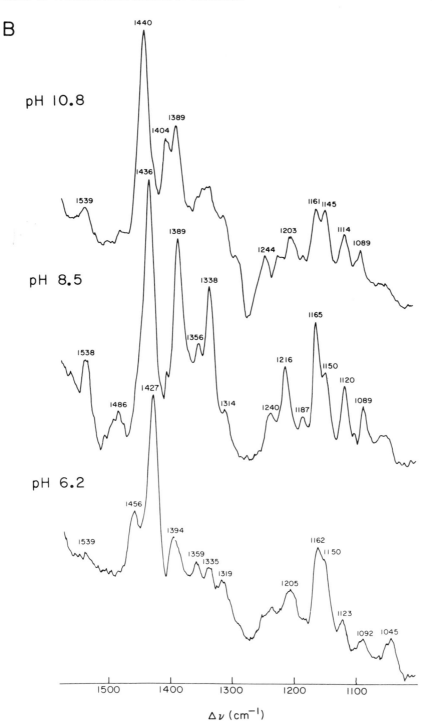

B. Resonance Raman Spectroscopic Studies
Of Arsanilazotyrosyl-248 Carboxypeptidase:
Structure of the Intramolecular Coordination Complex

The structural details of the azoTyr-284·Zn complex have
been examined by means of resonance Raman (rR) spectroscopy
(Scheule et al., 1977). The rR spectrum of arsanilazotyrosyl-
248 carboxypeptidase A is dominated by the vibrational bands
of its chromophoric azoTyr-248 residue and is uncomplicated by
background interference from either water or other components
of the protein. The spectrum contains multiple, discrete bands
that vary with pH, consistent with the interconversion of different azotyrosine species in solution (Figure 4). The identities of these species have been established by spectral analysis of model azophenols and the apoazoenzyme. All conclusions about the azoenzyme based on absorption spectroscopy,
including the apparent pK values for the interconversions, are
verified by the rR spectra. In addition, the properties of rR
bands that have been assigned to the motions of the specific
atoms of the complex on the basis of isotope substitutions
provide details of the interactions of those atoms with the
active site zinc atom, and hence, the presumable structure of
the intramolecular coordination complex.

C. Detection of Multiple Binding Modes at the Active Site
Of Carboxypeptidase A by Circular Dichroism –
Inhibitor Titrations

The azoenzyme has been titrated with a series of agents
known to inhibit native carboxypeptidase via different modes
(Johansen et al., 1976). Competitive inhibitors, e.g., L-ben-
zylsuccinate, L-Phe, L-phenyllactate (Figure 5), and the
pseudosubstrate, Gly-L-Tyr, generate one type of characteristic circular dichroic spectrum. The titration curves can be
fit with a single binding constant for each agent indicating
that only one molecule binds to the active center. Other
agents, e.g., β-phenylpropionate (Figure 6) and phenylacetate,
whose mixed modes of inhibition have previously been resolved
into competitive and noncompetitive components (Auld et al.,
1972), cause different spectral effects. Their circular di-
chroic-titration curves are consistent with two molecules of
inhibitor binding to the enzyme, as inferred from both equilibrium and kinetic studies. It is possible to differentiate
interactions leading to competitive and noncompetitive inhibition, respectively, by their characteristic effects on the extrema at 340 and 420 nm, reflecting azoTyr-248, and the one at
510 nm, due to the chelate with zinc. Noncompetitive inhibitors and modifiers induce yet additional spectral features,

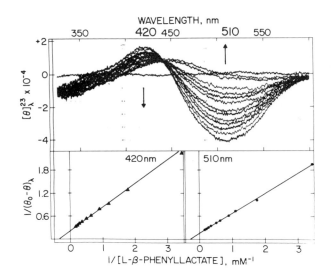

FIGURE 5. Upper Panel - Effect of L-phenyllactate on the circular dichroism spectrum of zinc azoTyr-248 carboxypeptidase, 0.05 M Tris-0.5 M NaCl, pH 8.5, 23°, uncorrected for dilution. L-Phenyllactate concentration is varied from 0 to 30 mM. Arrows indicate the direction of change at the corresponding wavelength set in larger ciphers.

Lower Panel - Double reciprocal plots ($1/\Theta_0 - \Theta)_\lambda$ vs. $1/$[L-phenyllactate]), on the right at 510 nm ($\bullet$) (K_{app}, 2.8 mM) and on the left at 420 nm ($\blacktriangle$), (K_{app}, 2.5 mM) both calculated from the circular dichroism spectral titrations in the upper panel. (From Johansen et al., 1976).

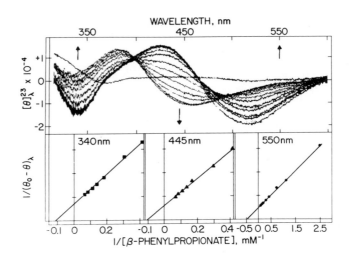

FIGURE 6.   Upper Panel - Effect of β-phenylpropionate on
the circular dichroism spectrum of zinc azoTyr-248 carboxypeptidase, 0.05 M Tris-0.05 M NaCl, pH 7.6, 23°.   β-Phenylpropionate concentration was varied from 0 to 91 mM.   The spectra
are not corrected for dilution, which is 18% for the final
spectrum.   Arrows indicate the direction of change at the corresponding wavelength, set in larger ciphers.

    Lower Panel - Double reciprocal plots of $1/(\Theta_o - \Theta_\lambda$ ver-
sus $1/[\beta$-phenylpropionate], right panel at 550 nm(●)(K_{app}, 2.4
mM), middle panel at 445 nm (▲) (K_{app}, 9.3 mM) and left panel
at 340 nm (■) (K_{app}, 9.4 mM) calculated from the circular
dichroic spectral titrations in the upper panel (From Johansen
et al., 1976).

each signaling changes in a particular active center environment. The general as well as the specific features of these
circular dichroic titrations are consistent with our mechanistic views (Vallee et al., 1968).

D. Monitoring Catalytic Events

AzoTyr-248 carboxypeptidase is eminently suitable for
mechanistic studies: It is enzymatically active toward both
peptides and esters; it contains a site-specific probe whose
spectra reflect both intramolecular metal complex and azoTyr-
248 itself; the probe responds dynamically to environmental
factors; and the response time of the probe is rapid enough
to measure the catalytic step. These combined kinetic and
spectral properties of the metal complex render it a potent
spectro-kinetic probe for observing microscopic details of the
catalytic process.

Thus, we have examined the kinetics of association and
dissociation of the azoTyr-248·Zn complex by stopped-flow pH
and temperature-jump methods (Harrison et al., 1975). The
rate constant for the dissociation process, 64,000 sec^{-1}, is
orders of magnitude greater than that for the catalytic step
itself (about 0.01-100 sec^{-1}). Rapidly hydrolyzed peptide and
ester substrates disrupt the azoTyr-248·Zn complex before
hydrolysis occurs. As a consequence, formation of uncomplexed
azoTyr-248, binding of substrate, and catalysis can all be
monitored independently. The results of these dynamic studies
specify a course of catalytic events, different from that postulated based on x-ray structure analysis. Tyr-248 is displaced away from the zinc, not inward as postulated on the
basis of x-ray studies (Lipscomb et al., 1968).

E. Monitoring Multiple Conformation States in Solution

The azoTyr-248·Zn complex also provides a means to detect
multiple conformations of the enzyme in solution by stopped-
flow pH and temperature-jump experiments (Harrision et al.,
1975). On mixing azoTyr-248 carboxypeptidase at pH 8.5 with pH
6.0 buffer, 97% of the red azoTyr-248·Zn complex is converted
to the yellow azophenol within 3 msec, the mixing time of the
stopped-flow instrument. The rate constant for this rapid
process, determined by temperature-jump analysis, is pH dependent and varies from 100,000 to 50,000 sec^{-1}. The remaining 3%
of the total absorbance change occurs slowly, with a first-
order rate constant of 0.4 sec^{-1}. This slow change implies a
relaxation process involving two different conformations of
yellow azocarboxypeptidase.

Table IV.   Metallocarboxypeptidase-Catalyzed Hydrolysis
Of Bz-(Gly)$_2$-L-Phe and Bz-(Gly)$_2$-L-OPhe[a]

	Bz-(Gly$_2$-L-Phe		Bz-(Gly)$_2$-L-OPhe	
Metal	k_{cat} (min^{-1})	$10^{-3}K_m$ (M^{-1})	$10^{-4}k_{cat}$ (min^{-1})	K_m^{-1} (M^{-1})
Cobalt	6000[b]	1.5[b]	3.9	3300
Zinc	1200	1.0	3.0	3000
Manganese	230[b]	2.8[b]	3.6	660
Cadmium	41	1.3	3.4	120

[a]Assays performed at 25°, pH 7.5, 1.0 M NaCl, and a buffer
concentration of 0.05 M Tris for peptide hydrolysis and
10^{-4}M Tris for ester hydrolysis. [b]Values are for carboxypeptidase A (Cox) (Auld and Vallee, 1970). All other
values are for carboxypeptidase A (Anson).

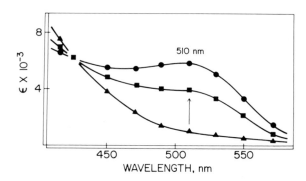

FIGURE 7.   Spectra of azoTyr-248-carboxypeptidase A, re-
constructed from stopped-flow pH jump experiments from pH 6.5
to 8.2 performed at 20 nm intervals.   The spectrum at zero
time (▲) changes in 3 msec to the spectrum of 60% of the azo-
Tyr-248·zinc complex (■) followed by a slower exponential
change (k = 2.2 sec^{-1}) to finally reach the equilibrium spectrum (●) of azoTyr-248·Zn complex.   Final conditions: enzyme,
25 μM, pH 8.2, 50 mM Hepes, 1 M NaCl, 25°.   (From Harrison et
al., 1975).

Jumping from pH 6 to 8.2 generates only about 60% of the final absorbance at 510 nm within the 3 msec mixing time of the experiment. The remaining 40% of the color appears with a first-order rate constant of 5.0 sec^{-1}. The reconstruction of the azoTyr-248 carboxypeptidase spectra in pH jump experiments from pH 6.5 to 8.2 identifies a time-dependent conformational step (Figure 7). This again suggests the presence of two yellow species, one that forms the metal complex readily on raising the pH and another that does not. By varying the initial pH of the jump experiments and keeping the final pH constant, it is possible to discern that the interconversion of the two yellow species is pH independent.

The mechanism governing these interrelationships involves two distinct processes. The first, $E_r + H^+ \rightleftharpoons E_y$, has an equilibrium constant, K_1, is extremely rapid ($k_{fast} \sim 10$ sec^{-1}), and reflects the pH <u>dependent</u> dissociation of the metal complex. The second, $E_y \rightleftharpoons E_{y'}$, is much slower ($k_{slow}$ = 5.0 sec^{-1}) and is due to the pH <u>independent</u> interconversion of two distinct populations of protein molecules, E_y and $E_{y'}$, in which the yellow azoTyr-248 has different conformations. These two separate processes can be recognized readily since the first is three to four orders of magnitude faster than the second. Even though both E_y and $E_{y'}$ are yellow and, hence, cannot be directly differentiated by their spectra, the amplitudes of the stopped-flow, pH-jump measurements are directly related to their concentrations. Therefore, the relaxation process becomes a precise gauge of their interconversion.

In a previous stopped-flow, pH-jump experiment, only a single process with a rate constant of about 6 sec^{-1} was observed (Quiocho et al., 1972). This study failed to detect the much more rapid process, and, as a consequence, led to a number of postulates which, obviously, are no longer pertinent.

The present studies point to the existence of rapidly equilibrating substructures of carboxypeptidase A, consistent with the view that in solution enzymes can adopt multiple, interconvertible, related conformations that either facilitate or impede catalysis. In crystals, rearrangements of molecular structures could be severely impaired or restricted, and crystallization might single out either active or inactive conformations. It is not surprising that in some cases crystals have greatly reduced activities or otherwise markedly altered catalytic behavior, as observed for carboxypeptidase A.

IV. SUMMARY

The organic and inorganic probes that we have employed are all intended to monitor the conformational and structural events coincident with catalysis. Information obtained through

the use of such probes provides insight into the dynamics that
can then be superimposed on the static picture provided by
x-ray crystallography, the only method currently available for
three-dimensional structural analysis. The resultant knowledge, when integrated with appropriate functional data, should
then allow the deduction of valid mechanisms of action.

ACKNOWLEDGMENTS

    This work was supported by Grant-in-Aid GM-15003 from the
National Institutes of Health of the Department of Health,
Education and Welfare.

Alter, G.M., Leussing, D.L., Neurath, H. & Vallee, B.L. (1977) Biochemistry 16, 3663-3668.
Anderson, W.F., Fletterick, R.J. & Steitz, T.A. (1974) J. Mol. Biol. 86, 261-269.
Auld, D.S. & Holmquist, B. (1974) Biochemistry 13, 4355-4361.
Auld, D.S., & Vallee, B.L. (1970) Biochemistry 9, 4352-4359.
Auld, D.S., Latt, S.A. & Vallee, B.L. (1972) Biochemistry 11, 4994-4999.
Bayne, S. & Ottesen, M. (1976) Carlsberg Res. Commun. 4, 211-216.
Blake, C.C.F. (1976) FEBS Letters Supplement 62, E30.
Coleman, J.E. & Vallee, B.L. (1962) Biochemistry 1, 1083-1092.
Coleman, J.E. & Vallee, B.L. (1964) Biochemistry 3, 1874-1879.
Förster, T. (1949) Ann. Phys. 2, 55.
Förster, T. (1965) in Modern Quantum Chemistry, Part III (Sinanoglu, O., ed.), p. 93, Academic Press, New York.
Harrison, L.W., Auld, D.S. & Vallee, B.L. (1975) Proc. Nat. Acad. Sci. USA 72, 4356-4360.
Holmquist, B. & Vallee, B.L. (1974) J. Biol. Chem. 249, 4601-4607.
Horrocks, W.DeW., Holmquist, B. & Vallee, B.L. (1975) Proc. Nat. Acad. Sci. USA 72, 4764-4768.
Johansen, J.T. & Vallee, B.L. (1971) Proc. Nat. Acad. Sci. USA 68, 2532-2535.
Johansen, J.T. & Vallee, B.L. (1973) Proc. Nat. Acad. Sci. USA 70, 2006-2010.
Johansen, J.T. & Vallee, B.L. (1975) Biochemistry 14, 649-660.
Johansen, J.T., Klyosov, A.A. & Vallee, B.L. (1976) Biochemistry 15, 296-303.
Kasvinsky, P.J. & Madsen, N.B. (1976) J. Biol. Chem. 251, 6852-6859.
Latt, S.A. & Vallee, B.L. (1971) Biochemistry 10, 4263-4269.
Latt, S.A., Auld, D.S. & Vallee, B.L. (1970) Proc. Nat. Acad. Sci. USA 67, 1383-1389.
Latt, S.A., Auld, D.S. & Vallee, B.L. (1972) Biochemistry 11, 3015-3022.
Linderstrøm-Lang, K. & Schellman, J.A. (1959) Enzymes 2nd ed., vol. 1, (Boyer, P.D., Lardy, H. & Myrbäck, K., eds), pp. 443-510, Academic Press, New York.
Lindskog, S. & Nyman, P.O. (1964) Biochim. Biophys. Acta 85, 462-474.
Lipscomb, W.N. (1973) Proc. Nat. Acad. Sci. USA 70, 3797-3801.
Lipscomb, W.N., Hartsuck, J.A., Reek, G.N., Quiocho, F.A., Bethge, P.H., Ludwig, M.L., Steitz, T.A., Muirhead, H. & Coppola, J.C. (1968) Brookhaven Symp. Biol. 21, 24-90.
Lumry, R. & Smith, E.L. (1955) Discuss. Faraday Soc. 20, 105-114.
Navon, G., Shulman, R.G., Wyluda, B.J. & Yamane, T. (1970) J. Mol. Biol. 51, 15-30.

Quiocho, F.A. & Richards, F.M. (1966) Biochemistry 5, 4062-4076.

Quiocho, F.A., McMurray, C.H. & Lipscomb, W.N. (1972) Proc. Nat. Acad. Sci. USA 69, 2850-2854.

Riordan, J.F. (1973) Biochemistry 12, 3915-3923.

Riordan, J.F., & Muszynska, G. (1974) Biochem. Biophys. Res. Commun. 57, 447-451.

Riordan, J.F. & Vallee, B.L. (1963) Biochemistry 2, 1460-1468.

Rossi, G.L. & Bernhard, S.A. (1970) J. Mol. Biol. 49, 85-91.

Sawyer, L. (1972) J. Mol. Biol. 71, 503-505.

Scheule, R.K., Van Wart, H.E., Vallee, B.L. & Scheraga, H.A. (1977) Proc. Nat. Acad. Sci. USA 74, 3273-3277.

Simpson, R.T. & Vallee, B.L. (1968) Biochemistry 7, 4343-4350.

Smith, E.L. (1949) Proc. Nat. Acad. Sci. USA 35, 80-90.

Spilburg, C.A., Bethune, J.L. & Vallee, B.L. (1974) Proc. Nat. Acad. Sci. USA 71, 3922-3926.

Spilburg, C.A., Bethune, J.L. & Vallee, B.L. (1977) Biochemistry 16, 1142-1150.

Sytkowski, A.J. & Vallee, B.L. (1976) Proc. Nat. Acad. Sci. USA 73, 344-348.

Vallee, B.L. & Riordan, J.F. (1968) Brookhaven Symp. Biol. 21, 91-119.

Vallee, B.L. & Williams, R.J.P. (1968) Proc. Nat. Acad. Sci. USA 59, 498-505.

Vallee, B.L. & Latt, S.A. (1970) in Structure-Function Relationships of Proteolytic Enzymes (Desnuelle, P., Neurath, H. & Ottesen, M., eds.), pp. 144-159, Academic Press, New York.

Vallee, B.L., Riordan, J.F. & Coleman, J.E. (1963) Proc. Nat. Acad. Sci. USA 49, 109-116.

Vallee, B.L., Riordan, J.F., Bethune, J.L., Coombs, T.L, Auld, D.S. & Sokolovsky, M. (1968) Biochemistry 7, 3547-3556.

Van Wart, H.E. & Vallee, B.L. (1977) Biochem. Biophys Res. Commun. 75, 732-738.

Waldschmidt-Leitz, E. (1931) Physiol. Rev. 11, 358-370.

Weber, G. (1975) Adv. Protein Chem. 29, 1-83.

Wells, M.A. & Bruice, T.C. (1977) J. Am. Chem. Soc. 99, 5341-5356.

STRUCTURE AND FUNCTION IN "MITOCHONDRIAL-TYPE" CYTOCHROMES $\underline{c}$[1]

Martin D. Kamen
Beverly J. Errede[2]

University of Southern California
Los Angeles, California
and
University of California at San Diego
La Jolla, California

INTRODUCTION

C-type cytochromes occur in all membrane systems which function in the bioenergetic apparatus of cells, whether aerobic or anaerobic (1,2). In the course of evolution, this apparatus has been packaged in various organelles, including the two extremes of evolutionary specialization in eukaryotes--the mito-chondrion and the chloroplast. Between these two extremes there is a continuum of vesicular and other variably differentiated membrane systems found in prokaryotes. All of these serve the same general function--conservation of redox energy through storage as ATP, associated with passage of reducing equivalents through electron transport chains. These chains exhibit similar patterns of components, the

---

[1] Researches on which this report is based were made possible by grants from the National Science Foundation (BMS 75-3708 and 75-13608) and the National Institutes of Health (GM-18528).
[2] Present address: Department of Radiation Biology, School of Medicine and Dentistry, University of Rochester, New York.

229

most accessible of which is some form of cytochrome
c, regardless of whether the overall metabolism is
functional for controlled reduction (uptake) of
oxygen (as in aerobic respiration), controlled reduction of inorganic or organic oxidants (as in anaerobic respiration), controlled evolution of oxygen
as in the anaerobic chloroplast system, or controlled photo-oxidation of inorganic or organic
substrates, as in bacterial photosyntheses.

In mitochondria, the cytochromes c have evolved
a unique structure--the so-called "cytochrome c
fold" (3,4). The heme prosthetic group is oriented
anomalously in a hydrophobic "cleft"--that is, its
polar propionyl side chains are pulled into the
non-polar low dielectric interior, rather than
appearing totally exposed to the environment as
they are in other heme proteins. This orientation
is maintained by covalent binding through thio-ether
linkage to the two cysteinyl residues (nos. 14 and
17 in the conventional numbering with the
N-Terminus as residue no. 1). In addition, a
complex set of hydrogen-bonding interactions is
operative between the propionyl carboxyls and
residues such as tryptophane (no. 59), tyrosine
(no. 48) and asparagine (no. 52). A genetic
burden is placed on biosynthesis in that the
primary structure must be such so that all these
interactions as well as the hydrophobic bindings
between many non-polar residues and the heme plane
come into play on folding to maintain this
anomalous orientation of the heme (5,6). This
folding occurs in all eukaryotic mitochondrial
cytochromes c, although variations in amino acid
composition may involve substitutions of a major
fraction of all the residues in the primary
structure (7). Only 28 residues remain invariant.
Thus, this class of cytochromes c comprises a
single restricted functional grouping despite this
great variability in primary structure. As we will
remark later, there is significant variation in
reactivity of eukaryotic cytochromes c, particularly protozoan forms, with the mitochondrial
redox systems of higher organisms.

At the other extreme of eukaryotic evolution,
the anaerobic chloroplast presents a
photochemically-coupled mechanism for photon energy
transduction with essentially the same bioenergetic
model--an electron transport chain like that of the
mitochondrion but modified so that the electron

donor is produced endogenously by photochemistry, and the terminal cytochrome c and cytochrome c oxidase are replaced by another c-type cytochrome-- cytochrome "f"--and a chlorophyll-protein complex. Cytochrome "f"--which exhibits a primary structure with apparently the same tertiary characteristic folding as mitochondrial cytochrome c--cannot replace mitochondrial cytochrome c in function as an electron donor to the mitochondrial oxidase, although the reverse is possible--that is, mito- chondrial cytochrome c can function in place of cytochrome "f" in the chloroplast.

In prokaryotes, a great variety of non- mitochondrial systems with a multiplicity of functions is found--still related in general to redox-coupled phosphate esterification. From an evolutionary standpoint, the cytochromes c of prokaryotes and those of protozoa may be related. However, our knowledge of these C-type cytochromes is still rather primitive, although we know already of at least five distinct classes among prokaryotic forms in which the "cytochrome c fold" is main- tained. These cytochromes c display a greatly extended range of primary structure and function, compared to the eukaryotic cytochromes c (8). One may note in particular, a class of small cytochromes c, (total residues ∿ 80-90) called "Pseudomonas cytochrome c551" which have been obtained from many species of the genus, *Pseudomonas* (9), as well as in two species from photosynthetic bacteria (10). A determination of a tertiary structure for one of these from a strain of *Ps. aeruginosa* (10) is available and shows clearly the same folding pattern and identical bonding ligands as in eukaryotic mitochondrial cytochrome c (4). However, no representative of this class functions as an electron donor or acceptor in the cytochrome c redox systems of mitochondria, nor is there any significant homology with mitochondrial cytochromes c.

An interesting duo of prokaryotic cytochromes c is found in green photosynthetic bacteria, one from the genus *Chlorobium* (sp. *limicola*, var. *thiosulfatophilum*) and the other from the genus *Prosthecochloris* (sp. *aestuarii*). These two cytochromes "c555" are homologous in primary structure and the former has the same 3-dimensional

structure as the pseudomonas cytochromes c_{551} (6,
11). However, as we will note below, the *Chlorobium* cytochrome c_{555} reacts significantly well with
mitochondrial cytochrome c oxidase and not at all
with the mitochondrial cytochrome c reductase;
the *P. aestuarii* cytochrome c_{555} reacts with neither
(12). These cytochromes c and those from the
pseudomonads appear to be somewhat homologous
with, and have structures similar to, the eukaryotic
cytochromes "f" with which, except for the *Chlorobium* cytochrome c_{555}, they share the same lack of
reactivity with the cytochrome c mitochondrial
redox system.

From the standpoint of structure, the most
remarkable group of prokaryotic cytochromes c are
those grouped under the term "cytochrome c_2".
Originally found in a strain of the non-sulfur
purple photosynthetic bacterium, *Rhodospirillum
rubrum* (13,14), the class now includes (15) repre-
sentatives from all of the "Bergey" species of
this family (*Rhodospirllaceae*), as well as a
representative of the nonphotosynthetic facultative
denitrifiers, *Paracoccus denitrificans*. In most
of these, amino acid sequences have been determined
(8,10). The tertiary structures of the *R. rubrum*
(16) and *P. denitrificans* (17) proteins have been
described and show remarkable homology to that
of the eukaryotic mitochondrial cytochromes c.
The primary structures indicate two subclasses of
cytochromes c. One of these (termed "IA" by Ambler
(8)) contains insertions in the invariant "core"
structure of the mitochondrial cytochrome c fold
in the regions corresponding to the "bottom" and
"left front surface", but still the "fold" is
retained. The second class ("IB") shows most
similarity to mitochondrial cytochrome c, there
being no insertions and only a minor deletion
occurring in the N-terminal helix portion. The
homologies displayed by two of the class IB
cytochromes c_2, those from *Rhodomicrobium vannielii*
and *Rhodopseudomonas viridis*--with eukaryotic
mitochondrial cytochromes c are as good or better
than those between some of the eukaryotic protozoan
cytochromes c and higher animal cytochromes c (18).

From these many variant forms of cytochrome c,
eukaryotic and prokaryotic, we can select proteins
which differ in details of surface topography,
particularly distribution of cationic groups which

cluster around the cleft containing the partially
exposed heme in a manner characteristic of
cytochromes c, while retaining the over-all three-
dimensional structure dictated by the "cytochrome
c" fold. By comparison of the reactivity of these
selected cytochromes c with that of beef heart
cytochrome c when they are substituted in the beef
heart reductase or oxidase mitochondrial complexes
we may hope to define structural parameters which
determine the redox function of the beef heart,
and other equivalent, eukaryotic mitochondrial
cytochromes c. This approach can also be extended
to other redox systems involving cytochromes c, by
use of appropriate oxidases, peroxidases and
reductases obtained from various eukaryotic and
prokaryotic sources (19,20).

Functional Assay of Cytochromes c
Reactivity--General Remarks

The ultimate goal of studies on cytochrome c
function is the clarification of nature of structural parameters at the molecular level which
determine reactivity of cytochrome c with its redox
partners in whatever electron transport chain such
cytochromes occur. In particular, one may focus
on the eukaryotic mitochondrial system with its
characteristic cytochrome c acting as the bridge
between the terminal positions of the redox chain
leading to reduction of molecular oxygen to water.
In this system, as well as other *in vivo* cases,
membrane organization is all-important and
constraints, more or less undefined, are introduced
in the contacts between cytochrome c and its
neighboring redox partners--the pyridine nucleotide-
linked cytochrome c reductase ("Complexes I and
III") and the oxidase ("Complex IV"). The identification of structural features which dictate
reactivity requires a detailed knowledge of
mitochondrial structure around the bound cytochrome
which is still unknown and likely to remain so for
some time. One may have recourse to kinetic
analyses to make an indirect approach but these
necessitate solubilisation of the reactive
complexes in as native form as possible as well as
preparation of soluble native cytochromes c. The
results of such analyses will have an undefined
relation to the mechanisms operative *in vivo*

because many more degrees of freedom are available
to the *in vitro* system, so that kinetic artifacts
are unavoidable. However, if one proposes only
to attempt a definition of structural parameters
in terms of how reactivity is affected by specific
changes in structural features, a kinetic approach
can be informative, provided it is adequately
performed--that is, a rate law is derived over a
wide range of substrate concentrations so that
inferences about mechanisms can be made. This
consideration has not been appreciated adequately
in the past and indeed is still ignored in many
on-going researches at present. Thus, a number
of surveys have been published showing relative
reactivities of cytochrome c from eukaryotic and
prokaryotic sources assayed with both beef heart
and bacterial oxidases but almost always at single
concentrations (21). Moreover, in most cases the
enzyme preparations employed have been ill defined.
Consequently an erroneous impression has been
generated over the years that all eukaryotic
mitochondrial cytochromes c are essentially equivalent in reactivity with mitochondrial oxidase or
reductase, regardless of source (22). Recently,
it has been reported that at very low concentrations of cytochromes c from baker's yeast, horse
heart and *Euglena gracilis*, large differences in
reactivity with beef heart oxidase are found (23).
As we will show, a similar result is found with the
cytochrome c from the protozoan, *Tetrahymena
pyriformis*, which is totally unreactive at any
concentration with beef heart reductase or oxidase.
One may conclude that the surveys made in the past
have been useful in indicating interesting cases
to investigate further but are essentially
uninterpretable in molecular terms. To proceed
further, it is clear kinetic studies in depth with
wide ranges of substrate concentrations, using
well-defined, reproducible preparations of highly
active, native cytochrome c reductases and
oxidases, are mandatory.

     There is a large literature on the reactions
of cytochrome oxidase (see ref. 24 for a review).
The researches by Smith and her associates defined
the general character of the reactions of cytochromes c with oxidase preparations (22). Likewise,
salient features of the reductase reactions were
established (25,26). However, insufficient ranges

of concentration were studied to permit deduction
of the needed general rate laws. Thus, it has not
been possible to proceed with the elaboration
of mechanisms that could be used to rationalize
the effect of different structural features of
cytochromes c on the oxidase- and reductase-cata-
lyzed cytochrome c reactions.

In the last five years, we have examined the
reactivity of a number of cytochromes c selected
for known structure modifications relative to
eukaryotic horse heart cytochrome c, as they interact with the beef heart oxidase and reductase
systems, reconstituted from the Complexes I, III,
and IV, first elaborated by Hatefi and his associates (27-29). A detailed discussion of the materials
and methods used is presented elsewhere (30,31).

Results of Kinetic Analyses
for the Oxidase-Catalyzed Reaction

It has been established previously (32) that
the reaction of the ferrocytochrome c were pseudo
first-order at all substrate concentrations
studies--that is, no saturation in reaction velocity occurred as cytochrome c concentration was
increased at constant concentration of oxidase.
We extended these observations to a range of
substrate concentrations covering some three orders
of magnitude, ranging from a few hundredths micromolar to several hundred micromolar. The rate
law obtained for all cytochromes c tested is of
the form:

$$\frac{-d\ (\text{ferrocyto c})}{dt} = \frac{\alpha_1 + \alpha_2 [C]}{1 + \beta_1 [C] + \beta_2 [C]^2} \ [\text{ferrocyto c}]\,(\text{oxidase})$$

(1)

where [C] is the total concentration of cytochrome
c, regardless of actual distribution between
oxidized and reduced forms. The reaction studied
was the oxidase-catalyzed oxidation of ferrocytochrome c by cytochrome a, conditions being
established in which oxygen tension was never
rate-limiting.

This rate law reflects the previous findings
of a first order rate constant, K', dependent in-
versely on initial total cytochrome c concentration

without saturation, but with addition of a quadratic
term in the function determining K', to account
for the fact the proportionality is not simply
inverse to cytochrome c concentration. Thus, one
must assume two, rather than just one, cytochrome
c substrate interaction is involved, especially
at high concentrations. The denominator terms,
as well as the linear numerator terms, represent
distribution of oxidase in various complexes and
in free form at steady state, the condition
established in these studies.

Any number of mechanisms consistent with this
general rate law can be devised. The choice
depends on criteria of economy in hypotheses and
plausibility in conceptual bases. Thus, mechanisms
which postulate productive complexes between substrate and enzyme, requiring first a binding step
which sets up a steady state process, appears
to biochemists as a plausible conceptual approach.
On the other hand, the lack of constraints inherent
in the *in vitro* system studied compared to the
structured *in vivo* system opens the possibility of
a diametrically opposed scheme in which most
collisions between the reactants produce non-
productive complexes and the forward reaction is
limited to occasional existence of transition
complexes, the concentration of which is determined
by various equilibria set up during the non-
productive complex formations and dissociations.
This is a type of "dead-end" mechanism kineticists
might naturally postulate. Variations on the
former type of mechanism--the "productive-complex"
scheme--can also be imagined. One example is based
on the assumption that independent sites of reaction are located in the enzyme surface, rather than
that there is only one primary site. If, as the
quadratic term in the general rate law implies,
two cytochromes c can react with one molecule of
oxidase, the reactions may be governed by sites
which are independent. Or one may assume reaction
of one cytochrome c at the reaction enzyme site
dictates the reaction of the second cytochrome c
at its site, or even the same site--a so-called
"dependent site" mechanism. To these two types
of productive-complex mechanisms, as well as the
"dead end" type one may add still another variation
in which reaction requires a permanently-bound
cytochrome c which interacts by electron exchange
with a free cytochrome c.

There are no definitive data for choosing among these possible schemes, or others that may be imagined, but one of the simplest and most plausible--"the dependent site" mechanism--provides a logical base for the comparative approach we have employed. It allows a distinction to be made between the two aspects of the rate behavior -- the effects of changes in binding constants and the comparative intrinsic catalytic efficiency of the complex--once binding in a productive manner has occurred. In Table I, we show the mechanistic scheme involved in the functional relations between observed and derived kinetic constants.

It is important to emphasize that the mechanism chosen in no way implies a claim to its operation *in vivo*. One might suppose, indeed, that in the mitochondrion, cytochrome c is bound more or less permanently to one or both of its redox reaction partners in a manner equivalent to "100% effective" collision. It is immaterial, however, which mechanism may best reflect the *in vivo* case. The purpose of the kinetic analysis based on the choice of the scheme shown in Table I is to determine what structural changes affect the binding equilibria which are dominant factors in this scheme. Once these have been evaluated by comparison of the kinetic behavior of the various cytochrome c structures studied, one may reason back to the topology that is likely to exist in the *in vivo* system. The mechanism shown in Table I is an extension of an earlier scheme due to Minnaert (33) which takes into account reaction of a second cytochrome c with the oxidase after the first cytochrome c has been bound and undergone reaction. This second reaction requires a certain concentration in complex formation between cytochrome c and oxidase. Elsewhere (30,31) we have presented detailed schemes for this and the other mechanisms mentioned with derivation of functions relating rate law parameters to reaction rate constants. The assumption involved in any of these schemes which are consistent with the first order kinetics observed are that "on" and "off" constant (or in the case of the "dead end" mechanism their ratio) be equal for the ferrocytochrome c and ferricytochrome c complex equilibria. Thus, in Table I, $k_1 = k_6$, $k_2 = k_5$, $k_7 = k_{12}$ and $k_8 = k_{11}$. In addition, reverse reaction rates (reduction of

TABLE I. "Dependent Site" Mechanism for Cytochrome $\underline{c}$-Cytochrome $\underline{c}$ Oxidase Reaction

"Site I"

$$\text{ferro c} \underset{K_2}{\overset{K_1}{\rightleftharpoons}} \text{[ferro c-ox]} \underset{K_4}{\overset{K_3}{\rightleftharpoons}} \text{[Ferri c-ox]} \underset{K_6}{\overset{K_5}{\rightleftharpoons}} \text{ferri c}$$
$$+ \qquad\qquad\qquad\qquad\qquad\qquad\qquad +$$
$$\text{ox} \qquad\qquad\qquad\qquad\qquad\qquad\qquad \text{ox}$$

"Site II"

$$\text{ferro c} \underset{K_8}{\overset{K_7}{\rightleftharpoons}} \text{[ferro c-ox-ferro c]} \underset{K_{10}}{\overset{K_9}{\rightleftharpoons}} \text{[ferri c-ox-ferro c]} \underset{K_{12}}{\overset{K_{11}}{\rightleftharpoons}} \text{ferri c}$$
$$+ \qquad\qquad\qquad\qquad\qquad\qquad\qquad\qquad\qquad\qquad +$$
$$\text{[ferro c-ox]} \qquad\qquad\qquad\qquad\qquad\qquad\qquad \text{[ferro c-ox]}$$

$$\text{ferro c} \underset{K_{14}}{\overset{K_{13}}{\rightleftharpoons}} \text{[ferro c-ox-ferri c]} \underset{K_{16}}{\overset{K_{15}}{\rightleftharpoons}} \text{[ferri c-ox- ferri c]} \underset{K_{18}}{\overset{K_{17}}{\rightleftharpoons}} \text{ferri c}$$
$$+ \qquad\qquad\qquad\qquad\qquad\qquad\qquad\qquad\qquad\qquad +$$
$$\text{[ferri c-ox]} \qquad\qquad\qquad\qquad\qquad\qquad\qquad \text{[ferri c-ox]}$$

TABLE I. (continued)

Rate Equation:

$$\text{Velocity} = \frac{K_1{}^\circ + K_1 K_1{}^\circ [c]}{1 + K_1[c] + K_1 K_2 [c]^2}$$

Definition of Constants:

Electron Transfer: "Site I": $K_1{}^\circ = \dfrac{K_1 K_3}{(K_2 + K_3)}$

                              "Site II": $K_2{}^\circ = \dfrac{K_7 K_9}{(K_8 + K_9)}$

Binding: "Site I": $K_1 = \dfrac{K_1}{K_2}$

                               "Site II": $K_2 = \dfrac{K_7}{K_8}$

product) are negligible ($k_4 = k_{10} = k_{16} = 0$) and
reactivities of substrate with either ferro- or
ferri- complexed oxidase are the same ($k_7 = k_{13}$,
$k_8 = k_{14}$; $k_9 = k_{15}$; $k_{11} = k_{17}$, and $k_{12} = k_{17}$).
This formidable collection of rate constants can
be tamed and an interpretation of observed reaction
rates and binding constants in terms of the α and β
parameters can be accomplished by determination
of these parameters at limiting concentrations,
as described in our previous reports (30,31).
Some rationalization of these assumptions, other
than that they work, is also provided.

The general rate law parameters are related to
certain constants which are derived from limit
value of K' at zero substrate concentrations and
binding constants observed at two extremes of the
concentration ranges studied (30,31). The values
of these kinetic constants, so derived for the
six selected cytochromes c shown, are given in
Table II. We have need to consider only binding
constants and rates for the "Site I" reactions at
low concentrations corresponding to the "high
affinity" sites mentioned, by others (23).

TABLE II. Values for the Kinetic Constants Defined
by the "Dependent Site" Mechanism

Cytochrome	I^a (mM)	K_1° ($M^{-1} s^{-1}$)	K_1 (M^{-1})
Horse heart c	44	2.50×10^8	6.21×10^6
	4	1.90×10^8	1.20×10^7
Cr. fasciculata c_{555}	44	7.94×10^7	1.40×10^6
C. thiosulfatophilum c_{555}	44	3.65×10^6	9.03×10^4
P. denitrificans c_{550}	44	6.05×10^5	1.29×10^5
R. rubrum c_2	44	9.70×10^4	2.56×10^4
	4	8.65×10^5	3.56×10^5
Rm. vannielii c_2	44	2.55×10^4	6.98×10^3

^{a} I = ionic strength

Taking the first three or the last three cytochromes c as separate groups, one notes that large
relative decreases in K_1 parallel decreases in $k_1°$.
That is, lowered binding correlates with lowered
reactivity, assuming that in the terms for $k_1°$ the
reaction rate constant k_3 is essentially equal for
each of the cytochromes c tested. This assumption
is not likely to be valid. However, a better basis
for the correlation between binding and reactivity
may be the data which compares rates $k_1°$ at different ionic strengths for two of these cytochromes
c (horse heart cytochrome c and $R.$ $rubrum$ cytochrome c_2), one from each group. We see that as
ionic strength drops from 44 mM to 4 mM, $k_1°$ and K_1
both increase in the case of $R.$ $rubrum$ cytochrome c_2.
This result is consistent with the many reports in
the literature which show that binding of cytochromes c to oxidase is electrostatic in character
(32,23,41). The corresponding data for the horse
heart cytochromes c are less definitive.

The break in correlation between $k_1°$ and K_1
which occurs between the $C.$ $limicola$ cytochrome
c_{555}, representing the least reactive cytochrome
c of the first group and $P.$ $denitrificans$
cytochrome c_{550}, the most reactive of the second
group (all cytochromes c_2) may reflect a marked
effect of cleft deformation on intrinsic catalytic
efficiency superimposed on defective binding. In
any case, the data available from these studies
as well as others (30,31) support the notion that
binding can be considered a dominant parameter in
determination of reactivity.

Deductions on Structural
Requirements for Reactivity

Granted that surface topology must affect
binding to the oxidase, one may proceed to examine
what is known about the structures of some cytochromes c for which relative binding data have
been obtained in the manner described for the
cytochromes c discussed in the preceding section.
These are the eukaryotic cytochromes c from a
higher animal (horse heart), a protozoan ($Crithidia$
$fasciculata$) and another protozoan ($Tetrahymena$
$pyriformis$). To these we add four prokaryotic
forms, the cytochromes c_2 from the photosynthetic

bacteria, *Rhodospirillum rubrum* and *Rhodomicrobium vannielii* and the non-photosynthetic denitrifier, *Paracoccus denitrificans*, and the small cytochrome c from the facultative nitrate reducer, *Pseudomonas aeruginosa*.

The three-dimensional structure of horse heart cytochrome c can be taken as identical with the highly resolved homologue from tuna, as determined by Swanson, Takano and their associates (34,35). Primary structures for the protozoan cytochrome c are available (36,37). High resolution three-dimensional structures for the prokaryotic cytochromes c_2 from *P. denitrificans* (38), *R. rubrum* (16), and *Ps. aeruginosa* (R. E. Dickerson, personal communication) have been published or are available. The primary structure of *Rm. vannielii* cytochrome c_2 (8) closely resembles that of the eukaryotic mitochondrial cytochromes c form.

In addition, a three-dimensional structure for the prokaryotic cytochrome c_{555} from the anaerobic green sulful photosynthetic bacterium, *Chlorobium limicola*, var. thiosulfatophilum has been determined recently (40). Regardless of how different the source, all the tertiary structures known show the same "cytochrome c fold" (4) and general over-all structural characteristics. Hence, one may assume the shape and folding character of those homologous cytochromes c for which the tertiary structures are not known will exhibit the same features as those cited. In Figure 1 we exhibit the seven cytochromes c chosen for binding and reactivity. The comparisons show, in particular, where certain charged residues may be located. The relative binding (taken to dominate reactivity) is shown in Table III, to be correlated with positions of the relevant cationic residues (mostly lysines) as deduced by homology.

As we have remarked, considerable evidence exists for the contention that the binding reaction *in vitro* is largely electrostatic in nature. The literature which is too extensive to quote presents studies on the effects of ionic strength, chemical derivatization of cationic groups, antibody blockage of certain specific areas, etc., all of which not only underscores the importance of the cationic residues present on the "front surface" of cytochrome c, encircling the heme cleft, but also implicating certain cationic groups as

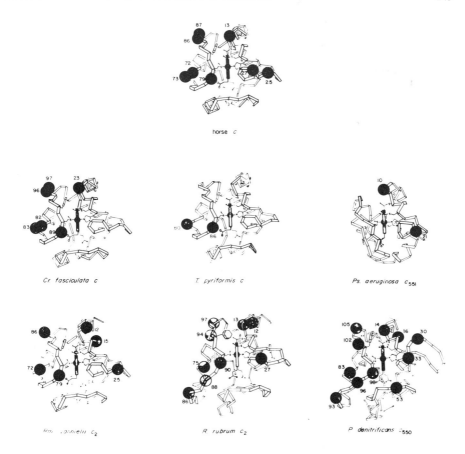

FIGURE 1.   Comparative locations of cationic
residues on the front surface of selected cyto-
chormes c.   Schematic representation of the "front"
view of the folded polypeptide chain with the edge
of the heme prosthetic group containing pyrrole
moieties II and II projecting toward the reader.
The binding extraplanar ligands, histidine
nitrogen and methionine sulfur, are shown.   Large
shaded circles indicate location of cationic
residues (all lysines for the cytochromes c shown).

TABLE III. Location of Cationic Residues in Selected
Cytochromes c (numbers refer to structurally analogous
positions on the "front" surface of Horse Heart
Cytochrome c)

Cytochrome	K_1^a	Positions							
		13	25	27	72	73	79	86	87
Horse Heart c	1.000	+	+	+	+	+	+	+	+
Cr. fasciculata c	0.200	+	−	−	+	+	+	+	+
P. denitrificans c_{550}	0.020	+	−	+	−	−	+	−	+
R. rubrum c_2	0.004	+	−	+	−	−	+	−	−
Rm. vannielii c_2	0.001	−	+	−	+	−	+	+	−
T. pyriformis c	< 0.001	−	−	−	−	+	+	−	+
Ps. aeruginosa c_{551}	< 0.001	+	+	−	−	−	−	−	−

^{a} Relative value with horse heart c taken as 1.000

particularly crucial in the interaction with the
oxidase (39). An example of studies demonstrating
that a given surface residue other than cationic
is not crucial for reactivity is that of Osheroff
et al. who find that iodination of tyrosine 74 in
horse heart cytochrome c does not influence reactivity with oxidase or succinate-linked reductase
(40). As to the importance of cationic groups, we
may cite the most recent, and apparently most
definitive study--that of Smith et al. (42)--in
which it is shown that modification of certain
specific highly conserved lysines located around
the cleft region in horse heart cytochrome c affect
activity with the oxidase. In these researches,
all derivatives are shown to retain structure and
physiochemical characteristics of native cytochrome
c. The modification procedures in this study involve derivatization of lysines 8, 13, 27, 72, 79 and
100, singly, by reaction of the protein with tri-
fluoromethyl(phenylisocyanate) under special
conditions which assure reactions of only
a limited number of the total lysines on the
surface. Only lysines 13, 27, 72, and 79-- those
immediately surrounding the emergent edge of the

heme in the crevice -- are revealed as needed in
effective binding for reactivity.

In any attempt to assess correlations of this
type, the tacit assumption as a "zero-order" approximation, is that binding variations in individual
cytochrome c interactions are assignable to single
residues rather than to an effect accumulated involving several surface residues. Moreover, the
effects of anionic charges which may chance to be
located near the cationic residue(s) in question
are difficult to evaluate. In any case, it is
instructive to attempt a correlation, however
simplistic.

At the front of the cytochrome c structure
in the horse heart mitochondrial protein, lysines
occur at positions 13, 25, 27, 72, 73, 79, 86,
and 87 (shown schematically in Fig. 1), forming
essentially a continuous ring of positive charge
at physiological pH around the opening of the heme
cleft from which the hydrophobic peripheral
portions or π - systems of heme pyrrole moieties
II and III protrude partially. In the protozoan
cytochrome c of $Cr.$ $fasciculata$, with a binding
constant K $\overline{1}/5$ that of horse heart c, the lysines
equivalent to positions 25 and 27 are absent so
that the right front surface is uncharged. Absence
of lysine at position 25 does not appear to effect
binding of yeast iso-cytochrome c (23) so that one
may suppose only the presence of positive charge
in the proper orientation can be achieved by
lysine 27. One may note that the relative efficiency of binding at the second site, K_2, is
reversed from that observed for the first site,
K_1, the former being four times greater. This
result can be rationalized by the suggestion that
prior binding at position 27 produces a structural
change in surface conformation which renders site
25 less accessible. In this connection, one notes
that the unreactive $T.$ $pyriformis$ cytochrome c
lacks positive charge completely on the right front
surface. This cytochrome c appears to function
in a wholly different oxidase system more akin
to the prokaryotic-type cytochrome "d" (43)--
originally called "cytochrome a_2"-- so that the
general absence of cationic charges on its front
surface may indicate that in this system hydrophobic bonding rather than electrostatic is
important. The $Ps.$ $aeruginosa$ cytochrome c_{551}
reveals a similar situation with a more hydrophobic

front surface compared to mitochondrial cytochromes c; it also functions with a prokaryotic
type cytochrome "d" oxidase.

   Comparison of presence of cationic residues
of the c_2-type cytochromes shows that the distribution on the front surface is significantly
different from that of horse heart cytochrome c,
or its eukaryotic analogues. The insertions and
deletions in *R. rubrum* and *P. denitrificans*
cytochromes c_2 may change, however slightly, the
orientation of the peptide backbond thereby
rendering otherwise homologous lysines structurally
non-equivalent for binding. Thus, the major
insertion on the left front surface causes the
lysine 75 sequentially analogous to horse heart
lysine 72 to be displaced backward. Likewise,
a single residue deletion beyond the ligand
methionine causes lysine 97 to be similarly
displaced relative to horse heart lysine 87.   The
three residue deletion in this region for *P.*
denitrificans cytochrome c_2, however, causes lysine
105 to assume an orientation like that of horse
heart lysine 87.   Thus, the increased reactivity
of the *P. denitrificans* protein compared to that
of *R. rubrum* can be rationalized as reflecting to
some extent this structural circumstance related
to lysine 87. Finally, the single deletion in
the helical region just above the crevice,
equivalent to horse heart cytochrome c valine
11 re-orients the helix for all three cytochromes
c_2. Hence it is likely that *Rm. vannielii* cytochrome c_2 lysine 12 is not equivalent in placement
to horse heart cytochrome c lysine 13. The very
marked unreactivity of this cytochrome c_2 may
reflect a very great importance for specific
binding attributable to precise placement of
cationic charge at the head of the crevice and
consequently the importance of lysine or arginine
13 location at this point on the surface. These
results based on comparative kinetic studies
agree with those of Smith *et al.* (42) derived from
chemical modification in implication of lysines
13, 27, 72 and/or 73, 86 and/or 87 as requisite
for binding to the oxidase. A model based on
charge complementarity between mitochondrial
cytochrome c and microsomal cytochrome b_5, as
proposed by Salemme (44), brings the invariant
lysines 13, 27, 72 and 79 of the former into

juxtaposition with complementary oppositely charged groups (aspartates 48, 60, and glutamate 44 and the most exposed propionate) of the latter. The distance between lysine terminal N atoms of cytochrome c and carboxylate 0 atoms of cytochrome b_5 is approximately 3 Å - consistent with an intermolecular complex in which bulk water is excluded at the interface and the closest distance of approach between the heterocyclic π-bonded atoms of the two heme planes, tilted at 15° to each other, is 8.4 Å.

Similar approaches to the solution of reductase bonding characteristics are as yet in a much less well defined status, primarily because kinetics based on the use of Complexes I and II as the cytochrome c, mediated redox system exhibit a complex two-phase kinetics, including initial zero order and consequent first-order reactions relative to the substrate ferricytochrome c. However, the first-order phase also shows lack of saturation at all substrate concentrations and a general rate law similar to that for the oxidase. A major difference is an apparent similarity in reaction rates for cytochromes c_2 compared to cytochromes c, whereas in the oxidase, reaction rates are so much lower for the former with mitochondrial oxidase. Thus, there is a greater similarity for binding of the various cytochromes c and cytochromes c_2 for the reductase than there is for the oxidase (see Table IV). The case of $T.$ $pyriformis$ cytochromes c is one of particular interest because it lacks all surface lysines except for residue 86, analogous to lys 79 in horse heart cytochrome c. This eukaryotic cytochrome c is as incapable of reaction with reductase as with oxidase. Hence, it is clear that more than lysine 79 is requisite for reductase binding. Likewise the value of K_1 for $Rm.$ $vannielii$ cytochrome c_2 is three orders of magnitude lower than that for horse heart cytochrome c in the oxidase reaction while it is only one order of magnitude lower for the reductase reaction (Table IV). Thus, the absence of lysine 13 is less important for reductase binding than it is for oxidase binding. A similar conclusion can be reached concerning lysine 27.

Thus, structural requirements for the reductase reaction are distinct from those for the oxidase

TABLE IV. Kinetic Constants (K°) and Binding Constants
(K) for the First Order Reaction Between Cytochromes
c and c_2 with Mitochondrial Reductase (Complex I and III)
after Errede (30)

Cytochrome	K° (sec^{-1} M^{-1})	K (M^{-1})
Horse Heart c	4.17×10^8	5.50×10^5
P. denitrificans c_{550}	5.53×10^7	1.84×10^5
R. rubrum c_2	1.08×10^8	9.24×10^4
Rm. vannielii c_2	8.50×10^7	5.91×10^4

in that they involve fewer of the cationic residues
surrounding the heme crevice, possibly only an overlap for reductase and oxidase binding in the left
front region of cytochromes c. Further comparisons
based on quantitative kinetic analyses for both
oxidase and reductase should permit specific assignment of crucial surface features for interaction
of cytochrome c with its redox partners in the mitochondrial reaction chain, and thus hasten the
eventual solutions of the general structural mechanisms operative in the redox processes involved.
Extension of such analyses to bacterial and protozoan oxidase and reductase systems is needed to
attain a general understanding of bio-energetics at
the molecular level which is based on cytochrome c
interactions with the various redox systems in
which it is functional.

ACKNOWLEDGMENTS

Structural data from which the schematic
representations of cytochrome c were adapted were
provided from published figures in the literature
cited, particularly those of Professor R. E.
Dickerson and associates, Professor F. R. Salemme
and Professor R. P. Ambler to whom we are indebted
for constructive discussions.

REFERENCES

1. Kamen, M. D., Proteins, Nucleic Acids, Enzymes (Japan) 18:753 (1972).
2. Kamen, M. D., Dus, K. M., Flatmark, T., and DeKlerk, H., in "Electron and Coupled Energy Transfers in Biological Systems," Vol. 1, Part A (T. E. King and M. Klingenberg, eds.) p. 243. M. Dekker, Inc., New York, 1972.
3. Dickerson, R. E., Sci. American, 226:58 (1972).
4. Dickerson, R. E., Timkovich, R., and Almassy, R. J., J. Mol. Biol. 100:473 (1976).
5. Salemme, F. R., Kraut, J., and Kamen, M. D., J. Biol. Chem. 243:7701 (1973).
6. Salemme, F. R., Ann. Revs. Biochem. 46:299 (1977).
7. Dickerson, R. E., and Timkovich, R., in "The Enzymes," 3rd edit., Vol. 11 (P. Boyer, ed.) p. 397, Academic Press, New York, 1975.
8. Ambler, R. P., in "Electron Transport in Micro-organisms" (J. Senez, J. LeGall, and H. Peck, eds.) C.N.R.S., Paris, in press, 1978.
9. Ambler, R. P., Biochem. J. 137:3 (1974).
10. Ambler, R. P., in "Int'l Sympos. on Evolution of Protein Molecules" (H. Matsubara and T. Yamanaka, eds.), Univ. Tokyo Press, Kobe, Japan, in press, 1978.
11. Korszun, Z. R., and Salemme, F. R., Proc. Nat. Acad. Sci. USA, in press (1978).
12. Davis, K. A., Hatefi, Y., Salemme, F. R., and Kamen, M. D., Biochem. Biophys. Res. Comm. 49:1139 (1972).
13. Vernon, L. P., Arch. Biochem. Biophys. 43:492 (1953).
14. Elsden, S. R., Kamen, M. D., and Vernon, L. P., J. Amer. Chem. Soc. 75:6347 (1953).
15. Bartsch, R. G., in "The Photosynthetic Bacteria" (R. K. Clayton and W. R. Sistrom, eds.), Plenum Press, New York, 1976.
16. Salemme, F. R., Freer, S. T., Nguyen Huu Xeong, Alden, R. A., and Kraut, J., J. Biol. Chem. 248:3910 (1973).
17. Timkovich, R., Dickerson, R. E., and Margoliash, E., J. Biol. Chem. 251:2197 (1976).
18. Ambler, R. P., Meyer, T. E., and Kamen, M. D., Proc. Nat. Acad. Sci. USA 73:472 (1976).
19. Kamen, M. D., Plant and Cell Physiology, Spec. Issue, p. 283, Japan (1977).

20. Errede, B., Kamen, M. D., and Meyer, T. E.,
    see reference 10.
21. Yamanaka, T., Adv. in Biophys. 3:227 (1972).
22. Smith, L., Nava, M. E., and Margoliash, E.,
    in "Oxidases and Related Systems," Vol. 2
    (T. E. King, H. S. Mason, and M. Morrison, eds.)
    p. 629, Univ. Park Press, Baltimore, Md., 1973.
23. Ferguson-Miller, S., Brautigan, D. L, and
    Margoliash, E., J. Biol. Chem. 251:1104 (1976).
24. Nicholls, P., Biochim. Biophys. Acta 346:261
    (1974).
25. Slater, E., Biochem. J. 46:499 (1950).
26. Smith, L., Davies, H. C., and Nava, M. E.,
    J. Biol. Chem. 249:2904 (1974).
27. Hatefi, Y., Haavik, A. G., and Jurshuk, P.,
    Biochim. Biophys. Acta 64:106 (1961).
28. Fowler, L. R., Richardson, S. H., and Hatefi,
    Y., Biochim. Biophys. Acta 64:170 (1962).
29. Errede, B., Kamen, M. D., and Hatefi, Y.,
    in "Manual of Enzymology" (S. Colowick and
    N. O. Kaplan, eds.), in press, 1978.
30. Errede, B., Doctoral Dissertation, Univ. of
    California at San Diego, La Jolla, Ca. (1977).
31. Errede, B., and Kamen, M. D., Biochemistry,
    in press (1978).
32. Smith, L., and Conrad, H. E., Arch. Biochem.
    Biophys. 63:403 (1956).
33. Minnaert, K., Biochem. Biophys. Acta 50:23
    (1961).
34. Swanson, R., Trus, B. L., Mandel, N., Mandel,
    G., Kallai, O. B., and Dickerson, R. E.
    J. Biol. Chem. 252:759 (1977).
35. Takano, T., Trus, B. L., Mandel, N., Mandel,
    G., Kallai, O. B., and Dickerson, R. E.,
    J. Biol. Chem. 252:776 (1977).
36. Hill, G. C., and Pettigrew, G. W., Eur. J.
    Biochem. 57:265 (1975).
37. Tarr, G. E., and Fitch, W. M., Biochem. J.
    159:193 (1976).
38. Timkovich, R., and Dickerson, R. E., J. Biol.
    Chem. 251:4033 (1976).
39. Ferguson-Miller, S., Brautigan, D. L., and
    Margoliash, E., in "The Porphyrins" (D. Dolphin,
    ed.), Academic Press, New York, in press (1978).
40. Osheroff, N., Feinberg, B. A., Margoliash, E.,
    and Morrison, M., J. Biol. Chem. 252:7743 (1977).

41. Van Gelder, B. F., Van Buuren, K.J.H., and
     Verboom, C. N., in "Electron Transfer Chains
     and Oxidative Phosphorylation" (E. Quagliari-
     ello *et al*.,eds.) pp. 63-*et seq*., North
     Holland Publishing Co., Amsterdam.
42. Smith, H. T., Staudenmayer, N., and Millet, F.,
     Biochemistry 16:4971 (1977).
43. Kilpatrick, L., and Ericinska, M., Biochim.
     Biophys. Acta 460:346 (1977).
44. Salemme, F. R., J. Mol. Biol. 102:563 (1976).

HUMAN LIVER ALCOHOL DEHYDROGENASE: ISOLATION AND PROPERTIES OF A NEW AND DISTINCTIVE MOLECULAR FORM[1]

William F. Bosron
Ting-Kai Li

Departments of Medicine and Biochemistry
Indiana University School of Medicine
Indianapolis, Indiana

Werner P. Dafeldecker
Bert L. Vallee

Biophysics Research Laboratory
Department of Biological Chemistry
Harvard Medical School
Boston, Massachusetts

It is generally accepted that the pharmacologic, addictive and pathologic consequences of alcohol consumption are directly related to the chemical properties of ethanol and/or its metabolic products. Hence, knowledge about the enzymes responsible for its elimination is fundamental to our understanding of the etiology and underlying mechanisms of alcoholism. Since these enzymes have not been available in suitable purity and quantity from human tissue until recently, such knowledge has had to be extrapolated from studies in other species, primarily the rat and horse. Moreover, a complex interrelationship between psychosocial and biologic factors exists in alcoholism which has been difficult to unravel. Perhaps owing to these shortcomings, a biochemical basis for this disorder has remained obscure. Nevertheless, there has appeared in recent years increasingly convincing evidence indicating a genetic predisposition not only for alcohol metabolizing capacity but also for alcohol drinking behavior and alcoholism in some individuals (1, 2). In this context, the study of the genetic variability of liver alcohol dehydrogenase (EC 1.1.1.1), ADH, is of particu-

[1]Supported by U.S. Public Health Service Grants AA 02342 and GM 15003.

lar interest, since it is the principal enzyme responsible
for the metabolic elimination of ethanol (3).

Human livers contain multiple molecular forms of ADH,
the number and amounts of which vary, seemingly dependent
upon the genetic background and the health of the donors.
Past studies have examined the genetic basis of this heterogeneity and have partially characterized the catalytic properties of some of these enzyme forms (3). Recently a new
molecular form of human ADH has been discovered whose kinetic properties are strikingly different from those of the
other molecular forms. Importantly, its Km for ethanol is
as much as 100 times that of the others and it is remarkably
insensitive to inhibition by pyrazole compounds, potent
inhibitors of all mammalian alcohol dehydrogenases thus far
studied (4). We have designated this new, pyrazole-insensitive, form of human liver alcohol dehydrogenase, Π-ADH (5).
This communication reviews those aspects of our past work
which led to the discovery of Π-ADH and describes its isolation, properties and functional role in hepatic ethanol
oxidation.

A. Molecular Heterogeneity of Human Liver ADH: Importance of Source of Material and the Discovery of the "Anodic Band"

It has been known since Vallee and coworkers first
attempted the isolation of human liver ADH that there are
multiple molecular forms of the enzyme (6). Subsequent work
in several laboratories confirmed this observation and
further demonstrated by electrophoresis on starch gels that
as many as 6-10 ADH molecular forms are present in some
liver homogenates (7-10). Interestingly, the number and the
amount of the individual molecular forms were found to vary
from liver to liver as did also total and specific enzymatic
activities. To account for these findings, a genetic model
for the formation of ADH molecular forms as isozymes was
proposed by Smith et al. (9, 11). According to this model,
there are 3 separate gene loci producing 3 different types
of polypeptide chains, α, β and γ, that can combine randomly
to form the active, dimeric ADH molecules. The complexity
and variability of the enzyme electrophoretic patterns arise
from individual differences in gene expression and genetic
polymorphism at the loci coding for the β and γ chains.
However, because the model was based almost entirely on
studies that employed crude homogenates and postmortem
specimens without regard to cause of death or the relative
stability of the different ADH forms, its validity could not

TABLE I.   Comparison of Human Liver ADH Activities in
Biopsy and Autopsy Specimens

Source	Number of Samples	Specific Activity+ (μmol product/min/mg protein) Range	Mean ± SD
A. Biopsy	16	0.053-0.104	0.078±0.016
B. Autopsy, Sudden Deaths	74	0.013-0.148	0.059±0.028
C. Autopsy, Disease-Related Deaths	47	0.006-0.067	0.025±0.014

+ Measured at pH 10.5 with 33 mM ethanol as substrate.

be accepted with certainty. Therefore, studies were performed to compare the activities (12, 13) and electrophoretic
patterns of livers (13) obtained by biopsy with those
obtained within 12 hours of death from individuals dying
from different causes.

The specific ADH activities of 16 biopsy samples
(group A), 74 autopsy samples from apparently healthy individuals who had died suddenly (group B), and 47 autopsy
samples from hospitalized patients who died of disease
(group C) are compared in Table I. The mean specific activity of the biopsy samples was higher than that of group B
which, in turn, was significantly higher than that of group
C. Thus, the health of the donor before death is a major
determinant of ADH activity in liver, although extended
storage at temperatures above -20° also resulted in loss of
activity.

In one of the above studies (13), the molecular forms
of ADH present in the liver extracts were simultaneously
examined by starch-gel electrophoresis. Importantly, it was
found that, in addition to all the molecular forms described
by Smith et al (11), there was present in the majority of
the livers a previously unidentified activity band whose
electrophoretic mobility was less than that of the αα enzyme form (Figure 1). In the studies of Smith et al (11),
the αα form, the major activity band in livers from premature
infants (sample Ap), was identified as the ADH isozyme with
the least electrophoretic mobility. Interestingly, the

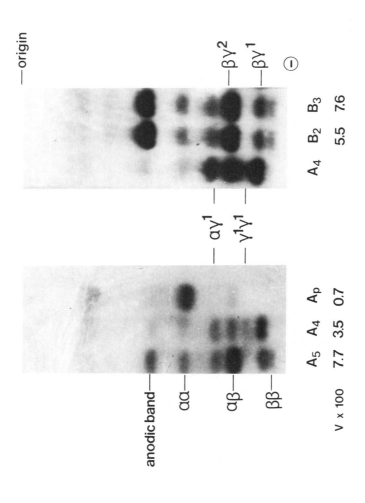

FIGURE 1. Starch gel electrophoresis of human liver ADH obtained from autopsy (A) and biopsy (B) specimens was performed at pH 7.7 (3). Enzyme activity is detected at pH 8.5 with ethanol as substrate. Specific enzymatic activity (V) of the samples is expressed as μmol of NADH formed per min per mg protein. Specimen A_p was from a premature infant. Specimen A_4 is phenotype ADH$3$1 and specimens A_5, B_2 and B_3 are ADH32-1, according to the nomenclature of Smith et al.

newly discovered enzyme form, initially called the "anodic band" (13), was very prominently visualized in all biopsy specimens (e.g. samples B_2 and B_3) and in the autopsy livers with high activity from the sudden-death group (sample A_5). It was only faintly visualized (sample A_4) or absent in autopsy livers from the disease-related group. Therefore, presumedly because this anodic enzyme form is more labile than the others both in vivo and in vitro, its existence had escaped consistent recognition in the past.

From the above data, it is now apparent that both the specific activity and degree of multiplicity of human liver ADH in autopsy livers depend critically upon the premortem history of the samples as well as storage conditions. Because the genetic model of ADH isozymes postulated by Smith et al was based primarily on studies of postmortem tissue from hospitalized patients, it does not account for the existence of the anodic band, present in biopsy liver specimens and the autopsy specimens exhibiting high activity. Hence, reevaluation of the molecular basis of the heterogeneity of ADH is necessary.

B. Purification of Human Liver ADH by Affinity Chromatography

Past studies utilized conventional techniques of salt fractionation and ion-exchange chromatography to isolate the molecular forms of ADH, either singly or collectively (6, 8, 10, 14). These approaches, for the most part, did not consistently yield sufficient quantities of pure enzyme to permit detailed physicochemical and kinetic characterization. Recently, a new and effective affinity chromatographic procedure has been developed that copurifies the molecular forms of ADH in bulk and high yield, free of contaminating proteins (15). This method employs a 4-substituted derivative of pyrazole, 4-[3-(N-6-aminocaproyl)-aminopropyl] pyrazole immobilized on Sepharose (CapGapp-Sepharose). Pyrazole and its 4-substituted analogs have been shown to be potent inhibitors of ADH by forming specific and tight dead-end ternary complexes with the enzyme and NAD^+. Thus in the presence of NAD^+, ADH is adsorbed onto the affinity column. It is then eluted from the column with ethanol through the formation of the productive ternary complexes enzyme·NAD^+· ethanol and enzyme·NADH·acetaldehyde. This dual selection process, based on the affinity of the inhibitor and substrate for the enzyme·coenzyme complex, imparts high specificity and, hence, a high degree of purification is attained. The method has been termed "double-ternary complex affinity chromatography."

The procedure has been applied successfully to the
isolation of ADH from human, horse, rat and rabbit livers
(15). In the case of human liver ADH, the total time required is only 6 hours in contrast to the several days
needed with more conventional methods. The yield is high,
65%, and the preparations are free of contaminating proteins
as determined by SDS polyacrylamide gel electrophoresis, gel
filtration and ultracentrifugation. They contain all the
pyrazole-sensitive molecular forms of ADH originally present
in the liver extracts (5, 15). The physicochemical and
kinetic properties of such preparations have been studied
previously by Lange et al (16).

C. Isolation of a Pyrazole-Insensitive Enzyme Form,
Π-ADH, and Proof of its Identity with the Anodic Band

The relationship of the anodic band to high specific
ADH activity suggested that it must contribute significantly
to total ADH activity in liver. Hence its isolation was
attempted with the use of CapGapp-Sepharose affinity chromatography. Homogenate supernatants of human autopsy livers
exhibiting high specific enzymatic activity were first
partially purified on DEAE-cellulose and then applied to
CapGapp-Sepharose in the presence of NAD$^+$ (15). As shown in
Figure 2, a significant portion of activity did not bind to
the affinity resin but eluted in the void fraction from the
column with the bulk of the protein. In accord with previous studies (15), the majority of the activity bound to the
resin and was eluted with 0.5 M ethanol.
        The failure of a substantial part of ADH activity to
bind to CapGapp-Sepharose suggested that some of the molecular
forms are less susceptible to inhibition by pyrazole compounds
than others. Therefore, the enzyme electrophoretic patterns
of the liver homogenate and the fractions which bound and
did not bind to the resin were examined by staining for
ethanol oxidizing activity in the presence and absence of 4-
methylpyrazole (Figure 3). In the homogenate, the band with
the least electrophoretic mobility was identified as the
anodic band. The other bands corresponded to those enzyme
forms characteristic of phenotype ADH$_3$2, according to the
nomenclature of Smith et al (11). Interestingly, the enzyme
fraction that did not bind to CapGapp-Sepharose contained
only the anodic enzyme form, whereas the fraction which
bound to the resin contained all the other molecular forms.
As might be expected from this result, the anodic enzyme
form was found not to be inhibited by 4-methylpyrazole at
concentrations which completely inhibited the activities of
all other molecular forms (Figure 3). This relative in-

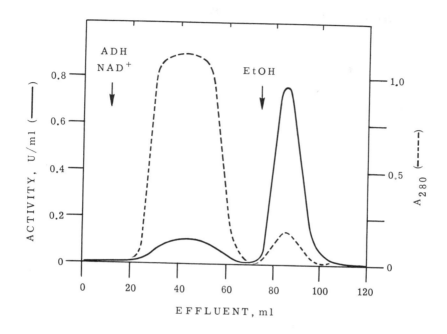

FIGURE 2.   CapGapp-Sepharose affinity chromatography of human liver ADH.   The supernatant from a liver homogenate was first purified on DEAE-cellulose and then applied to a 0.9 X 35 cm column of CapGapp-Sepharose in 50 mM Na Pi and 3 mM NAD$^+$ at pH 7.5.   ADH activity was determined with 33 mM ethanol, 2.4 mM NAD$^+$ in 0.1 M glycine-NaOH at pH 10.0.

sensitivity of the anodic enzyme form to inhibition by pyrazole accounts for its ease of separation from the other pyrazole-sensitive molecular forms of ADH and, on the basis of this distinctive property, it was renamed, Π-ADH (5).

Π-ADH was then purified by affinity chromatography on AMP-Agarose.   Virtually all the activity bound to the resin and was subsequently eluted with a linear gradient of NADH

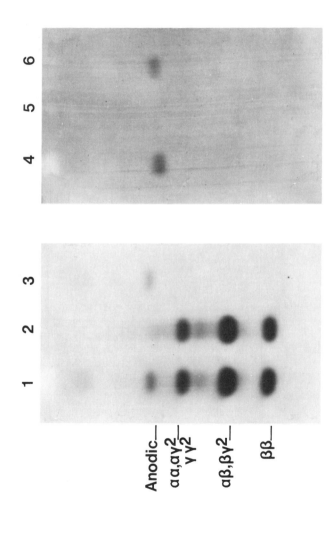

FIGURE 3. Identification of the ADH molecular forms in liver homogenate and the enzyme fractions separated by CapGapp-Sepharose. Starch gel electrophoresis of the homogenate supernatant, samples 1 and 4, the enzyme fraction which bound, samples 2 and 5, and did not bind, samples 3 and 6, to the affinity resin was performed at pH 7.7. Gels were stained in the absence, samples 1-3, or presence, samples 4-6, of 2 mM 4-methylpyrazole (5).

(Figure 4). Purified in this manner, II-ADH was homogeneous as evidenced by SDS-gel electrophoresis and analytical ultracentrifugation (17).

D. Molecular and Catalytic Properties of II-ADH

The inability to detect II-ADH in certain autopsy liver specimens as opposed to those obtained at biopsy suggested that this molecular form of the enzyme is more labile than the others in vivo (13). Consistent with this conclusion, purified II-ADH was also very unstable when stored in vitro. Approximately 50% of activity was lost within 24 h at pH 7.5 and 4°. However, addition of 10^{-2} M ethanol effectively stabilized enzymatic activity for up to two weeks (17).

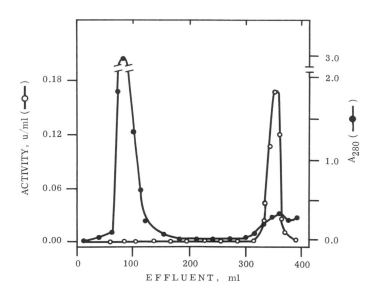

FIGURE 4. AMP-Agarose affinity chromatography of II-ADH. The enzyme fraction that did not bind to CapGapp-Sepharose was precipitated with 75% ammonium sulfate and gel filtered on Bio-Gel P-6 in order to remove NAD^{+}. Enzyme was then chromatographed on a 2.5 X 30 cm column of agarose-hexane-AMP (PL Biochemicals, Type II) in 0.1 M Tris-Cl, pH 8.6 at 4°. II-ADH was eluted with a linear gradient of 0 to 7×10^{-5} M NADH beginning at 300 ml of effluent. II-ADH activity was determined as described in Figure 2.

Purified Π-ADH has many physical and kinetic properties
in common with horse-liver ADH (4) and with the other enzyme
forms isolated previously from human liver (3). Π-ADH is a
dimer composed of 42,000 dalton subunits as determined by
SDS-gel electrophoresis and analytical ultracentrifugation
(17). Its amino acid composition and zinc content, 4 g-
atoms/mole, are similar to other, pyrazole-sensitive, human
ADH molecular forms (18). Such similarities suggest that Π-
ADH and the other enzyme forms may be genetically related.

Detailed kinetic studies with Π-ADH indicate that it
obeys an ordered BiBi mechanism for the oxidation of ethanol
and reduction of acetaldehyde (19). The pH optimum for the
maximal velocity of ethanol oxidation (V_M) with Π-ADH is
above 10.0. Similar to other human ADH molecular forms, it
is specific for NAD^+ and NADH as coenzymes with Michaelis
constants (Km) of 14 µM and 16 µM, respectively. The enzyme
exhibits broad substrate specificity for alcohol oxidation
(Table II). Increasing the primary alcohol chain length from
2 to 5 carbons decreases the Km from 34 to .036 mM while V_M
remains relatively constant at 0.48 to 0.50 µmol/min/mg.
The relationship between log 1/Km and log octanol-water
partition coefficient for these alcohols was found to be
linear. This suggests that hydrophobic binding energies
play an important role in Π-ADH substrate specificity (20).

Importantly, Π-ADH exhibits certain kinetic properties
that are strikingly different from those described for
previous preparations of the human enzyme (6, 10, 14, 16,
21). It does not oxidize methanol or ethylene glycol, even
at a concentration of 100 mM (Table II). The Km values for
both ethanol, 34 mM, and acetaldehyde, 30 mM, are as much as
100 times greater than those reported previously for impure
preparations or for mixtures of the pyrazole-sensitive forms.

TABLE II. Substrate Specificity of Π-ADH

Alcohol	V_M	K_M
	(µmol/min/mg)	(mM)
Methanol	0	--
Ethanol	0.50	34
Butanol	0.50	0.14
Pentanol	0.48	0.036
3-Pyridylcarbinol	0.49	0.24
Ethylene Glycol	0	--

0.1 M Pi, pH 7.5, 2.4 mM NAD^+, 25°.

As demonstrated both by spectrophotometric assay and by activity staining of starch gels, Π-ADH is insensitive to inhibition by 4-methylpyrazole at concentrations that completely inhibit the other forms (5, 17). However, if the pyrazole or 4-methylpyrazole concentrations are increased approximately 1000 fold above the K_I previously reported for preparations of horse and human ADH (14, 22), Π-ADH is eventually inhibited. The mode of inhibition, under these circumstances, is no longer competitive with respect to ethanol in the oxidative direction, but is now competitive with respect to acetaldehyde in the reductive direction (Figure 5). K_I values calculated from the increase in slope of the reciprocal plots for pyrazole and 4-methylpyrazole are 30 and 1.4 mM, respectively. By contrast, the corresponding K_I values for the pyrazole-sensitive forms of human liver ADH are 2.6 and 0.21 μM (14).

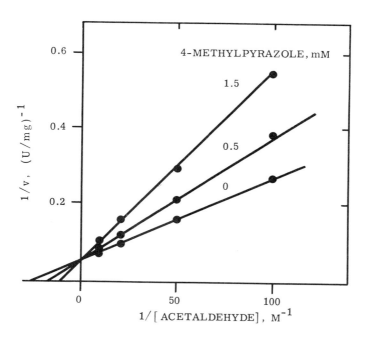

FIGURE 5. 4-Methylpyrazole inhibition of acetaldehyde reduction by Π-ADH. Enzymatic activity was determined with 0.2 mM NADH in 0.1 M Na Pi at pH 7.5 and 25° in the absence and presence of 4-methylpyrazole. The data were analyzed by plotting the reciprocal of the velocity versus the reciprocal of the acetaldehyde concentration.

E. Quantitation of the Functional Role of Π-ADH in Hepatic Ethanol Oxidation

These distinctive kinetic properties of Π-ADH, i.e. its high Km for ethanol and insensitivity to inhibition by 4-methylpyrazole, have enabled the elucidation of its content in human liver samples and its potential role in hepatic ethanol oxidation. When assayed at pH 10.5 with 33 mM ethanol, 2.4 mM NAD$^+$ and 33 μM 4-methylpyrazole, the residual activity of liver homogenate supernatant can almost entirely be attributed to Π-ADH. We found that the mean specific activity of Π-ADH in 56 of the samples from the autopsied, sudden-death group in Table I was 0.011 μmol/min/mg protein. By contrast, the Π-ADH activity in 34 samples from the autopsied, disease-related group was 5.5 times lower, 0.002 μmol/min/mg protein. These findings, therefore, confirm the earlier observations by electrophoresis, indicating that the anodic band or Π-ADH activity is preferentially lost in disease states (13).

The high Km for ethanol of Π-ADH suggested that it may play an important role in ethanol elimination at concentrations that are saturating for other enzyme forms. Hence the activity in liver homogenate supernatants was measured as a function of ethanol concentration at pH 7.5 (Figure 6). In the absence of 4-methylpyrazole, activity increased progressively over the range of 0.3 to 100 mM ethanol. However, in the presence of 0.2 mM 4-methylpyrazole, activity appeared only when ethanol concentration exceeded 3 mM and increased thereafter in parallel with that observed in the absence of 4-methylpyrazole. Therefore, Π-ADH contributes substantially to total activity at high ethanol concentrations in accord with its high Km for ethanol (17). Importantly, at concentrations of ethanol that produce moderate to severe intoxication, i.e. 30 to 100 mM, Π-ADH represented as much as 40% of the total alcohol oxidizing activity. In 10 other homogenate supernatants similarly studied, the contribution of Π-ADH at an ethanol concentration of 60 mM ranged from 17% to 39% with an average of 27% (5). This degree of variation suggests that, as with total ADH activity, there may exist an inherent biologic variation in Π-ADH activity. However, to what extent this may have been modified by postmortem change or other environmental factors is presently unknown.

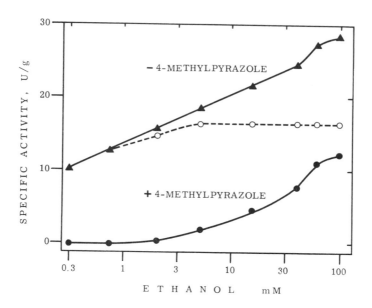

FIGURE 6. Contribution of pyrazole-sensitive and -insensitive (Ⅱ-ADH) activities to total ADH activity in a liver homogenate. Activity in the homogenate supernatant was determined at 0.3 to 100 mM ethanol with 2.4 mM NAD$^+$ in 0.1 M Na Pi, pH 7.5. Ⅱ-ADH activity was determined in the presence of 0.2 mM 4-methylpyrazole and total activity in the absence of 4-methylpyrazole. The difference between total and Ⅱ-ADH activity is calculated to be the pyrazole-sensitive activity, the dashed line (5).

CONCLUSION

The discovery of Ⅱ-ADH as a functionally distinct enzyme form should bear importantly upon our understanding of normal human alcohol metabolism and its pathological derangement. For example: 1) It is a widely held belief that ethanol elimination rates become maximal when ethanol concentrations exceed 5 mM. The studies here shown, however, indicate that oxidation rates <u>in vivo</u> should increase when blood ethanol concentrations rise to intoxicating levels in some individuals. 2) The failure of pyrazole compounds to inhibit ethanol oxidation, especially at high concentrations of ethanol, has been commonly viewed as evidence for pathways of ethanol metabolism other than that catalyzed by ADH.

The pyrazole insensitivity of Ⅱ-ADH indicates that such alternate pathways or their lack in humans cannot be inferred exclusively from the effects of these compounds.

As already noted, both the molecular heterogeneity of liver ADH and alcoholism in some individuals appear to be under genetic control. These considerations raise the provocative question whether the presence or absence of Ⅱ-ADH or of any of the other enzyme molecular forms may prove to be biochemical links to alcoholism. Whether chronic alcohol consumption, malnutrition and disease alter the relative distribution and amount of Ⅱ-ADH and other enzyme forms are particularly pertinent questions. The elucidation of these interrelationships provides a realistic experimental basis from which to advance the understanding of the biochemical determinants and mechanisms in alcoholism and alcohol-related diseases.

REFERENCES

1. Partanen, J., Brunn, K. & Markkanen, T. (1966) Inheritance of Drinking Behavior (Rutgers University Center of Alcohol Studies, New Brunswick, NJ).

2. Goodwin, D.W., Schulsinger, F., Moller, N., Hermansen, L., Winokur, G. & Guze, S.B. (1974) Arch. Gen. Psychiat. 31, 164-169.

3. Li, T.-K. (1977) Adv. Enzymol. 45, 427-483.

4. Branden, C.-I., Jornvall, H., Eklund, H. & Furugren, B. (1975) in The Enzymes, ed. Boyer, P.D. (Academic Press, New York), 3rd ed., Vol. XI, Part A, pp. 104-190.

5. Li, T.-K., Bosron, W.F., Dafeldecker, W.P., Lange, L.G. & Vallee, B.L. (1977) Proc. Nat. Acad. Sci. 74, 4378-4381.

6. Blair, A.H. & Vallee, B.L. (1966) Biochemistry 5, 2026-2034.

7. von Wartburg, J.-P., Bethune, J.L. & Vallee, B.L. (1964) Biochemistry 3, 1775-1782.

8. Schenker, T.M., Teeple, L.J. & von Wartburg, J.-P. (1971) Eur. J. Biochem. 24, 271-279.

9. Smith, M., Hopkinson, D.A. & Harris, H. (1971)
    Ann. Hum. Genet. 34, 251-271.

10. Pietruszko, R., Theorell, H. & de Zalenski,
    C. (1972) Arch. Biochem. Biophys. 153, 279-
    293.

11. Smith, M., Hopkinson, D.A. & Harris, H. (1972)
    Ann. Hum. Genet. 35, 243-253.

12. Azevedo, E., Smith, M., Hopkinson, D.A. &
    Harris, H. (1974) Ann. Hum. Genet. 38, 31
    37.

13. Li, T.-K. & Magnes, L.J. (1975) Biochem.
    Biophys. Res. Commun. 63, 202-208.

14. Li, T.-K. & Theorell, H. (1969) Acta Chem.
    Scand. 23, 892-902.

15. Lange, L.G. & Vallee, B.L. (1976) Biochemistry
    15, 4681-4686.

16. Lange, L.G., Sytkowski, A.J. & Vallee, B.L.
    (1976) Biochemistry 15, 4687-4693.

17. Bosron, W.F., Li, T.-K., Lange, L.G.,
    Dafeldecker, W.P. & Vallee, B.L. (1977)
    Biochem. Biophys. Res. Commun. 74, 85-91.

18. Dafeldecker, W.P., Vallee, B.L., Bosron, W.F. &
    Li, T.-K. (unpublished observations).

19. Bosron, W.F., Li, T.-K., Dafeldecker, W.P. &
    Vallee, B.L. (manuscript in preparation).

20. Hansch, C., Schaeffer, J. & Kerley, R. (1972)
    J. Biol. Chem. 247, 4703-4710.

21. Pietruszko, R. (1975) Biochem. Pharmacol. 24,
    1603-1607.

22. Theorell, H., Yonetani, T. & Sjoberg, B.
    (1969) Acta Chem. Scand. 23, 255-260.

SYNTHETIC ANTIGENS WITH PEPTIDE
DETERMINANTS OF DEFINED SEQUENCE

Michael Sela[1]
Michal Schwartz
Edna Mozes

Department of Chemical Immunology
The Weizmann Institute of Science
Rehovot, Israel

INTRODUCTION

Synthetic polymers of amino acids have been used extensively in many biological studies (1,2). Ever since it has been shown that such polymers may be immunogenic (3), they have been used increasingly to define the molecular requirements for antigenicity as well as to elucidate the molecular parameters of a variety of immunological phenomena (4).

The first synthetic antigen investigated in our laboratory was (T,G)-A--L, a multichain polymer in which poly-L-lysine served as a backbone and the poly-DL-alanyl side chains, attached to the ε-amino groups of the lysine residues, were in turn elongated with peptides containing tyrosine and glutamic acid (5,6). These peptides were prepared by polymeric techniques, and thus both the detailed amino acid composition and their sequence varied from one polymeric side-chain to another.

(T,G)-A--L was chosen in many laboratories as

---

[1]Michael Sela is an Established Investigator of the Chief Scientist's Bureau, Ministry of Health, Israel.

a model antigen for immunochemical and immuno-
biological studies as well as for studies on the
genetic control of immuno response in mice (7-10),
rats (11), chicken (12), etc. In view of the
intensity of studies with this polymer, it was of
interest to better define its antigenic determinants, to have a better notion of the nature of
the antibodies produced in terms of their structure,
specificity and affinity, as well as to better understand the fine specificity of the genetic control
of its immune response. For this purpose we have
synthesized several tetrapeptides composed of two
tyrosine and two glutamic acid residues, attached
them to multichain poly-DL-alanine and investigated
their immunochemical properties.

Synthesis and Immunological Characterization

The polymer (T,G)-A--L is prepared by polymeric
techniques throughout, i.e., the specific determinants of this immunogen are prepared from multichain poly-DL-alanine serving as a polyvalent
initiator for the polymerization of N-carboxyanhy-
drides of tyrosine and glutamic acid which are
still reversibly blocked in their reactive
functions (6). After deblocking, the final product
contains chains of tyrosine and glutamic acid which
may differ in the ratio of the two amino acids, in
the length of the peptides, as well as in its
precise amino acid sequence. For this reason we
always assumed that (T,G)-A--L, which is less
complex than a protein, still possesses a hetero-
geneous mixture of antigenic determinants.
In order to elucidate this problem we have
prepared by stepwise method of peptide synthesis
(Fig. 1, (13)) several tetrapeptides, each containing two tyrosine and two glutamic acid residues
(13,14). These peptides, appropriately and reversibly blocked, were attached to A--L. The resulting
immunogens were tested for their capacity to induce
antibody formation in several inbred mouse
strains (14,15).
Out of the four tetrapeptides tested only one,
namely (Tyr-Tyr-Glu-Glu)-poly(DLAla)--poly(LLys),
[(T-T-G-G)-A--L], was found to be similar to the
random (T,G)-A--L. The similarity was expressed
both in the pattern of immune response of the

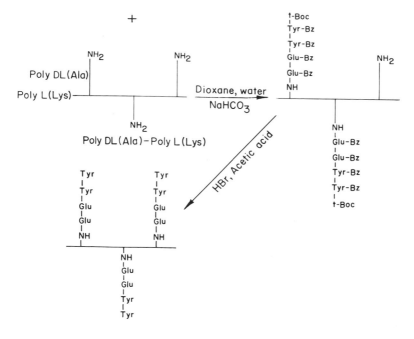

N-tertbutyl-oxycarbonyl-(o-benzyltyrosyl)$_2$-(γ-benzylglutamyl)$_2$- hydroxy succinimide ester

FIGURE 1. Synthesis of the ordered synthetic polypeptide.

TABLE I. Antibody response to (T-T-G-G)-A--L

| Mouse strains | Log_2 hemagglutination[a] | |
	tested with (T-T-G-G)-A--L	tested with (T,G)-A--L
C3H.SW	9	5.6
C57BL/6J	7	5.7
AKR/Cu	2	2
C3H/HeJ	2	2
SJL/J	2	N.D.
CWB	6.5	N.D.
CKB	1.77	N.D.

[a]The data present average of Log_2 of hemagglutination titers of at least 15 individuals in each experimental group as was tested with tanned red blood cells coated with (T-T-G-G)-A--L or with (T,G)-A--L.

TABLE II. Specificity of anti-(T,G)-A--L antibodies.

Test antigen	Log_2 of hemmaglutination titer[a]
(T,G)-A--L	8
(T,G)-Pro--L	7
(T-T-G-G)-Pro--L	7
(T-G-T-T)-Pro-L	3
(G-T-T-G)-Pro--L	2

[a]Log_2 of hemagglutination titers of sera pooled from 10 C3H.SW mice immunized with (T,G)-A--L (14).

different mouse strains (16) and in the cross-reactivity of antibodies elicited to (T-T-G-G)-A--L with the random (T,G)-A--L (Table I (14)). Furthermore, antibodies elicited upon immunization with the random (T,G)-A--L cross-reacted very well with the ordered (T-T-G-G)-A--L (Table II (14)). Thus, it was suggested that T-T-G-G is the major determinant of the random (T,G)-A--L.

Genetic Control of Immune Response

The gene controlling the ability to respond to the random polypeptide (T,G)-A--L was found to be linked to the major histocompatibility (H-2) locus of the mouse (16). Thus, mice possessing the H-2^b allele are high responders to (T,G)-A--L, whereas those carrying the H-2^k antigenic specificity are low responders to this immunogen (17). In order to find out whether the response potential to (T-T-G-G)-A--L is also linked to H-2, inbred mice of C3H.SW and C3H/HeJ strains, their F$_1$ hybrids and backcross offsprings were tested for their immune response to (T-T-G-G)-A--L. The results of the genetic analysis indicated that the immune response to (T-T-G-G)-A--L is governed by a dominant gene(s) which is closely linked to H-2 (Fig. 2 (15)).

Since (T-T-G-G) was shown to be the immunodominant determinant in (T,G)-A--L, it could be used in studies aimed at establishing the nature of the genetic defect in low responder mice to (T,G)-A--L by performing experiments such as isoelectric focusing (IEF) analyses of specific antibodies, and by affinity measurements.

It was shown that it is possible to restore the response of low responder mice to (T,G)-A--L by immunizing them with a complex of the antigen with methylated bovine serum albumin (MBSA) (18). However, the mechanism of this effect remained unknown, because it could not be ruled out that the antibodies produced by MBSA-reconstituted low responder mice might be directed against different determinants of either the (T,G)-A--L-MBSA complex or to "minor" determinants, which might exist in the (T,G)-A--L (18). Using the antigenically restricted (T-T-G-G)-A--L and its MBSA complex,

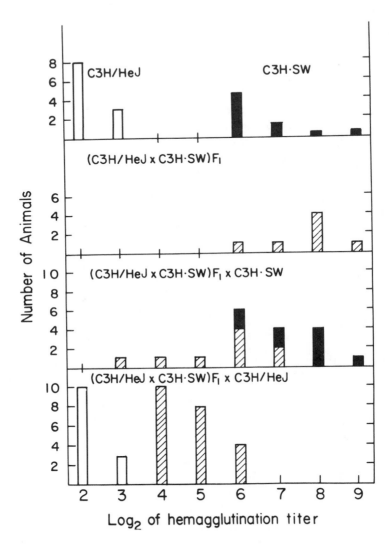

FIGURE 2. Linkage of the immune response
potential to (T-T-G-G)-A--L with H-2. The immune
response of mice immunized with (T-T-G-G)-A--L and
titrated with (T-T-G-G)-A--L-coated SRBC using the
passive hemagglutination assay. (□) animals carrying the H-2^k allele donated by the C3H/HeJ mice;
(■) animals carrying the H-2^b allele donated by the
C3H.SW mice (▨) animals carrying the H-2$^{b/k}$
alleles (15).

it was shown that immunizations of the low
responder mouse strains CKB and C3H/HeJ with the
ordered copolymer (T-T-G-G)-A--L complexed to
MBSA also resulted in a marked increase in specific
antibody titers and binding capacities. The
highly significant differences between high and
low responders were no longer detected if the
complexed antigen was used (Table III (19)). Since
the ordered (T-T-G-G)-A--L possesses mainly one
determinant, such enhancement by itself suggests
that the antibodies produced against the random
(T,G)-A--L by high responder and reconstituted low
responder mice are directed to the same determinant,
namely the immunodominant T-T-G-G (Table IV (19)).

Characterization of Antibodies
in High and Low Responder Strains

Further characterization of the properties of
the specific antibodies to both the random and the
ordered polypeptides which were obtained under
these various immunizing conditions, was achieved
utilizing an independent method. For the analysis
of specific IgG antibodies by IEF, a method was
used (20) which is slightly more complicated than
simple antigen overlay techniques (21). Since
both (T,G)-A--L and (T-T-G-G)-A--L have a high
molecular weight and because of their physical
properties extensive washing of unbound material
was required, focused antibodies were insolu-
bilized by cross-linking with glutaraldehyde. By
this, background radioactivity could be reduced to
a minimum and no binding was detectable in normal
sera. In both complexed antigens the sequence
T-T-G-G contributes a very similar and predominant
portion to the specificity of the response in
MBSA reconstituted low and high responder mice.
All the IEF data show that there are hardly any
differences in the degree of heterogeneity of the
antibody populations raised against both
polypeptides (19). Under these experimental
conditions, as far as the high responder is
concerned, there is no difference in the quality
of the IEF spectrotypes, whether MBSA is (Fig. 3
(19)) or is not (Fig. 4 (19)) used together with
the antigen. Upon reconstitution with MBSA,

TABLE III. Enhancement of antibody response to (T-T-G-G)-A--L in low responder strains upon immunization with a complex of the antigen with MBSA.

| | Immunization with | | | |
| | (T-T-G-G)-A--L | | (T-T-G-G)-A--L + MBSA | |
Mouse strains	Log$_2$ of hemagglutination[a]	Binding[b] (%)	Log$_2$ of hemagglutination	Binding (%)
CWB	6.58 ± 0.25[c]	62	7.53 ± 0.54	64
CKB	1.47 ± 0.19	2	6.05 ± 0.48	62
C3H.SW	5.37 ± 0.24	68.5	7.23 ± 0.34	75
C3H/HeJ	1.40 ± 0.30	7	7.09 ± 0.34	71

[a] SRBC were coated with (T-T-G-G)-A--L.
[b] Antigen used in the assay: ^{125}I-(T-T-G-G)-A--L. Assays were done on pools of sera in each group.
[c] Geometric means of tests of 10-30 individual antisera are given with standard deviations (19).

TABLE IV. Titration of antibodies raised against (T,G)-A--L and (T,G)-A--L + MBSA in high and low responder mice.

IMMUNIZATION WITH

Mouse strains	(T,G)-A--L			(T,G)-A--L + MBSA		
	Log$_2$ of hemagglutination[a] with (T-T-G-G)-A--L	with (T,G)-A--L	Binding[b] of (T-T-G-G)-A--L (%)	Log$_2$ of hemmagglutination with (T-T-G-G)-A--L	with (T,G)-A--L	Binding of with (T-T-G-G)-A--L (%):
CWB	6.22 ± 0.36[c]	5.5 ± 0.5	58	5.84 ± 0.38	6.5 ± 0.5	54
CKB	1.48 ± 0.22	2.8 ± 0.04	7	5.93 ± 0.46	5.5 ± 0.5	56

a SRBC coated with either (T-T-G-G)-A--L or (T,G)-A--L were used in the hemagglutination assay.
b ^{125}I-(T-T-G-G)-A--L was used for antigen binding assays on pooled antisera.
c Geometric means and standard deviations as in Table III (19).

277

immun.:(T-T-G-G)-A--L +MBSA

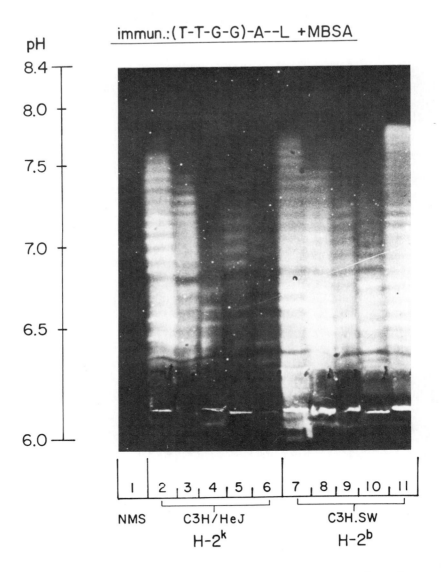

FIGURE 3.   IEF spectra of representative immune sera of C3H/HeJ (H-2^k) and C3H.SW (H-2^b) mice immunized with (T-T-G-G)-A--L + MBSA.   Antigen used for developments ^{131}I-(T-T-G-G)-A--L.   Exposure time to film: 24 h (19).

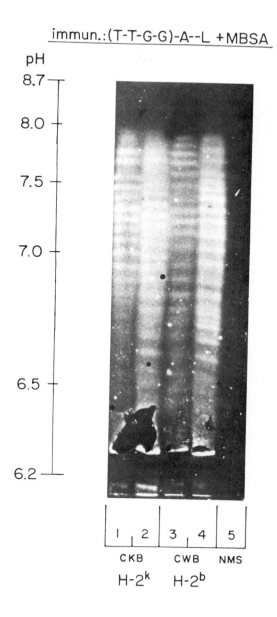

immun.:(T-T-G-G)-A--L +MBSA

pH

8.7 —

8.0

7.5

7.0

6.5

6.2 —

| 1 | 2 | 3 | 4 | 5 |

CKB   CWB   NMS

H-2^k H-2^b

FIGURE 4. IEF patterns of immune sera against (T-T-G-G)-A--L + MBSA produced by CKB (H-2^k, Ig-1^b) and CWB (H-2^b, Ig-1^b) mice. Sera were developed with ^{131}I-(T-T-G-G)-A--L, exposure time: 18 h (19).

however, the low responder IEF spectra are also
characterized by discrete bands in all immune sera
tested (Fig. 3 (19)). As can be seen in Fig. 3,
under these conditions the degree of complexity
of the high and low responder IEF spectra is very
similar (Fig. 3 (19)).

The above mentioned study was extended to a
comparison of association constants of antibodies
elicited by high and low responder mice specific
to the (T-T-G-G) determinant, obtained either by
immunization with (T,G)-A--L and its MBSA complex,
or with (T-T-G-G)-A--L and its complex (22). The
availability of the free tetrapeptide enable for
the first time measurements of the affinity of
anti-(T,G)-A--L antibodies. Measurements were
performed by antigen binding assay, using iodinated
(T-T-G-G)-A--L, and by equilibrium dialysis using
the tetrapeptide Tyr-Tyr-Glu-Glu to which ^{14}C-DL-
alanine was attached at the C-terminus. The association constants determined by equilibrium
dialysis were of the order of magnitude of $10^4 M^{-1}$
in all cases, showing similarity between the antibodies produced in high and low responders in the
presence of MBSA or in high responders immunized
with the antigen alone.

When the measurements were performed by the
antigen binding assay with (T-T-G-G)-A--L, the
resulting association constants were two orders of
magnitude higher than those obtained with the
hapten (Fig. 5 (22)). Calculations were performed
in an attempt to understand these differences,
based on probability analysis of the kind employed
by Berzofsky et al. (23). A valency of 40-80 was
obtained for the polypeptide antigen, if only
probability calculations were taken into consideration. This value seems to be too high for
an antigen of molecular weight of 200,000, which
interacts with the antibody through the Fab
Fragment. Therefore, it is possible that the
probability effect only partially accounts for the
observed high apparent Ko for the antigen.

Although the values obtained by the two techniques were different, both indicated that antibodies produced by high responder mice immunized
either with the antigen alone or with its MBSA
complex do not differ according to this criterion.
These experiments indicated also that the intrinsic
association constants of antibodies elicited in

low responders to the complex of (T-T-G-G)-A--L
with MBSA are of the same order of magnitude as
those produced by high responders either to the
complex or to the antigen (T-T-G-G)-A--L
itself (Fig. 5 (22)).

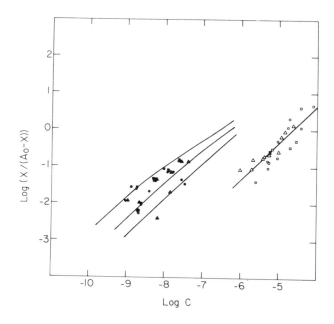

FIGURE 5. Sips plots of the binding data
where c and x are respectively the free and bound
concentrations of hapten or antigen (molecules)
and Ao is total antibody sites concentration.
Circles and squares denote respectively antibodies elicited in high responder C3H.SW mice
against (T-T-G-G)-A--L and its MBSA complex
triangles denote antibodies against the latter
antigen elicited by low responder C3H/HeJ mice.
Open figures represent equilibrium dialysis
experiments and black signs stand for antigen
binding capacity experiments. The line for
equilibrium dialysis is obtained by least
squares calculation. The lines for antigen
binding capacity are calculated with different
values of f, the apparent valence of the antigen
as described in the experimental section (22).

The avidity (defined as the apparent binding
constant derived from measurements of the interaction of the antibody with macromolecular
antigen) of antibodies elicited against the random
antigen (T,G)-A--L or its MBSA complex is also
similar to that of the ordered antigen (T-T-G-G)-
A--L (Fig. 6 (22)). This is in line with the
observation that T-T-G-G constitutes the main
antigenic determinant of (T,G)-A--L.

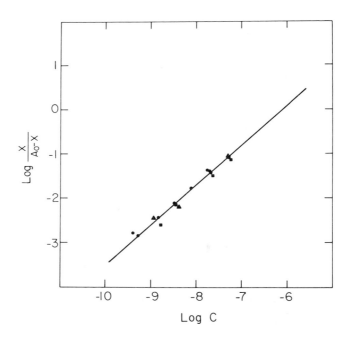

FIGURE 6.   Sips plots of the binding data,
where c and x are respectively the free and
the bound concentrations of antigen (molecules)
and Ao is the total antibody sites concentration.
Circles and squares denote respectively antibodies
elicited in high responder C3H.SW mice against
(T,G)-A--L and its MBSA complex.   Triangles
denote antibodies against the latter antigen
elicited by low responder C3H/HeJ mice (22).

The above study demonstrated that H-2^k low responder mice were capable of responding to the random (T,G)-A--L (18) and to the ordered (T-T-G-G)-A--L (22), similarly to the high responder mice.

Specificity of Humoral Responses to Ordered Peptide Antigens

Two additional tetrapeptides which were tested, Tyr-Glu-Tyr-Glu (T-G-T-G) and Glu-Tyr-Tyr-Glu (G-T-T-G), appeared to play a minor role as determinants in the random (T,G)-A--L (14). Furthermore, the genetic control of the immune response to one of them, coupled to multichain poly-DL-alanine, namely (T-G-T-G)-A--L, was found to be different from that observed for (T-T-G-G)-A--L (Fig. 7 (15)). Thus, the two ordered tetrapeptides (T-T-G-G)-A--L and (T-G-T-G)-A--L, which are composed of the same amino acid residues and differ only in their order, were found to be under qualitatively different genetic controls, indicating the high degree of discrimination of the genes involved in the immune response.

The cross reactivity between the random (T,G)-A--L and the ordered polypeptides was checked only on the level of their immunogenicity but also on the level of their capability to induce specific tolerance and the effect of such tolerance on the ability to mount an immune response to the heterologous antigens. Preliminary data indicated that (T-T-G-G) is the major tolerogenic determinant in the random (T,G)-A--L as was checked by the capability of tolerant mice to (T,G)-A--L or to (T-T-G-G)-A--L to respond to (T-T-G-G)-A--L (Table V). These mice responded well to (T-G-T-G)-A--L. On the other hand specific tolerance to (T-G-T-G)-A--L did not affect the capability to respond to (T-T-G-G) when the immunizing antigen was wither (T-T-G-G)-A--L or (T,G)-A--L.

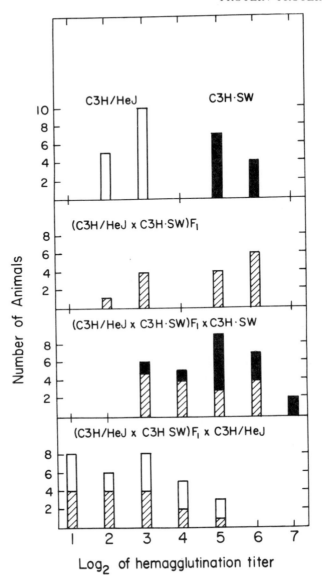

FIGURE 7. The immune response of mice immunized with (T-G-T-G)-A--L and titrated with (T-G-T-G)-A--L coated SRBC using the passive microhemagglutination assay. (□) animals carrying the H-2^k allele denoted by the C3H/HeJ mice. (■) animals carrying the H-2^b allele denoted by the C3H.SW mice. (▨) animals carrying the H-2$^{b/k}$ allele (15).

Need for Cell Cooperation

Further analysis showed that the two poly-peptides (T-T-G-G) and (T-G-T-G) differ also in their need for T-B cell cooperation in the process of antibody production toward them (24). (T-T-G-G)-A--L was shown to be a T-dependent immunogen (24), as expected from its similarity to the random (T,G)-A--L (25). In contrast, (T-G-T-G)-A--L can activate B cells to efficiently elicit antibodies without the help of thymocytes or T cells (Table VI, VII (24)), as was shown by determination of the humoral immune response to this antigen in T-depleted mice.

It has been previously shown that repeating antigenic determinants (26) and slow metabolism are among the prerequisites for T-independency (27). The two ordered peptide antigens tested (T-T-G-G)-A--L and (T-G-T-G)-A--L , were found to differ in their T-B cell cooperation in the process of anti-body production towards them, although the chemical differences between them are very limited, as both of them possess repeating antigenic determinants composed of L-amino acids only. However the two ordered peptide antigens might differ in other physicochemical properties such as their tendency to aggregate.

Mode of Interaction with Macrophages

The different modes of responses towards the above mentioned antigens may result from a long persistance in the body, either free or bound to the surface of cells such as macrophages. Thus, it is possible that an immunogen may persist in a macromolecular form and thus be T-independent and slowly metabolized because its availability to digestion by proteolytic enzymes is limited due to interaction with cell surface.

We have shown that the T-independent (T-G-T-G)-A--L was bound to the macrophage membrane to a higher extent than the T-dependent (T-T-G-G)-A--L (Fig. 8) under two experimental conditions, at 37°C in the presence of sodium azide, and at 4°C. On the other hand, the uptake of (T-G-T-G)-A--L by macrophages was lower by a factor of two when compared to the other antigen, and this difference is even bigger when its high

TABLE V.   Specificity of the humoral response
to ordered peptide antigens

	Tolerance to		
	(T-T-G-G)- A--L	(T,G)- A--L	(T-G-T-G)- A--L
Immunogen and assaying antigen	Hemagglutination titers		
(T-T-G-G)-A--L	2	<2	4 ± 1.5
(T-G-T-G)-A--L	3.4 ± 0.73	5 ± 0.7	<2
(T,G)-A--L	N.D.	<2	5.2 ± 0.4

a)   Tolerance was induced by i.p. injection
of 2 mg. of antigen, aggregate free.

b)   Antibody titers were determined by the
passive microhemagglutination assay. Results are
expressed as means of $\log_2$ of hemagglutination
titer ± S.E.

TABLE VI.   The immune response to (T-G-T-G)-
A--L of irradiated reconstituted CKB mice

Cells transferred	$\log_2 HA$[a]
Intact mice	3.5 ± 0.40
Spleen cells (3×10^7)	4.0 ± 0.34
Spleen cells treated with antiserum against Θ (2×10^7)	3.6 ± 0.23

Spleen cells treated with
  antiserum against
  Θ (2×10^7) + thymocytes
  1×10^8)
	2.7 ± 0.22

[a] Average of $\log_2$ hemagglutination of 5-6
animals in each group. The titers of control sera
from irradiated, spleen repopulated, nonimmunized
mice was lower than 2 (24).

TABLE VII. Immune response of CKB "B mice" to (T-G-T-G)-A--L

Animals	Log₂ Hemagglutination[a]			
	(T,G)-Pro--L[b]	(T-G-T-G)-A--L[b]	(T-G-T-G)-A--L[c]	SRBC[d]
CKB "B mice"	1.6 ± 1.0	3.9 ± 0.9	4.5 ± 0.28	3.3 ± 0.35
CKB intact mice	5.7 ± 0.87	3.8 ± 0.85	4.4 ± 0.48	6.8 ± 0.32

[a] Mice were injected with 10 μg of antigen with complete Freund's adjuvant and boosted with the same dose of antigen in aqueous solution.

[b] Sera obtained 11 days after booster injection.

[c] Sera obtained 14 days after booster injection.

[d] Titers of the same mice immunized 17 days after booster injection with 10⁸ SRBC and bled 5 days later (24).

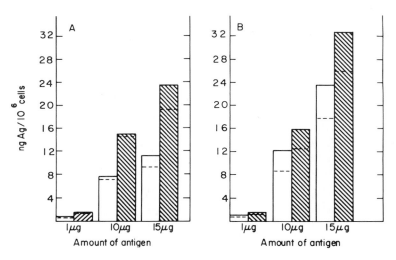

FIGURE 8.   Binding of labeled antigens on macrophage surface following incubation with 1, 10, 15 μg
of antigen/ml/10⁶ cells.   A- at 4°C, B- at 37°C in
the presence of NaN₃ 10⁻¹ M. ☐ binding of ¹²⁵I-
(T-T-G-G)-A--L and ▨ binding of ¹²⁵I-(T-G-T-G)-A--L.
The broken lines represent the amount of antigen
removed by trypsin.

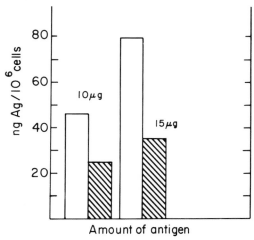

FIGURE 9.   Uptake of radiolabeled antigens by
nonstimulated macrophages (10 and 15 μg/ml/10⁶
cells) following 1 hr incubation at 37°C. ▨ uptake
of ¹²⁵I-(T-G-T-G)-A--L and ☐ uptake of ¹²⁵I-
(T-T-G-G)-A--L.

binding to the membrane is taken into account
(Fig. 9). It is of interest that susceptibility
of both antigens to hydrolytic enzymes, once taken
by the cells, was identical (Fig. 10). If indeed
direct stimulation of B cells is possible when the
polymeric antigen is preserved *in vivo* for a long
time it could be achieved either with antigens
which are unsusceptible to enzymatic degradation
as was found for polymers composed of D amino acids
(27), or alternatively when the polymeric antigen
is specifically bound to the surface of cells as
might be the case with (T-G-T-G)-A--L. Thus, the
preferential binding of (T-G-T-G)-A--L to macrophage surface might be related to the direct B cell
triggering. The exact mechanism which determines
the preferential binding of (T-G-T-G)-A--L to
macrophage surface has not been established.

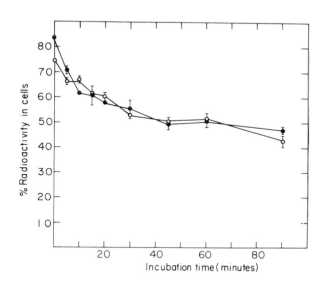

FIGURE 10. Degradation of radiolabeled
antigen by thioglycolate stimulated macrophages
followed by uptake of liposomes in which the
antigens were entrapped. Results demonstrate cell
associated radio-activity at different time intervals. O - (T-T-G-G)-A--L ● - (T-G-T-G)-A--L.

However, it involves most probably both specific features of the macrophage membrane and some unique physicochemical properties of the antigen.

Concluding Remarks

It is apparent from this study that a minimal change - a reversion of the sequence of the two interior amino acid residues within the tetrapeptide which serves as the antigenic determinant - causes a series of major differences in the biological properties of the immunogens of which they represent the major building blocks. These properties include the genetic control of immune responsiveness and the need for cell cooperation in order to mount an efficient immune response. It is most likely that the difference in the three-dimensional structure of the related tetrapeptides must be a significant one, and worth investigating, as this might lead to a better understanding of the above biological phenomena.

REFERENCES

1. Katchalski, E., Sela, M., Silman, H. I., and Berger, A., Polyamino Acids as Protein Models. In "The Proteins" (H. Neurath, ed.) Vol. 2, p. 405. Acad. Press, Inc., New York (1964).
2. Sela, M., in "Peptides, Polypeptides and Proteins" (E. R. Blout, F. A. Bovey, M. Goodman and N. Lotan, eds.) p. 570. John Wiley and Sons, Inc., New York (1974).
3. Sela, M., Adv. Immunol. 5:29 (1966).
4. Sela, M., Science 166: 1365 (1969).
5. Sela, M., and Arnon, R., Biochim. Biophys. Acta 40: 382 (1960).
6. Sela, M., Fuchs, S., and Arnon, R., Biochem. J. 85: 223 (1962).
7. McDevitt, H. O., and Benacerraf, B., Adv. Immunol. 11: 31 (1969).
8. Benacerraf, B., and McDevitt, H. O., Science 175: 273 (1972).
9. Mozes, E., and Shearer, G. M., Curr. Top. Microbiol. Immunol. 259: 167 (1972).
10. Mozes, E., Immunogenetics 2: 397 (1975).
11. Günther, E., Rüde, E., and Stark, O., Eur. J. Immunol. 2: 151 (1972).

12. Günther, E., Balcarová, J., Hálá, K., Rüde, E., and Hraba, T., Eur. J. Immunol. 4:548 (1974).

13. Ramachandran, J., Berger, A., and Katchalski, E., Biopolymers 10:1829 (1971).

14. Mozes, E., Schwartz, M., and Sela, M., J. Exp. Med. 140:349 (1974).

15. Schwartz, M., Mozes, E., and Sela, M., Eur. J. Immunol. 5:871 (1975).

16. McDevitt, H. O., and Tyan, M. L., J. Exp. Med. 128:1 (1968).

17. McDevitt, H. O., and Chinitz, A., Science 163:1207 (1969).

18. McDevitt, H. O., J. Immunol. 100:485 (1968).

19. Cramer, M., Schwartz, M., Mozes, E., and Sela, M., Eur. J. Immunol. 6:618 (1976).

20. Keck, K., Grossberg, A. L., and Pressman, D., Eur. J. Immunol. 3:99 (1973).

21. Cramer, M., and Braun, D. G., J. Exp. Med. 139:1513 (1974).

22. Schwartz, M., Lancet, D., Mozes, E., Sela, M., Immunology, submitted.

23. Berzofsky, J. A., Curd, J. G., and Schechter, A. N., Biochemistry 15:2113 (1976).

24. Schwartz, M., Hooghe, R. J., Mozes, E., and Sela, M., Proc. Nat. Acad. Sci. USA 73:4184 (1976).

25. Lichtenberg, L., Mozes, E., Shearer, G. M., and Sela, M., Eur. J. Immunol. 4:430 (1974).

26. Mozes, E., and Sela, M., Transpl. Rev. 23:189 (1975).

27. Sela, M., Mozes, E., and Shearer, J. M., Proc. Nat. Acad. Sci. USA 64:2696 (1972).

Three-dimensional Structure of an Antibody Molecule and Several Fragments

J. Deisenhofer, M. Matsushima, P. M. Colman, M. Marquart and R. Huber

Max-Planck-Institut für Biochemie and Physikalisch-Chemisches Institut der Technischen Universität, D-8033 Martinsried bei München, Germany

and W. Palm

Institut für Medizinische Biochemie der Universität, A-8010 Graz, Austria

Antibody molecules serve a dual function: to recognise foreign cells and macromolecules and to trigger the events leading to their elimination. Specific recognition requires surface structures complementary to the antigen and hence a huge variety of antibody molecules. In contrast the effector functions would most economically be performed if all antibody molecules were identical. Crystal structure analyses have provided a detailed view of the domain-wise construction scheme of antibody molecules and the folding of these domains. The polypeptide chain in all immunoglobulin domains is arranged in two pleated sheets with variations in detail in the various light and heavy chain domains. Structural studies of hapten binding have shown that preformed spatial complementarity forms the basis of hapten-antibody interaction. No detailed information is at hand about interaction of antibodies with the macromolecules of the complement system or with Fc receptors.

There is strong, extended lateral interaction between V_L-V_H, C_L-C_H1 and C_H3-C_H3 domains, but not between the C_H2 domains. Here, the potential interacting surface is covered by the carbohydrate moiety. The longitudinal interactions of the domains along the chain (pearls on a string) appear weak. This is demonstrated by variations of the elbow angle (angle between local diads of V and C modules) and the disorder of the Fc part relative to the Fab arms in an intact IgG antibody. The absence of longitudinal interactions suggests a highly flexible IgG molecule. Flexibility might be advantageous for binding to hapentic sites of various distances, but a flexible molecule does not allow cooperative phenomena between antigen binding at the V module

293

and effector functions located at the Fc part. It
is unclear whether a rigid conformer plays a role.
Structural studies of specific antibodies and their
complexes with antigen as well as complexes between
antibodies and specifically interacting proteins
might provide a clue.

1) Colman, P.M., J. Deisenhofer, R. Huber and
   W. Palm, J. Mol. Biol. 100 (1976) 257-282.

2) Davies, D.R., E.A. Padlan and D.M. Segal,
   Rev. Biochem. 44 (1975) 639-667.

3) Deisenhofer, J., P.M. Colman, O. Epp and
   R. Huber, Hoppe-Seyler's Z. Physiol. Chem. 357
   (1976) 1421-1434.

4) Epp, O., E.E. Lattman, M. Schiffer, R. Huber
   and W. Palm, Biochemistry 14 (1975), 4943-4952.

5) Fehlhammer, H., M. Schiffer, O. Epp, P.M. Colman,
   E.E. Lattman, P. Schwager, W. Steigemann and
   H.J. Schramm, Biophys. Struct. Mechanism 1
   (1975) 139-146.

6) Huber, R., J. Deisenhofer, P.M. Colman,
   M. Matsushima and W. Palm, Nature 264 (1976)
   415-420.

7) Matsushima, M., M. Marquart, T.A. Jones,
   P.M. Colman, R. Huber and W. Palm,
   J. Mol. Biol. (1978) in press.

8) Silverton, E.W., M.A. Navia, D.R. Davies,
   Proc. Natn. Acad. Sci. U.S.A. 74 (1977)
   5140-5144.

STRUCTURE AND FUNCTION OF THE ANTIBODY COMBINING SITE

David Givol

Department of Chemical Immunology
The Weizmann Institute of Science
Rehovot, Israel

I. INTRODUCTION

If we consider antibodies and enzymes as a collection of recognition molecules, we will observe several noticeable differences between them. Enzymes are designed to bind ligands and catalyze a chemical reaction, the binding and catalytic sites being parts of an active site. Antibodies have the property of reversible binding of ligands; they are not capable of the formation or breaking of covalent bonds. Enzymes are produced in all cells; antibodies are the unique products of specific cells in the immune system, i.e. lymphocytes and plasma cells. It is interesting that lymphocytes can make more different combining sites, as antibodies, than all the other cells of the body *in toto*, since antibodies can be made against any particular foreign protein, including antibodies themselves. The structure of enzymes is widely variable for different enzymes. There is no similarity between an enzyme that acts on carbohydrates (e.g. lysozyme) and an enzyme that acts on polypeptides (e.g. trypsin). On the other hand, antibodies to carbohydrates and polypeptides cannot be resolved by any physico-chemical criteria, their general structure is very similar and they differ only in their combining site. The number of combining sites in an enzyme molecule may differ in different enzymes but generally one combining site is formed by one peptide chain. In antibodies there are always two combining sites per molecule (ten in IgM which is the pentamer of the basic subunit) and each combining site is formed by two dissimilar peptide chains. Random pairing between these two chains could give rise to a

295

large number (n^2) of different combining sites with a much
smaller number (n) of the two chains. This reduced the gene
load necessary to code for such a large variety of proteins.
In antibodies the peptide chains can be divided into a constant
(C) region whose amino acid sequence is identical for all antibodies of one class, and a variable (V) region whose sequences
differ from one antibody to another.

A. Location of the Combining Site

Direct proof that the combining site is entirely formed
from the variable region has been obtained by us several years
ago (1). The mouse IgA protein, produced by myeloma MOPC 315,
with high affinity for nitrophenyl ligands (e.g. dinitrophenyl)
was used. The Fab'fragment was digested with pepsin at pH 3.7
and fractionated on a Dnp-lysine-Sepharose column. The bound
material was eluted specifically with Dnp-glycine and fractionated on Sephadex G-75 column. The major fraction, named Fv, has
a molecular weight of 25000 daltons and was shown to be composed of two polypeptide chains of approximately 12000 daltons,
which were separated on DEAE-cellulose in 8M-urea. One chain
was shown to have the same N-terminal sequence as the intact
heavy chain, whereas the other chain was blocked as is the
light chain. It was concluded that Fv, which retains full binding capacity, is formed by V_L and V_H domains. Recently another
Fv was obtained also from mouse myeloma protein which is produced by the mouse myeloma XRPC 25 (2). The properties of various antibody fragments is given in Table I.

TABLE I. A Comparison of Some Properties of Protein 315
          and Its Fab' and Fv Fragments

	Protein 315	Fab'	Fv
$S_{20.w}$	7 S	3.7 S	2.6 S
Mol wt	153,000	59,000	25,000
Isoelectric point	pH 4.5	pH 4.7	pH 5.8
Peptide chains	H, L	Fd, L	V_H, V_L
N terminals	PCA,Asp	PCA,Asp	PCA,Asp
Binding sites	2	1	1
Association constant (M^{-1})	2.4×10^6	2.4×10^6	2.8×10^6

B. The Folding and Association of Domains in the Fv Fragments

In 8 M urea Fv dissociated into V_L and V_H that can be separated on DEAE-cellulose in urea. When a 1:1 mixture of the
isolated chains in urea was diluted into PBS, binding activity
was fully regained (3).

The kinetics of the association process between V_L and V_H
to form an active Fv was followed by monitoring their fluorescence. There is a marked increase in the fluorescence intensity
of Fv in 8 M urea, along with a red shift of 11 nm. Fv reconstituted from urea by dilution into PBS has the same emission
spectrum as native Fv. Thus the renaturation of Fv (or a mixture of V_L and V_H) from 8 M urea can be followed by the decrease in protein fluorescence. The results of such studies, at
different protein concentrations, showed that at a relatively
high protein concentration(2×10^{-6}), the renaturation of Fv is
composed of two distinct steps: (1) a rapid step which cannot
be resolved by the regular manual mixing and comprises about
30% of the regain of native fluorescence and (2) a slow process
which takes place in about 15 minutes. As the protein concentration decreases, the rapid step becomes more prominent, and,
at 1.6×10^{-8} M, the entire renaturation process is due to the
rapid process only. Since it is very likely that the rate of
the bimolecular association between V_L and V_H will increase
with increasing protein concentration, despite our finding that
the native conformation at higher concentrations is regained
more slowly, we assume that at higher protein concentrations
many of the interactions between the chains lead to inactive,
scrambled associates which can be converted slowly to the correct, more stable, native species. At low protein concentration, the high affinity between the correctly folded V_L and V_H
is the major force that governs the chain-chain interactions,
and the entire process of renaturation takes place in less than
30 seconds (the time required for mixing and manual measuring
of fluorescence).

The kinetics of regain of binding activity was analyzed in
these reactions by a rapid titration with equivalent amounts
of DNP-lysine during the renaturation, measuring the fluorescence quenching due to the binding of DNP-lysine to Fv. At all
protein concentrations, the kinetics of regain of native fluorescence was identical with that of regain of binding activity.
In addition, we found that the presence of the hapten during
the renaturation process did not increase the rate of renaturation, indicating that the formation of the combining site is
not the rate-limiting step in the folding and association process. The data also imply that the association constant of V_L
and V_H is greater than 10^8 M^{-1} since full regain of active Fv

was obtained at 10^{-8} M concentrations of V_L and V_H (4).

C. Oxidation and Renaturation of Completely Reduced Fv

Fv was reduced in 8 M urea, 0.1 M Tris-Cl pH 8.2 and 0.1 M
β-mercaptoethanol at 37°C for 1 hour. Alkylation of a portion
of this solution with iodoacetamide, followed by amino acid
analysis, showed four carboxymethylcysteines per mole of Fv.
Various attempts to oxidize and renature the completely reduced Fv, by removing the urea and mercaptoethanol on Sephadex
G-25 or by dilution and dialysis, did not result in a good
yield of an active Fv. Since either V_L or V_H each contain only
one disulfide bond, we anticipated that if oxidation of the
reduced Fv were to be performed in 8 M urea, no "wrong" disulfide bonds coul be generated, and only the single intrachain
disulfide bond would be formed. Reduced Fv (3 mg/ml) was dialyzed against 8 M urea to remove β-mercaptoethanol and during
this time it was also reoxidized. Subsequent dilution of this
Fv into 0.15 M NaCl - 0.01 M phosphate buffer pH 7.4 (PBS)
resulted in a complete regain of binding activity by the re-
natured Fv. Similar experiments were also performed with reduced V_L which was reoxidized in urea and combined with V_H in
8 M urea. Dilution of this mixture into PBS yielded completely
active Fv (4).
These results reinforce the previously published evidence
that the folding and specificity of the antibody combining
site are entirely dependent on the amino acid sequence (5) and
are in line with the thermodynamic hypothesis of protein folding (6). In our case, 100% of the binding activity was regained
by reoxidation of completely reduced V domains of a homogeneous antibody. The complete renaturation of fully reduced Fv
also provides strong support to the domain hypothesis of
immunoglobulin structure (7). This study demonstrates that the
V domains not only exist in immunoglobulin structure but can
also fold independently of the rest of the peptide chain to
form the native structure with full binding activity. Moreover,
the folding of V_L and V_H are completely independent of each
other.

D. Chemical Mapping of the Combining Site

This was performed by affinity labeling with DNP derivatives which contain bromoacetyl group as an active chemical
group (X) which forms a covalent bond with amino acid residue
at or near the combining site. Since the ligand is rigidly

held in the site affinity labeling reagents of different
length can be used as yardsticks to map the site. The results
(8,9) demonstrated that Tyr 34 on V_L and Lys 52 on V_H are
being labeled by reagents of different length (Table II).

TABLE II. Affinity Labeling of Protein 315*

Reagent		% Labeled residue and chain	
		Tyr 34L	Lys 52H
BADE	DNP-NH-CH$_2$-CH$_2$-NH-X	96	4
BADB	DNP-NH-CH$_2$-CH$_2$-CH-NH-X 　　　　　　　　| 　　　　　　　COOH	87	13
BADO	DNP-NH-CH$_2$-CH$_2$-CH$_2$-CH-NH-X 　　　　　　　　　　　| 　　　　　　　　　COOH	66	34
BADL	DNP-NH-CH$_2$-CH$_2$-CH$_2$-CH$_2$-CH-NH-X 　　　　　　　　　　　　　| 　　　　　　　　　　COOH	5	95

* X = COCH$_2$Br

If, indeed, the specificity of the chemical reaction is a
function of the positioning of the bromoacetyl group, the
foregoing results suggest that the distance between the labeled lysine on the H chain and the labeled tyrosine on L chain
is approximately equal to the difference in the length of BADE
and BADL. Hence, a bifunctional reagent with two bromoacetyl
groups separated by a distance equal to the difference in
length between BADE and BADL would react simultaneously with
both lysine and tyrosine and thus would cross-link the heavy
and light chains (10). Indeed the bifunctional reagent, Dnp-
HN-(CH$_2$)$_2$CH(HNCOCH$_2$Br)CO(NH)$_2$COCH$_2$Br, covalently cross-linked
the H and L chains to yield a molecule with a molecular weight
of 72,000. The cross-linked H-L contained labeled Tyr and Lys
in a ratio of 1:1, indicating that Tyr (on L) and Lys (on H)
were involved in the cross-linking. This affinity-directed
cross-linking of two peptide chains demonstrates the close
spatial arrangement of H and L chains in the construction of
the antigen-binding site.

E. Physical Mapping of subsites in the antibody combining site

The combining site of protein 315 shows strong dependence of the association constant on the size and character of various DNP ligands. It was possible therefore to map the combining site and to subdivide it into subsites of interaction with various portions of the hapten (Table III).

TABLE III.   Association Constant of Protein 315 for Ligands with Various Side Chains

	Hapten	$K_A \times 10^{-5} (M^{-1})$
1	$DNPCH_2CH_3$	19.5
2	$DNPCH_2COO^-$	1.27
3	$DNPCH(CH_2)COO^-$	3.60
4	$DNPCH_2CH_2CH_3$	27.0
5	$DNPCH_2CH_2OH$	6.75
6	$DNPCH_2CH_2COO^-$	6.40
7	$DNPCH_2CH_2NH_3^+$	0.89
8	$DNPCH_2CH_2CH_2CH_3$	32.0
9	$DNPCH_2CH_2CH_2COO^-$	84.0
10	$DNPCH_2CH_2CH_2NH_3^+$	0.98
11	$DNPCH_2CH_2CH_2CH_2COO^-$	22.0

Based on these data (11) the combining site was subdivided into: a) The DNP subsite which interacts with the dinitrobenzene ring and contributes approximately 50% of the binding energy, b) The hydrophobic subsite which is close to the DNP ring and favors hydrophobic moiety in the ligand at this position (e.g. ligand 1 vs. 2 or 4 vs. 5 in Table III), c) The positive subsite which favors negatively charged group at approximately 9 Å from the center of the DNP ring (e.g. ligands 9 vs. 10 in Table IV). The differences between the K_A of ligands of identical size but different chemical character clearly show the possibility of defining subsites in the antibody combining site.

F. Location of Subsites in the
peptide chains of Fv

Since V_L and V_H can be isolated from Fv (3) it is possible to compare the binding properties of Fv ($V_L + V_H$) to that of V_L or V_H. It is shown (Table IV) that V_L but not V_H contains the binding site to DNP ligands (12).

TABLE IV.   Comparison of Some Properties of the Subunits of Fv 315

	V_L	V_H	Fv
MW (NaDodSO$_4$)	12,000	13,500	12,000 + 13,500
MW (PBS)	24,000	Aggregate	25,000
$s_{20,w}$	2.18	Aggregate	2.6
Dnp binding sites	2	0	1
K_A x 10^{-3}	2.3	-	2,000
Binding to Dnp-Lys-Sepharose	+(90%)	-(6%)	+(90%)

V_L which exists as a dimer binds 2 molecules of DNP lysine with approximately 1000 fold lower K_A than that of Fv. However when a homologous series of DNP ligands was used to map the site of V_L dimer it appeared that the combining site of V_L dimer binds the DNP ring per se (e.g. DNPOH) almost as strong as Fv (Table V).

Some significant features of the combining site of protein 315 are absent in that of V_L dimer. The increase in affinity with the increase in length of ligand side chain disappears almost completely. A remarkable feature of the results presented in Table V is the small difference in the association constant for DNPOH of $(V_L)_2$ and protein 315. For example, the affinity of protein 315 for DNP-lysine or DNP-aminocaproate is respectively 708 and 1636 greater than that of $(V_L)_2$, whereas the affinity of protein 315 for DNPOH is only 6.5 fold greater than that of $(V_L)_2$. This indicates that the predominant binding energy of the DNP ring is provided by V_L. It is also noteworthy that in $(V_L)_2$ there is almost no increase in K_I between DNPOH and other ligands with longer side chains, whereas, in protein 315 combining site which is composed of both V_L and V_H, the affinity for DNP-lysine is 567-fold greater than for DNPOH (Table V). Moreover, different side chains of the ligand can sense special features of protein 315 combining site. Thus

DNP-aminobutyric acid binds about 6-fold better than DNP-aminopropylamine due to a "positive subsite" in protein 315. This feature is also absent in $(V_L)_2$. It is suggested that a significant part of the interactions with side chains of DNP ligands are due to contributions from the heavy chain, whereas the DNP ring is bound predominantly to the light chain. It is therefore a clear cut illustration of the subdivision of the antibody combining site into defined subsites which can in this particular case be identified with one of the chains.

TABLE V. Binding of Various Ligands to Protein 315 IgA
          and Its V_L Subunits

Ligand	$K_I (M^{-1} \times 10^{-3})$		$K_I (IgA)$
	V_L	IgA	$\overline{K_I (V_L)}$
DNP-OH	0.46	3	6.5
DNP-NHCH$_2$COOH	1.60	50	31
DNP-NHCH$_2$CH$_2$CH$_2$COOH	1.40	5200	3714
DNP-NHCH$_2$CH$_2$CH$_2$NH$_2$	1.20	870	725
DNP-NHCH$_2$CH$_2$CH$_2$CH(NH$_2$)COOH	2.0	3400	1700
DNP-NHCH$_2$CH$_2$CH$_2$CH$_2$CH$_2$COOH	2.2	3600	1636
DNP-NHCH$_2$CH$_2$CH$_2$CH$_2$CH(NH$_2$)COOH	2.4	1700	708

G. Model Building Study of Fv
and its Combining Site

Recent progress in X-ray analysis of Fab' fragments of homogeneous myeloma proteins has led to the elucidation of the three-dimensional structure of the binding site of two anti-bodies: the human protein New which binds vitamin K and the mouse protein McPC 603 which binds phosphorylcholine (13, 14). From these studies it is clear that the residues forming the sites that are complementary to the antigen are contributed by the hypervariable regions (15) (three segments of 5-10 residues each, in both light and heavy chain termed L1, L2, L3 and H1, H2, H3 respectively) and that replacements of amino acid residues in these segments will generate binding sites with new specificities. Such replacement will not disturb the "immuno-globulin fold" of the variable domains which remain similar in all antibodies. Thus on a background of a common three-dimensional structure a vast number of specificities can be generated. This diversity in recognition is the essence of the immune system and therefore a comparative analysis of a large number of combining sites is required in order to provide the details of molecular recognition by antibodies. A quantitative comparison of the tertiary structure of human and mouse variable domains has demonstrated that the immunoglobulin fold is very similar in different variable domains, while structural differences occur predominantly in the hypervariable loops which comprise the combining site (16). Hence the variable region can be regarded as consisting of a rigid framework to which are attached the hypervariable loops. This provides a basis for comparative analysis of antibody combining sites by model building, which then uses the coordinates of the framework from X-ray analysis data of known immunoglobulins and the sequence of the immunoglobulin which is to be analysed.

As soon as a tentative model was completed, even before any adjustments were made to fit hapten, several features of the binding site became apparent (17). The middle of the hypervariable surface was dominated by a rather pronounced cavity. Compared to protein 603, the 315 cavity contained substantially more aromatic residues. It seemed most likely that the DNP moiety would be bound within this cavity.

The hapten-binding site is bounded by L1 on one side and by H1 and H2 on the other. The floor is formed by L3 and the roof is formed primarily by H3. The hypervariable residues that project into the hapten-binding cavity include Phe 34, Asn 36, Asp 99, and Leu 103 of the heavy chain and Tyr 34, Asn 36, Trp 93 and Phe 98 of the light chain. Ser 32 and Asn 96 of the light chain and Phe 50, Lys 52, and Asp 101 of the heavy chain ring the opening of the pocket. As in protein 603, L2 is screened from the hapten-binding site by L1 and H3, and no residues from L2 project into the binding cavity.

The model so built (17), immediately suggested the likely site for binding DNP. This was based on H-bonding of the NO_2 group of DNP to protein residues (Asn 36L or Asn 36H), the stacking with a tryptophan residue from V_L (Trp 93L) and further adjustment to ensure the feasibility of affinity labeling of Tyr 33L and Lys 52H by BADE and BADL (9) respectively. The possibility for cross linking by the bifunctional reagent previously described (10) is obvious. The model also confirmed the location of the DNP ring subsite to V_L (12) since Trp 93L was the major candidate for interaction with the DNP ring. The model was compatible with many of the binding data of the various ligands to protein 315 (11). The next important step was to use the model as a basis for structural

analysis by magnetic resonance and electron spin resonance.
On the basis of the model it was possible by these methods to
determine the general dimension of the site, its polarity and
assymetry, to assign contact residues with the ligand, and to
determine their correct orientation with respect to the hapten
(18). The positioning of Trp 93L, Phe 34H, Tyr 34L, Asn 36L
and Arg 95L and other residues allows understanding of the
specificity of binding DNP ligand to protein 315.

Several interesting points emerge from the comparison of
the combining site of protein 315 with those of protein New
and 603. In Fab' New-vitamin K_IOH complex, the naphthoquinone
ring of vitamin K_IOH is in close contact with Tyr 90L (13).
Sequence homology shows that Trp 93L in protein 315 which
interacts with the DNP is homologous to Tyr 90L in New, and
therefore this provides an interesting example for a change in
specificity on replacing one residue in the site. Note that
protein 315 also binds menadione in lower affinity than DNP
but protein New does not bind DNP ligands. Charge interactions
play a dominant part in the binding of protein 603 to the
hapten phosphoryl choline (14). On the other hand, another
model building study on rabbit anti-type III polysaccharide
(BS-5) suggest clustering of the hydroxyamino acids Tyr, Ser
and Thr and also Asn in the combining site which can form H
bonds with hydroxyls of the sugar ligand (19). It is clearly
seen that some pattern of changes in the antibody binding site
can be deduced from such comparative studies.

In conclusion, two types of structural parameters will
influence antibody specificity: 1) size differences of the
hypervariable loops. This is particularly pronounced in L1 and
H3 and will affect the dimension of the site and its depth;
2) replacements in the hypervariable loops which will provide
different contact residues with the ligand.

Since the complementarity determining residues are all
present in loops which extend into the solution and do not
influence the framework of the domain, there is great flexibility in replacing residues in these loops. Thus different
antibodies may display a unique antigen-complementary site
although they share a common three-dimensional structure. This
can be compared with the known enzyme structures. Models of
various enzymes have shown that such sites are clefts of
grooves lined with 10-20 amino acids. Just how easily the
specificity of binding of small molecules by proteins can be
changed was demonstrated by Hartley who compared the sites of
chymotrypsin, trypsin and elastase (20). These sites differ in
that serine 189 in the site of chymotrypsin is changed to
aspartic acid in the active site of trypsin. Such a change in
one amino acid will change the specificity of the combining

site from the binding of an aromatic residue to the binding
of a positive charged residue. If valine replaced glycine at
position 216, as is the case in elastase, again the binding
specificity will be dramatically changed.

There is no difficulty, therefore, in envisaging how such
replacements in hypervariable segments will change antibody
specificity, and the number of 15-20 hypervariable residues in
each chain seems enough to account for as many variants as
antibody diversity demands. It is, however, likely that
elucidation of the details of structural variations between
complementarity determining residues will help understanding
the mechanism that generates antibody diversity.

REFERENCES

1. Inbar, D., Hochman, J. and Givol, D., Proc. Natl. Acad.
    Sci. 69:2659 (1972).
2. Sharon, J. and Givol, D., Biochemistry 15:1591 (1976).
3. Hochman, J., Inbar, D. and Givol, D., Biochemistry 12:1130
    (1973).
4. Hochman, J., Gavish, M., Inbar, D. and Givol, D.,
    Biochemistry 15:2706 (1976).
5. Haber, E., Proc. Natl. Acad. Sci. 52:1099 (1964).
6. Anfinsen, C.B. (1973) Science 181:223 (1973).
7. Edelman, G.M. and Gall, W.E., Ann. Rev. Biochem. 38:415
    (1969).
8. Haimovich, J., Givol, D. and Eisen, H.N., Proc. Natl. Acad.
    Sci. 67:1656 (1970).
9. Haimovich, J., Eisen, H.N., Hurwitz, E. and Givol, D.
    Biochemistry 11:2389 (1972).
10. Givol, D., Strausbach, P.H., Hurwitz, E., Wilchek, M.,
    Haimovich, J. and Eisen, H.N., Biochemistry 10:3461 (1971).
11. Haselkorn, D., Friedman, S., Givol, D. and Pecht, I.
    Biochemistry 13:2210 (1974).
12. Gavish, M., Dwek, R.A. and Givol, D. Biochemistry 16:3154
    (1977).
13. Amzel, L.M., Poljak, R.J., Saul, F., Varga, J.M. and
    Richards, F.F., Proc. Natl. Acad. Sci. 71:1427 (1974).
14. Segal, D.M., Padlan, E.A., Cohen, G.H., Rudikoff, S.,
    Potter, M. and Davies, D.R., Proc. Natl. Acad. Sci.
    71:4298 (1974).
15. Wu, T.T. and Kabat, E.A., J. Exp. Med. 132:211 (1970).
16. Padlan, E.A. and Davies, D.R., Proc. Natl. Acad. Sci.
    72:819 (1975).

17. Padlan, E.A., Davies, D.R., Pecht, I., Givol, D. and
     Wright, C., Cold Spring Harbor Symp. Quant. Biol. XL1:627
     (1977).
18. Dwek, R.A., Wain-Hobson, S., Dower, S., Gettins, P.,
     Sutton, B., Perkins, S.J. and Givol, D., Nature 266:31
     (1977).
19. Davies, D.R. and Padlan, E.A., in Antibodies in Human
     Diagnosis and Therapy (ed. Haber, E. and Krause, R.M.)
     Raven Press, New York p. 119 (1977).
20. Hartley, B.S., Phil. Trans. Roy. Soc. Lon. Ser. B.
     257:77 (1970).

PROTEIN INHIBITORS OF SERINE
PROTEINASES; CONVERGENT EVOLUTION,
MULTIPLE DOMAINS AND HYPERVARIABILITY
OF REACTIVE SITES*

Michael Laskowski, Jr.
Ikunoshin Kato
William J. Kohr

Department of Chemistry
Purdue University
West Lafayette, Indiana

The current interest in protein proteinase inhibitors
owes its largest impetus to the present appreciation that
proteinases excercise a plethora of control functions in the
organism (1). Inhibitors, in turn, control those controllers
and thus are deemed worthy of intensive study. However,
there are other motivations for the study of inhibitors.
Enzyme-inhibitor association is probably the best understood
example of protein-protein association. The study of enzyme-
inhibitor interaction has already contributed greatly to the
understanding of the mechanism and specificity of proteinases
and offers prospects of contributing more. Some inhibitors
since they are small, stable and well characterized, have
become favorite objects of study of physical protein chemists.
In this article we would like to convince the reader that
the evolution of proteinase inhibitors is also a particularly
useful area of study.

Protein proteinase inhibitors are ubiquitous. They
have been found essentially everywhere where they were looked
for, often in very large amounts and very frequently as
mixtures of many different inhibitors. We would estimate that

*Supported by grant GM 10831 from National Institutes of
Health

307

TABLE I.   Phenomenological Characteristics of the Standard
            Mechanism

1. Incubation of the inhibitor with catalytic quantities
   of the enzyme it inhibits leads to specific hydrolysis of one peptide bond--the reactive site peptide
   bond--in the inhibitor molecule, thus converting
   the inhibitor from virgin (peptide bond intact)
   to modified (peptide bond hydrolyzed).
2. The two fragments of modified inhibitor do not
   dissociate but are held together. In all cases
   studied thus far the fragments are held together
   by one or more disulfide bridges. The equilibrium
   constant for virgin to modified inhibitor conversion
   is close to unity at neutral pH.
3. Both virgin and modified inhibitors are active but
   modified inhibitor reacts with the enzyme it inhibits
   much more slowly than virgin inhibitor. The stable
   enzyme-inhibitor complex is the same chemical
   substance whether formed from virgin or from modified inhibitor.
4. Kinetically controlled dissociation of enzyme-
   inhibitor complex produces predominantly virgin
   inhibitor.
5. Removal of either the newly formed COOH terminal
   amino acid residue or the newly formed NH_2 terminal
   amino acid residue from modified inhibitor causes
   loss of activity. Modifications or replacements
   of the COOH terminal amino acid side chain frequently cause predictable changes in specificity
   or loss of activity. Any blockage of the NH_2
   terminal amino acid acid residue preventing reformation of the reactive site peptide bond causes loss of
   activity.
6. The complex involves extremely close fit between the
   reactive site of the inhibitor and the active site
   of the enzyme. The carbonyl of the reactive site
   of the inhibitor is tetrahedral and forms long
   covalent bond with oxygen O^γ of the catalytic serine
   of the enzyme (2,3).
7. The conformation of the residues in the inhibitor
   interacting with the enzyme is closely similar in
   various inhibitors even though the inhibitors themselves are not conformationally similar (2,3).

a single mammal or bird must contain at least 100 different
inhibitors. Thus the problem of talking about protein
proteinase inhibitors in general is formidable. We must
delimit the field.

Proteinases can be divided into four mechanistic classes--
serine proteinases, sulfhydryl proteinases, acid proteinases
and metaloproteinases. Protein inhibitors for each class
of enzymes were described. However, we are not aware of any
well documented case based on experiments on highly purified
inhibitors and enzymes where one inhibitor inhibits enzymes
belonging to more than one mechanistic class. If this is so
we can divide the inhibitors into the same four classes.
From now on we will discuss only inhibitors of serine
proteinases.

In the last 13 years the study of enzyme-inhibitor interaction produced a remarkable concensus. Majority, but not
all, of well studied inhibitors appear to interact with the
enzymes they inhibit according to a common mechanism, the
characteristics of which are listed in Table I. The
exceptions appear to be most inhibitors from mammalian sera
and inhibitors of thrombin but in both cases there is little
general agreement about what mechanism is obeyed.

The number of inhibitors for which the mechanism has
been demonstrated is extremely large. Rather surprisingly
such inhibitors are not all homologous but rather form
a large set of homologous families. Table II is a list of

TABLE II.   Families of Inhibitors Obeying the Standard
             Mechanism

I	Kunitz BPTI Family
II	Kazal Secretory Family
III	Trypsin Inhibitor from Ascaris Family
IV	Chymotrypsin Inhibitor from Ascaris Family
V	Bowman-Birk Legume Family
VI	Kunitz STI Family
VII	Potato I Family
VIII	Potato II Family
IX	Unassigned Families

The list of references to papers showing that these inhibitors obey the standard mechanism is too formidably long to
be included here.

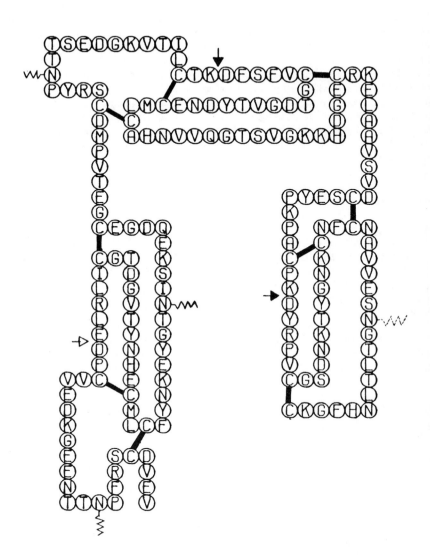

FIGURE 1. Preliminary amino acid sequence of Japanese quail (Coturnix coturnix japonica) ovomucoid (8). (residue 158 was now assigned as N and residues 65 and 67 were exchanged). ⌁ indicates carbohydrate attachment position. ➤ indicate the two trypsin reactive sites in domains 2 and 3. ▷ indicates the putative reactive site in domain 1.

currently known families but it is almost certain that when some already known members of group IX are sequenced more families will have to be established. Of course, lack of homology is an extremely difficult characteristic to prove and it is still possible that after more thorough examination a distant relationship between some of the families will become apparent. At the present level of our ability to detect such relationships it appears that the ability to inhibit serine proteinases according to the standard mechanism arose independently in each family as a result of convergent evolution. We claim (with some hesitancy) that the families are analogous and not homologous to one another.

Each family in turn probably has a great many homologous members. This was already amply demonstrated for Families I, II and V. Table III is the known listing of members of family II--inhibitors homologous to pancreatic secretory trypsin inhibitor (Kazal)

We have sequenced ovomucoids (see Fig. 1) from several birds. The striking feature of this sequence is the presence of three separate tandem domains. Each domain has three intradomain and non-interdomain disulfide bridges and each has a single actual or potential reactive site for inhibition of serine proteinases. In some cases (for example the first domain of Japanese quail ovomucoid shown in Fig. 1) we have not as yet found an enzyme which is inhibited by it and thus

TABLE III. Kazal Type Inhibitors

	# of Domains
Pancreatic Secretory Trypsin Inhibitor (Kazal) (4)	1
Seminal Plasma Inhibitor (5)	1
Bdellin (6)	1
Dog Submandibular Inhibitor (7)	2
Ovomucoid (8)	3
Ovoinhibitor - proteinase Inhibitor from avian serum (9)	6
S-SI Inhibitor (?) (10)	1 (?)

we refer to its reactive site (assigned by homology to the
actual reactive sites) as a potential reactive site. The
question whether any enzyme will ever be found that is inhibited by this and several related domains is most intriguing
but at the moment not answerable.

In favorable cases the domains can be separated by
splitting the connecting peptide by limited proteolysis (this
has now been successful for all 20 species of birds tried
between the second and third domains) or by cyanogen bromide
cleavage of the connecting peptide between first and second
domains. In those cases where the separation technique use
does not additionally cause internal "nicks" in the separated
domains they are independently active as inhibitors.
Limited physicochemical studies suggest that in the intact
ovomucoid molecule the separated domains do not interact
appreciably.

An even more striking result comes from our most recent
sequencing studies on Japanese quail and chicken ovoinhibitor,
a minor inhibitory component of some (possibly all) avian egg
whites. We do not yet have a complete amino acid sequence,
but we have strong evidence that the scheme shown in Fig. 2
is the correct description of the molecule. The ovoinhibitor
molecule consists of six tandem domains, each homologous
to the pancreatic secretory inhibitor (Kazal) but particularly

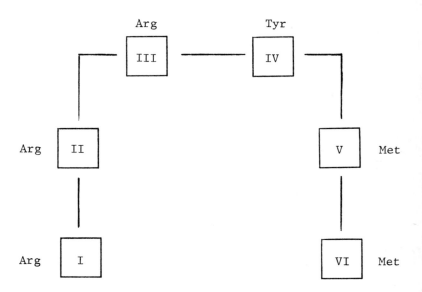

FIGURE 2.   Schematic diagram of Japanese quail ovoinhibitor
molecule (9). P_1 residue indicated for each domain.

JQ OI	Ile Glu Val Asp Cys Ser Gln Tyr Ser Ser Gly Ile Ser Lys Asp Gly
Chicken OI	Ile Glu Val Asn Cys Ser Leu Tyr Ala Ser Gly Ile Gly Lys Asp Gly
Chicken serum PI	Ile Glu Val Asn Cys Ser Leu Tyr Ala Ser Gly Ile Gly Lys Asp Gly

JQ OI	Thr Ser Trp Val Ala Cys Pro Arg Asn Leu Lys Pro Val Cys Gly...
Chicken OI	Thr Ser Trp Val Ala Cys Pro Arg Asn Leu Lys Pro Val Cys Gly...
Chicken serum PI	Thr Ser Trp Val Ala Cys Pro Arg Asn Leu Lys Pro Val Cys Gly...

FIGURE 3. Amino terminal sequences of Japanese quail (JQ) and chicken ovoinhibitors (OI) and of a proteinase inhibitor (PI) from chicken serum (9).

In both chicken proteins Asn[4] is glycosylated. The sequences of chicken ovoinhibitors and of chicken serum proteinase inhibitor are identical, the differences between those sequences and the Japanese quail ovoinhibitor sequence are underlined.

TABLE IV.   Amino Acid Sequences Surrounding the Reactive
            Sites of the Six Domains of Japanese Quail
            Ovoinhibitor (9)

↓

I	...Val Ala Cys Pro Arg Asn Leu Lys Pro Val Cys...
II	...Val Ala Cys Pro Arg Asn Met Lys Pro Val Cys...
III	...Val Ala Cys Pro Arg Asn Leu Lys Pro Val Cys...
IV	...Ala Ala Cys Pro Tyr Ile Leu His Glu Ile Cys...
V	...Met Ala Cys Thr Met Ile Tyr Asp Pro Val Cys...
VI	...Pro Val Cys Thr Met Glu Tyr Ile Pro His Cys...

strongly homologous to the ovomucoid domains. Table IV shows
the amino acid sequences at the six reactive sites of the
ovoinhibitor molecule. We did not as yet carry out definitive
studies to show which of the domains are actually active.
It is, however, quite striking that in the first three the
P_1 amino acid is Arg and thus they are potentially tryptic
sites, which the remaining three Tyr and Met are potentially
chymotrypsin-elastase sites. Similar pattern appears to hold
for chicken ovoinhibitor as well. In ovoinhibitors as in
ovomucoids we have some evidence that separated domains are
separately active.

In 1974 Barrett has isolated a new proteinase inhibitor
from chicken plasma (11). He has shown that this inhibitor is
closely similar to but not identical with chicken ovoinhibitor
and shares with it reaction of immunological identity. We have
isolated this inhibitor as well and in Fig. 3 we compare the
NH_2 terminal sequences of this inhibitor with the NH_2 terminal
sequences of Japanese quail and of chicken ovoinhibitor. It
is seen there that the sequences of the two chicken proteins
share an identical sequence for the first 30 residues. We
therefore assume that the two proteins have the same amino
sequence throughout the molecule and that the minor differences
are the cosequence of posttranslational modification, most
likely the extent of glycosylation, as already suggested by
Barrett.

This finding sheds some light on the question--why so
many independent domains? It is well known that plasma
proteins must have appreciably large molecular weights in
order to be retained in circulation. Gene elongation by
repeated gene duplication is an efficient way of increasing
molecular weight without losing a great deal of biological

activity by adding some long neutral peptide. However, we do not have as yet a similar explanation for the three domains in ovomucoids or for the two in dog submandibular inhibitor.

The tendency to acquire more than one reactive site on a single polypeptide chain by gene elongation/duplication events is not limited to inhibitors in the Kazal homology group. Hochstrasser has shown that the COOH terminus of the very large inter α trypsin inhibitor in mammalian sera consists of two tandem domains homologous to Kunitz bovine pancreatic trypsin inhibitor. A variety of inhibitors from legumes, homologous to Bowman-Birk soybean inhibitor all consist of two homology regions each with a reactive site for a serine proteinase. However, unlike in ovomucoids, ovoinhibitors, dog submandibular inhibitor and inter-α-trypsin inhibitors the two homology regions do not form separate, intradomain disulfide-bridged domains. Instead the disulfide bridges cross link the entire molecule into a single domain (12). Thus gene elongation by duplication appears far more common among inhibitors than among other biologically active molecules.

However, the most striking contrast between inhibitors of serine proteinases and most other proteins is the tolerance of the residues comprising the reactive site towards replacement by other maino acid residue. In most proteins replacement of residues in the active site even by closely related ones leads to either complete loss or dramatic decrease in biological activity. Similarly amino acid residues comprising the active site of most proteins are among the evolutionarily most conserved residues. Neither of these two statements appears to be valid for protein proteinase inhibitors.

Some time ago (Fig. 4) we succeeded in replacing the reactive site Arg[63] in soybean trypsin inhibitor (Kunitz) by Lys[63] (13). The "mutant" inhibitor is fully active in fact a slightly stronger inhibitor of trypsin than the original molecule. Later Leary in our laboratory succeeded in replacing Arg[63] by Trp[63] and thus converted a strong trypsin/weak chymotrypsin inhibitor to an inhibitor that inhibits chymotrypsin very slowly and trypsin not all (14-15). These findings were extended to Kunitz bovine pancreatic trypsin inhibitor by Jering and Tschesche (16). Again the "mutant" inhibitors were quite active and had the expected inhibitory specificity.

The replacements of P[1] residues in the reactive sites of inhibitors are not confined to the laboratory. They frequently occur in nature in various inhibitor families and again they often lead to changes in inhibitory specificity. By far the most dramatic example of variability of the P[1] residues are

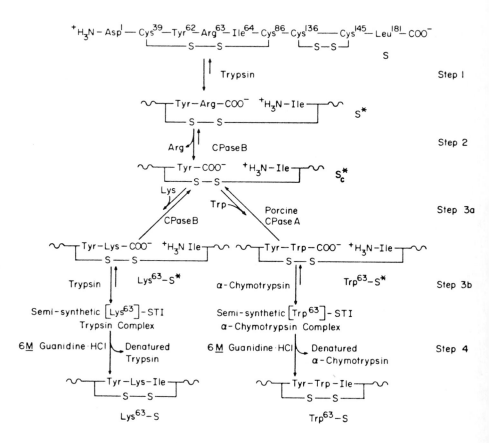

FIGURE 4.   The sequence of reactions employed in replacements of Arg[63] in soybean trypsin inhibitor (Kunitz) by Lys[63]
(on the left (13)) and by Trp[63] (on the right (14,15)).

the third domains (see Fig. 1) of avian ovomucoids. We have obtained by limited proteolysis ovomucoid third domains from 20 species of birds and determined their amino acid sequences. Startlingly when the twenty sequences are aligned it turns out that the most varied residue in the set is the P_1 residue of the reactive site. The reactive sequences of these third domains are listed below (Fig. 5). As expected the inhibition spectra of the various domains differ strikingly confirming and greatly extending the early observation of Feeney and coworkers (18) that (entire) ovomucoids from different closely related species have strikingly different inhibition spectra.

The physiological function of ovomucoids is not known but it is tempting to speculate that the differences between ovomucoids in closely related species may in fact contribute to morphological differences between these species. However, this conclusion is highly speculative.

	↓
JAPANESE QUAIL	....Cys Pro Lys Asp Tyr Arg Pro....
OSTRICH	....Cys Pro Leu Asp Tyr Met Pro....
EMU	....Cys Ser Leu Glu Tyr Met Pro....
PHEASANTS (5 SPECIES)	....Cys Thr Met Glu Tyr Arg Pro....
DUCK	....Cys Thr Met Glu Tyr Met Pro....
TURKEY, CHUKAR: CALIFORNIA, GAMBEL'S AND SCALED QUAILS	....Cys Thr Leu Glu Tyr Arg Pro....
BOBWHITE QUAIL	....Cys Met Ser Glu Tyr Arg Pro....
GOOSE	....Cys Thr Val Glu Tyr Met Pro....
CHICKEN	....Cys Thr Ala Glu Asp Arg Pro....
CHACHALACA	....Cys Leu Gln Glu Gln Lys Pro....

FIGURE 5. Sequences surrounding the reactive sites of third Domains of ovomucoids from various birds (17).

REFERENCES

1. Neurath, H. and Walsh, K. A. in "Regulatory Proteolytic
    Enzymes" Proceedings 11th FEBS Meeting (1978).
2. Huber, R., Kukla, D., Bode, W., Schwager, P., Bartels, K.,
    Diesenhofer, J. and Steigmann, W., J. Mol. Biol. 89 73
    (1974).
3. Sweet, R. M., Wright, H. T., Janin, J., Chothia, C. H.
    and Blow, D. M., Biochemistry 13, 4212 (1974).
4. Greene, L. J. and Bartelt, D. C., J. Biol. Chem. 244 2646
    (1969).
5. Tschesche, H., Kupfer, S., Klauser, R., Fink, E. and Fritz,
    H., Protides Biol. Fluids 23, 255 (1976).
6. Krejci, K. and Frita, H. FEBS Lett. 64, 152 (1976).
7. Hochstrasser, K. and Fritz, H., Z. physiol. Chem. 365,
    1659 (1975).
8. Kato, I., Schrode, J., Wilson, K. A. and Laskowski, M. Jr.,
    Protides Biol. Fluids 23, 235 (1976).
9. Kato, I. and Kohr, W. J., Fed. Proc. 37, 000, (1978).
10. Ikenaka, T., Odani, S., Sakai, M., Nabeshima, Y., Sato, S.
    and Murao, S., J. Biochem 76, 1191 (1974).
11. Barrett, A. J., Biochim. Biophys. Acta 371, 52 (1974).
12. Ikenaka, T. and Odani, S. in "Regulatory Proteolytic Enzyme
    Proceedings of 11th FEBS Meeting, 1978.
13. Sealock, R. W. and Laskowski, M. Jr., Biochemistry 8, 3703
    (1969).
14. Leary, T. R. and Laskowski, M. Jr., Fed Proc. 32, 465 (1973
15. Kowalski, D., Leary, T. R., McKee, R. E., Sealock, R. W.,
    Wang, D. and Laskowski, M. Jr., Bayer Symposium 5, 597
    (1974).
16. Jering, H. and Tschesche, H., Eur. J. Biochem. 61, 453
    (1976).
17. Kato, I., Kohr, W. J. and Laskowski, M. Jr. in "Regulatory
    Proteolytic Enzymes" Proceedings of 11th FEBS Meeting
    (1978).
18. Rhodes, M. B., Bennet, N. and Feeney, R. E., J. Biol. Chem.
    235, 1686 (1960).

RIBOSOMES AND PROTEIN BIOSYNTHESIS

H. G. WITTMANN

Max-Planck-Institut für Molekulare Genetik
Berlin-Dahlem, Germany

The best studied ribosome is that of Escherichia coli. It consists of a 30S subunit (with 21 proteins and 16S RNA) and a 50S subunit (with 34 proteins as well as 5S and 23S RNAs). All components have been isolated in the presence of urea and acetic acid as well as under non-denaturing conditions. They have been extensively characterized by chemical, physical and immunological methods. Furthermore, the primary structures of 42 proteins, with a total of more than 6000 amino acids, have so far been determined. The protein-chemical and the immunological results demonstrate that there are no homologous structures among the proteins studied, with the exception of two pairs of proteins: L7-L12 and S20-L26. The secondary structures of the proteins have been studied by circular dichroism and have also been predicted from their amino acid sequences using four different programmes.

The architecture of the ribosomal subunits is being studied by the following methods: cross-linking between adjacent proteins or between RNA and proteins; localization of the protein binding sites on the three ribosomal RNAs; fluorescence spectroscopy; neutron scattering; immune electron microscopy. The application of these techniques, especially the latter one, has resulted in a rather detailed model for the contours of the subunits and for the spatial arrangement of the proteins and RNAs within the ribosomal particle. The cross-linking studies have progressed to the point where the position of the cross-linked amino acids within a protein pair, and the distance between them can be determined. The same is true for the cross-linked amino acid and the nucleotide in a protein-RNA cross-link.

The involvement of the ribosomal components in the various steps of protein biosynthesis and the molecular mechanisms of antibiotics are being studied by several methods. The most

319

efficient of these are affinity labeling as well as partial
and total reconstitution. Some examples are given.

Comparative studies by immunological and electrophoretic
methods as well as by primary structure determinations of ribosomal proteins and of 5S RNAs isolated from pro- and eukaryotic species allow conclusions to be drawn about the evolution
of ribosomes. The structure of these organelles has changed
drastically during evolution, whereas their function has remained the same.

I. INTRODUCTION

Ribosomes are small cell organelles on which the biosynthesis of proteins occurs. Since proteins are essential components of living cells, ribosomes must be present in all organisms. They consist of two subunits of unequal size, and
protein biosynthesis can only take place when the two subunits
are associated and combined with messenger-RNA as well as with
other non-ribosomal components, such as transfer-RNAs and protein factors.

Ribosomal particles present in the cytoplasm of eukaryotic cells, regardless whether from plants or animals, are somewhat bigger and sediment faster (80S) than those from prokaryotes (70S) or from chloroplasts and mitochondria. However, no
essential difference has been found between the functions of
ribosomes isolated from different classes. Much more is known
about the components, the topography and the function of ribosomes from Escherichia coli than from any other organism and
in the following our present knowledge of E. coli ribosomes
is briefly described.

II. COMPONENTS

Ribosomes from E. coli and other bacteria sediment with
a coefficient of 70S and consist of a small (30S) and a large
(50S) subunit. There are two RNA molecules (5S and 23S) in the
large and one RNA molecule (16S) in the small subunit. The primary structure of the 5S RNA, which has 120 nucleotides, has
been determined (1), whereas work on the 16S RNA with 1,600
nucleotides is still in progress (2-6).

In order to determine the exact number of proteins in
each of the subunits, a procedure for two-dimensional polyacrylamide gel electrophoresis was developed (7); the method
separates all E. coli ribosomal proteins. It was shown that

the small subunit contains 21 proteins, designated S1-S21, and the large subunit 34 proteins, L1-L34 (8).

The proteins of both subunits were isolated in a pure form and on a large scale by a combination of column chromatography on carboxymethyl cellulose in the presence of urea and by gel filtration. While the mixture of the 21 proteins from the small subunit was separated without a prefractionation step, the proteins of the large subunit were fractionated into several groups, each of which was resolved by column chromatography and gel filtration (9). Isolation of 50S proteins in relatively small amounts was accomplished without a prefractionation step (10).

The pure proteins were characterized by chemical, physical, and immunological methods. The molecular weights of the proteins, as determined by SDS electrophoresis and sedimentation equilibrium, range from 9,000 to 65,000 (9). However, sequence analysis proved that the molecular weights, especially of the smallest and most basic proteins, were somewhat overestimated. Apparently, SDS gel electrophoresis of small, basic proteins gives molecular weights that are too high.

As expected from their electrophoretic migration, the ribosomal proteins are very basic. This is reflected in their high content of arginine and lysine residues which can amount up to 35% (11-13), and in the high isoelectric points of most proteins (14). Only three of them (S6, L7 and L12) have isoelectric points near pH 5 whereas those of the other proteins range from pH 7 to 12.

The extent of secondary structure is rather different among the various ribosomal proteins as predicted from their amino acid sequences (15-18), and as determined experimentally (19-20). The α-helix content ranges from 7 to 65% and is highest for proteins L7/L12 from the large, as well as S20 and S21 from the small subunit. Shape determinations by hydrodynamic methods and by small angle X-ray diffraction revealed that most of the proteins studied so far are not spherical but have an elongated shape (reviewed in ref. 21). These results with isolated proteins fully agree with immune electron microscopic studies (see below) which give information about the shape of the proteins in situ.

Very rapid progress has been made during the last several years in the determination of the primary structure of the ribosomal proteins. The sequence is now known for more than 6,000 of the 8,000 amino acids present in the E. coli ribosome. The primary structure has so far been established for 17 proteins from the small subunit and 25 proteins from the large subunit (reviewed in ref. 21-23). The amino acid sequences of additional proteins can be expected in the near future.

Knowledge of the primary structure of 42 ribosomal proteins with more than 6,000 amino acids allows one to determine

whether sequence homologies exist among these proteins. A
first answer to this question can be given from immunological
studies. As soon as pure ribosomal proteins became available
they were injected into rabbits or sheep, and antibodies were
raised. No cross-reaction was found between the ribosomal proteins indicating no extensive homologies among them (24,25).
However, there were two exceptions: antibodies against protein L7 cross-reacted with protein L12 and vice versa. A similar result was obtained for proteins S20 and L26. Sequence
analyses of these proteins showed that proteins L7 and L12
differ only in the presence or absence of an acetyl group at
the N-terminus and that proteins S20 and L26 are identical.
    A more direct and quantitative answer to the question of
homologies among ribosomal proteins is based on the comparison
of the known amino acid sequences of the proteins. Although
many short peptides with up to six amino acids are identical
in two or more ribosomal proteins (26), the frequency of their
occurrence does not significantly differ from the corresponding occurrence in computer-simulated proteins of the same
lengths and compositions as the ribosomal proteins but with
randomized amino acid sequences (27). In other words: homologous structures do not occur more frequently than can be
expected on a random basis.

                           III. TOPOGRAPHY

    The spatial arrangement of the numerous components in
the E. coli ribosomal particle is being studied by a number
of chemical, physical and immunological methods which have
already produced a rather detailed knowledge of ribosome topography. Several of these techniques, e.g. fluorescence
spectroscopy (28), low-angle scattering of neutrons (29-32),
and cross-linking of proteins with bifunctional reagents of
known length (33-37), have already given ample information
about the distances between the components in the ribosome.
    Fluorescence spectroscopy and neutron scattering can
give results about relatively long distances in the particle,
whereas bifunctional reagents identify proteins that are so
close that they can be regarded as neighbors. More detailed
knowledge of the spatial structure of the subunits can be obtained by determining the amino acids that are cross-linked
and by positioning them in the known primary sequence of the
neighboring proteins. This information is especially valuable
for proteins that have elongated shapes.
    Cross-links can be made not only between ribosomal pro-
teins but also between ribosomal RNA and proteins. This has

been accomplished by means of ultraviolet light or by chemical reagents. In this way the cross-link between the 3'-end of the 16S RNA and protein S1 and S21 (38) as well as that between ribosomal proteins and nucleotide stretches of the 16S and 23S RNAs have been demonstrated (39-41). Experiments are now in progress to identify the peptide and the oligonucleotide involved in the cross-link. Besides these studies with intact subunits other attempts have been made to cross-link RNA and protein in complexes made in vitro by binding a single ribosomal protein, e.g., S4, to 16S RNA and to identify the peptides cross-linked to the RNA(42). A prerequisite for these studies is that the protein binds directly and independently to the RNA.

Three proteins (L5, L18 and L25) bind to the 5S RNA, seven proteins (S4, S7, S8, (S13), S15, (S17) and S20) bind to the 16S RNA and ten proteins (L1, L2, L3, L4, L6, L13, L16, L20, L23 and L24) to the 23S RNA (for reviews see refs 43-45). The RNA binding sites of many of these proteins were determined in the following way: A single protein is bound to the RNA in conditions that minimize unspecific binding. Then those RNA regions that are not protected by the bound protein are digested by RNase treatment and the RNA-protein complex is isolated. After removing the protein with phenol, the RNA fragment is sequenced. Since the primary structures of the three ribosomal RNAs are completely or at least partly known (see above), it is possible to determine to which region of the RNA chain the RNA fragment that is protected by the binding protein corresponds. In this way, it was shown that for instance proteins S4 and S20 bind to the 5'-region of the 16S RNA, proteins S8 and S15 to the middle region, and protein S7 to the 3'-region. The sizes of the RNA binding sites vary considerably, e.g., from 40-50 nucleotides for protein S8 to approximately 500 nucleotides for protein S4. It has recently been shown that the number of proteins binding in vitro to the 16S or 23S RNAs depends on the isolation methods for both the proteins and the RNAs (46-48).

Another way to determine to which RNA region a given protein binds is to isolate rather large RNA fragments after mild digestion of the RNA and to test which protein binds to which fragment (43, 49-51). The results from this procedure are in good agreement with those of the experiments just mentioned and they permit the localization of a number of proteins along the 16S and 23S RNA in relatively great detail. They find further support in results from studies on specific ribonucleoprotein fragments that are obtained by mild treatment of ribosomal subunits with nuclease (52-54). Each of these fragments contains several proteins and an RNA region. The proteins in a given fragment are considered to be relatively close to each other and to the RNA region. Proteins S4, S16, S17, and S20

were found together with the 5'-proximal region of the 16S
RNA, proteins S6, S8, S15, and S18 with the central region,
and proteins S7, S9, S13, and S19 with the 3'-proximal region
(55,56). Similar information is available for the binding
sites of five 50S proteins on the 23S RNA (57). The advantage
of this approach is that it allows the localization of those
proteins that are not primary binding proteins, i.e., that do
not bind singly and directly to the RNA. A second subset of
proteins can only bind after the primary binding proteins are
in position, a third set only after the second, and so on.
This is illustrated by an "assembly map" that reflects the cooperative interactions between the ribosomal components (58).

    Valuable information about the topography of ribosomal
subunits has come from the use of antibodies directed specifically against individual ribosomal proteins. As mentioned
above there is no cross-reaction between any of the proteins
in the E. coli ribosome, except proteins L7/L12 and S20/L26.
Furthermore, each of the 30S proteins and many of the 50S proteins are accessible for their cognate antibodies within the
intact particle (59,60), i.e., at least part of the protein
chain is located at the ribosomal surface. This provides the
possibility of locating the proteins in situ by immune electron microscopy. This technique allows the direct electron
microscopic visualization of an antibody attached to the ribosome at that point where the protein is located against which
the antibody is directed. Because of the bivalency of an antibody molecule, "dimers" are formed that consist of two ribosomal subunits connected by one antibody molecule. Since both
the 30S and the 50S subunits have characteristic shapes with
some "landmarks" that are readily recognizable in the electron
microscope, the attachment points can be unambiguously located
for most antibodies.

    The immune electron microscopic technique was first
applied to the localization of a few proteins on the 50S subunit (61) and then widely extended to the localization of all
30S proteins and about 60% of the 50S proteins (62,63). It has
been used independently to locate several proteins on the small
subunit (64,65) and proteins L7 and L12 on the large one (66,
67). Although the results from the various groups still differ
to some extent, especially with regard to the shape of the
small subunit and the orientation of the subunits within the
70S ribosome (reviewed in ref. 23), the immune electron microscopic method has led to the most detailed description of the
topography of the E. coli ribosome. It gives information not
only about the location of the various proteins within the
particle but also about their shape in situ. The results
suggest that many ribosomal proteins are elongated. In many
cases antibodies directed against a single protein attach
to distinct regions of the particle, and these regions are

more distant from each other than can be explained if a protein of that molecular weight were globular. The conclusion that numerous ribosomal proteins have an elongated shape is supported by hydrodynamic measurements (see above).

The topographical model of the two ribosomal subunits and the 70S ribosome as derived from immune electron microscopy as well as the correlation between this model and the functional data on ribosomal components have recently been discussed in detail elsewhere (23), and the reader is referred to that review for further information.

IV. FUNCTION

When it became clear that the E. coli ribosome contains more than 50 components, the finding occasioned some surprise. It had not been expected that the main function of the ribosome, namely the polymerization of amino acids into a protein chain, needed such a complex structure. However, it is now clear that protein biosynthesis is accomplished by numerous coordinated and complicated events in which the ribosomal proteins and the RNAs as well as several protein factors are involved. This process consists of three main steps, namely, initiation, elongation, and termination, and each can be further divided into several sub-steps.

Since several recent reviews deal with the various steps in protein biosynthesis and the involvement of the ribosomal components in these processes, (44,68,69), in the following the emphasis will be on the approach that has been used for the elucidation of the function of the ribosomal components. The question was as to how many and which of the numerous components are involved in each of the various steps in protein biosynthesis, e.g., in binding of the messenger-RNA or the aminoacyl-tRNA, in peptidyl transferase activity, in translocation, and so forth. A solution of this problem is especially difficult in view of the functional interdependence and cooperativity among the ribosomal components. It was therefore necessary to use several methods for a successful approach to this problem. The two most efficient of them are affinity labeling and reconstitution.

A. Affinity Labeling

This method allows the identification of ribosomal proteins or RNA located at or near the binding sites of non-ribosomal components such as messenger-RNA, aminoacyl-tRNA or GTP
which are essential for protein biosynthesis, and also of
various antibiotics. The component to be studied is, in its
radioactive form, chemically modified in such a way that it
binds covalently to the ribosome or its appropriate subunit.
An elegant way to achieve this is by photoaffinity labeling:
the radioactive component is bound to the ribosome but it can
only covalently react after irradiation at the appropriate
wavelength. It is very important to show that the binding is
specific and occurs at the same site as that of the corresponding non-modified component.

In order to identify the ribosomal protein(s) or RNA to
which the chemically modified and radioactive component has
been bound, the ribosome is disrupted and the complex between
the ribosomal and the radioactive component is isolated by gel
electrophoresis or column chromatography. A sensitive and unambiguous way for identification of the ribosomal protein(s)
in the complex is by means of specific antibodies.

The affinity labeling technique has already given a great
deal of information about the involvement of ribosomal components in active sites, such as the peptidyl transferase center
with the A- and P-sites, the binding sites for mRNA and tRNA,
the ribosomal GTPase, etc (reviewed in refs 70-72).

B. Reconstitution

This technique has been used for functional studies in
the following way: By stepwise treatment of ribosomal subunits
with increasing concentrations of salt, e.g., LiCl, an   increasing number of proteins are removed from the subunits resulting in a series of "split proteins" and of protein-defi-
cient "core" particles. The loss of a functional activity by
the various cores, e.g., the ability to form peptide bonds
(peptidyl transferase) or to bind aminoacyl-tRNA or antibiotics,
is then tested. Addition of the corresponding group of split
proteins to an inactive core often leads to restoration of the
activity. This means, that one (or more of the) split protein(s)
is (are) directly or indirectly involved in the activity under
investigation. In order to identify the protein, the appropriate group of split proteins is separated by column chromatography and/or gel filtration into the single components and
each of them is added singly to the inactive core. The reconstituted particle is then tested for activity. In this way a
correlation between a structural element and a given ribosomal

function can be revealed (see refs 73-76 for more information).

Under special conditions it is possible to remove only proteins L7 and L12 from the 50S subunit leaving all other proteins on the particle (77). The removal of these two proteins, which are almost identical, leads to a drastic decrease of several ribosomal functions, e.g., binding of elongation and termination factors, whereas other functions, such as peptidyl transferase, are not impaired. Readdition of proteins L7 and L12 restores the full activity of the 50S subunits (reviewed in ref 78).

In the procedures of partial reconstitution just described, only several (and not all) of the ribosomal proteins are removed from the subunit and are then added back to the core particle. It is, however, possible to completely separate the ribosomal RNA and protein moieties and reconstitute them under appropriate conditions to fully active particles. This total reconstitution has been achieved for both the small (79) and the large (80) subunit of the E. coli ribosomes, and it was used to analyze the assembly process of both subunits (81,82) and the involvement of ribosomal proteins (82) and the 5S RNA (83) in the various steps during protein biosynthesis. For instance, the specific reconstitution of the peptidyl transferase center consisting of 23S RNA and only a very limited number of 50S proteins has recently been achieved (84). In this way it is possible to eliminate many proteins from being involved in the peptidyl transferase activity and to restrict the number of candidates for this enzymatic activity to a few 50S proteins. It has further been shown that at least one of the 50S ribosomal proteins, namely L24, is not involved in a ribosomal function although it is very important for the process of ribosome assembly (85).

V. COMPARISON WITH OTHER ORGANISMS

Extensive comparisons have been made of the structure of ribosomes from E. coli and those of other organisms. Two-dimensional gel electrophoresis showed very similar protein patterns among the Enterobacteriaceae, a bacterial family to which E. coli belongs. As expected, the differences increase as more distantly related species of bacteria are compared (86,87); finally, there are no detectable similarities between E. coli ribosomes and those of eukaryotes, e.g., yeast, mammals or plants. The same general conclusion can be drawn from immunological studies using specific antibodies against individual E. coli ribosomal proteins. However, there are a few proteins, especially L7 and L12, that seem to be conserved during evolution. In addition to immunological cross-reaction between these two proteins from E. coli and their analogues from eukaryotes, e.g., yeast and rat, it was found that protein L7/L12 from E. coli can replace their eukaryotic equivalents in functional studies (reviewed in refs 78,88).

More quantitative information about the degree of homology can be obtained from comparative studies of the primary structure of ribosomal components isolated from different species. Using the data on the N-terminal regions of ribosomal proteins from various bacteria, relatively strong homology has been found between E. coli and various Bacillus species (89,90, and S. Osawa personal communication). This finding is consistent with the possibility of reconstituting hybrid 30S ribosomal subunits from E. coli RNA and Bacillus proteins and vice versa (91). It is possible also to reconstitute functionally active 30S ribosomes in which an individual Bacillus protein is replaced by its E. coli equivalent (92).

From the immunological, protein-chemical and reconstitution experiments it can be concluded that the fundamental organization of bacterial ribosomes is similar and has been conserved. This is also true for the thermophilic Bacillus stearothermophilus, whereas the ribosomes from the halophilic Halobacterium cutirubrum, which are only stable in near-saturated salt conditions, differ from those of all other bacteria that have been studied in that they contain acidic instead of very basic proteins (93). It is interesting that ribosomes from Halobacterium also contain a protein whose sequence is homologous to that of E. coli proteins L7/L12 (94), which seem to be highly conserved throughout nature (see above).

Similar comparative studies, as summarized for the ribosomal proteins, have also been carried out with ribosomal 5S RNAs. They have been reviewed in detail (45). Recently, it was concluded from reconstitution experiments that 5S RNA from E. coli is homologous to 5.8S RNA (and not to 5S RNA) from eukaryotic ribosomes (95).

From the structural studies on both the ribosomal proteins and RNAs of organisms belonging to different classes it is clear that the ribosomal components have changed during evolution to such an extent that there is only little homology between ribosomes from pro- and eukaryotes. On the other hand, no essential difference has been found in the way in which pro- and eukaryotic ribosomes function in protein biosynthesis. This apparent discrepancy raises the question as to how the components in such a complex structure as the ribosome, in which numerous proteins and several RNAs interact with each other, can change so drastically without impairing the vital function. To answer this question, more information is necessary, especially on ribosomes other than those of E. coli. No doubt, ribosomes are an ideal system to study this and many other problems of great scientific interest.

REFERENCES

1. Brownlee, G. G., Sanger, F., and Barrell, B. G., J. Mol. Biol.34: 379 (1968).
2. Ehresmann, C., Stiegler, P., Mackie, G., Zimmermann, R. A., Ebel, J. P., and Fellner, P., Nucl. Acid. Res. 2: 265 (1975).
3. Magrum, L., Zablen, L., Stahl, D., and Woese, C., Nature 257: 423 (1975).
4. Ehresmann, C., Stiegler, P., Carbon, P., and Ebel, J. P., FEBS Lett. 84: 337 (1977).
5. Branlant, C., Sriwidada, J., Krol, A., and Ebel, J. P., Eur. J. Biochem. 74: 155 (1977).
6. Ross, A., and Brimacombe, R., Nucl. Acid. Res., in press (1978).
7. Kaltschmidt, E., and Wittmann, H. G., Anal. Biochem. 36: 401 (1970).
8. Kaltschmidt, E., and Wittmann H. G., Proc. Natl. Acad. Sci. USA 67: 1276 (1970).
9. Wittmann, H. G., in "Ribosomes" (M. Nomura, P. Lengyel, and A. Tissières, ed.), Cold Spring Harbor Laboratory Press, New York, p. 93, 1974.
10. Zimmermann, R. A., and Stöffler, G., Biochemistry 15: 2007 (1976).
11. Craven, G. R., Voynow, P., Hardy, S. J. S., and Kurland, C. G., Biochemistry 8: 2908 (1969).
12. Kaltschmidt, E., Dzionara, M., and Wittmann, H. G., Mol. Gen. Genet. 109: 292 (1970).
13. Mora, G., Donner, D., Thammana, T., Lutter, L., and Kurland, C. G., Mol. Gen. Genet. 112: 229 (1971).
14. Kaltschmidt, E., Anal. Biochem. 43: 25 (1971).
15. Wittmann-Liebold, B., Robinson, S. M. L., and Dzionara, M., FEBS Lett. 77: 301 (1977).
16. Dzionara, M., Robinson, S. M. L., and Wittmann-Liebold, B., Hoppe-Seyler's Zeitschr. Physiol. Chem. 358: 1003 (1977).
17. Wittmann-Liebold, B., Robinson, S. M. L., and Dzionara, M., FEBS Lett. 81: 204 (1977).
18. Dzionara, M., Robinson, S. M. L., and Wittmann-Liebold, B., J. Supramol. Struct., in press (1978).
19. Heiland, I., Dzionara, M., and Suatzke, G., manuscript in preparation.

20. Littlechild, J., Dijk, J., and Morrison, C. A., manuscript in preparation.
21. Brimacombe, R., Stöffler, G., and Wittmann, H. G., Ann. Rev. Biochem. 47: 271 (1978).
22. Wittmann, H. G., and Wittmann-Liebold, B., in "Ribosomes" (M. Nomura, P. Lengyel, and A. Tissières, ed.), Cold Spring Harbor Laboratory Press, New York,p. 115, 1974.
23. Stöffler, G., and Wittmann, H. G., in "Molecular Mechanisms of Protein Biosynthesis" (H. Weissbach and S. Pestka, ed.), Acad. Press, New York, p. 117, 1977.
24. Stöffler, G., and Wittmann, H. G., Proc. Nat. Acad. Sci USA 68: 2283 (1971).
25. Stöffler, G., and Wittmann, H. G., J. Mol. Biol. 62: 407 (1971).
26. Dzionara, M., and Wittmann-Liebold, B., FEBS Lett. 61: 14 (1976).
27. Dzionara, M., and Wittmann-Liebold, B., FEBS Lett. 65: 281 (1976).
28. Cantor, C. R., Huang, K. H., and Fairclough, R., in "Ribosomes" (M. Nomura, P. Lengyel, and A. Tissières, ed.), Cold Spring Harbor Laboratory Press, New York, p. 587, 1974.
29. Moore, P. B., Engelmann, D. M., and Schoenborn, B. P., in "Ribosomes" (M. Nomura, P. Lengyel, and A. Tissières, ed.), Cold Spring Harbor Laboratory Press, New York, p. 601, 1974.
30. Hoppe, W., May, R., Stöckel, P., Lorenz, S., Erdmann, V. A., Wittmann, H. G., Crespi, H. L., Katz, J. J., and Ibel, K., Brookhaven Symp. Biol. 27: IV 38 (1975).
31. Stuhrmann, H. B., Haas, J., Ibel, K., De Wolf, B., Koch, M. H. J., Parfait, R., and Crichton, R. R., Proc. Natl. Acad. Sci. USA 73: 2379 (1976).
32. Moore, P. B., Engelmann, D. M., and Schoenborn, B. P., J. Mol. Biol. 112: 199 (1977).
33. Traut, R. R., Heimark, R. L., Sun, T. T., Hershey, J. W. B., and Bollen, A., in "Ribosomes" (M. Nomura, P. Lengyel, and A. Tissières, ed.), Cold Spring Harbor Laboratory Press, New York, p. 271, 1974.
34. Lutter, L. C., Ortandl, F., and Fasold, H., FEBS Lett. 48: 288 (1974).
35. Barritault, D., Expert-Bezançon, A., Milet, M., and Hayes, D. H., FEBS Lett. 50: 114 (1975).
36. Lutter, L. C., Kurland, C. G., and Stöffler, G., FEBS Lett. 54: 144 (1975).
37. Sommer, A., and Traut, R. A., J. Mol. Biol. 106: 995 (1976).
38. Czernilofsky, A. P., Kurland, C. G., and Stöffler, G., FEBS Lett. 58: 281 (1975).
39. Möller, K., and Brimacombe, R., Mol. Gen. Genet. 141: 343 (1975).

40. Möller, K., Rinke, J., Ross, A., Buddle, G., and Brima-
    combe, R., Eur. J. Biochem. 76: 175 (1977).

41. Ulmer, E., Meinke, M., Ross, A., Fink, G., and Brimacombe,
    R., Mol. Gen. Genet., in press (1978).

42. Ehresmann, B., Reinbolt, J., and Ebel, J. P., FEBS Lett.
    58: 106 (1975).

43. Zimmermann, R. A., in "Ribosomes" (M. Nomura, P. Lengyel,
    and A. Tissières, ed.), Cold Spring Harbor Laboratory
    Press, New York, p. 225, 1974.

44. Brimacombe, R., Nierhaus, K. H., Garrett, R. A., and
    Wittmann, H. G., Prog. Nucleic Acid Res. Mol. Biol. 18: 1
    (1976).

45. Erdmann, V. A., Prog. Nucleic Acid Res. Mol. Biol. 18: 45
    (1976).

46. Hochkeppel, H. K., and Craven, G. R., Mol. Gen. Genet.
    153: 325 (1977).

47. Hochkeppel, H. K., and Craven, G. R., J. Mol. Biol. 113:
    623 (1977).

48. Littlechild, J., Dijk, J., and Garrett, R. A., FEBS Lett.
    74: 292 (1977).

49. Zimmermann, R. A., Mackie, G. A., Muto, A., Garrett, R. A.,
    Ungewickell, E., Ehresmann, C., Stiegler, P., Ebel, J. P.,
    and Fellner, P., Nucleic Acid Res. 2: 279 (1975).

50. Chen-Schmeisser, U., and Garrett, R. A., Eur. J. Biochem.
    69: 410 (1976).

51. Spierer, P., and Zimmermann, R. A., J. Mol. Biol. 103: 647
    (1976).

52. Morgan, J., and Brimacombe, R., Eur. J. Biochem. 37: 472
    (1973).

53. Roth, H. E., and Nierhaus, K. H., FEBS Lett. 31: 35 (1973).

54. Spierer, P., Zimmermann, R., and Mackie, G. A., Eur. J.
    Biochem. 52: 459 (1975).

55. Rinke, J., Yuki, A., and Brimacombe, R., Eur. J. Biochem.
    64: 77 (1976).

56. Yuki, A., and Brimacombe, R., Eur. J. Biochem. 56: 23
    (1975).

57. Branlant, C., Krol, A., Sriwidada, J., and Brimacombe,
    R., Eur. J. Biochem. 70: 483 (1977).

58. Mizushima, S., and Nomura, M., Nature (London) 266: 1214
    (1970).

59. Stöffler, G., in "Ribosomes" (M. Nomura, P. Lengyel, and
    A. Tissières, ed.), Cold Spring Harbor Laboratory Press,
    New York, p. 615, 1974.

60. Stöffler, G., Hasenbank, R., Lütgehaus, M., Maschler, R.,
    Morrison, C. A., Zeichhardt, H., and Garrett, R. A., Mol.
    Gen. Genet. 127: 89 (1973).

61. Wabl, M. R. (Ph.D. Thesis), Berlin: Freie Universität
    (1973).

62. Tischendorf, G. W., Zeichhardt, H., and Stöffler, G.,
    Mol. Gen. Genet. 134: 209 (1974).
63. Tischendorf, G. W., Zeichhardt, H., and Stöffler, G.,
    Proc. Natl. Acad. Sci. USA 72: 4820 (1975).
64. Lake, J. A., and Kahan, L., J. Mol. Biol. 99: 631 (1975).
65. Lake, J. A., Pendergast, M., Kahan, L., and Nomura, M.,
    Proc. Natl. Acad. Sci. USA 71: 4688 (1974).
66. Boublik, M., Hellmann, W., and Roth, H. E., J. Mol. Biol.
    107: 479 (1976).
67. Lake, J. A., Proc. Natl. Acad. Sci. USA 74: 1904 (1977).
68. Lengyel, P., in "Ribosomes" (M. Nomura, P. Lengyel, and
    A. Tissières, ed.), Cold Spring Harbor Laboratory Press,
    New York, p. 13, 1974.
69. Pongs, O., Nierhaus, K. H., Erdmann, V. A., and Wittmann,
    H. G., FEBS Lett. 40: Suppl. 28 (1974).
70. Cantor, C. R., Pellegrini, M., and Oen, H., in "Ribosomes"
    (M. Nomura, P. Lengyel, and A. Tissières, ed.), Cold
    Spring Harbor Laboratory Press, New York, p. 573, 1974.
71. Cooperman, B. S., Bioorg. Chem., in press.
72. Kuechler, E., Angew. Chem. 88: 555 (1976).
73. Maglott, D., and Staehelin, T., Methods Enzymol. 20: 408
    (1971).
74. Nierhaus, D., and Nierhaus, K. H., Proc. Natl. Acad. Sci.
    USA 70: 2224 (1973).
75. Nierhaus, K. H., and Montejo, V., Proc. Natl. Acad. Sci.
    USA 70: 1931 (1973).
76. Sander, G., Marsh, R. C., and Parmeggiani, A., FEBS
    Lett. 33: 132 (1973).
77. Hamel, E., Koka, M., and Nakamoto, T., J. Biol. Chem. 247:
    805 (1972).
78. Möller, W., in "Ribosomes" (M. Nomura, P. Lengyel, and
    A. Tissières, ed.), Cold Spring Harbor Laboratory Press,
    New York, p. 711, 1974.
79. Traub, P., and Nomura, M., Proc. Natl. Acad. Sci. USA
    59: 777 (1968).
80. Nierhaus, K. H., and Dohme, F., Proc. Natl. Acad. Sci.
    USA 71: 4713 (1974).
81. Dohme, F., and Nierhaus, K. H., J. Mol. Biol. 107: 585
    (1976).
82. Nomura, M., and Held, W. A., in "Ribosomes" (M. Nomura,
    P. Lengyel, and A. Tissières, ed.), Cold Spring Harbor
    Laboratory Press, New York, p. 193, 1974.
83. Dohme, F., and Nierhaus, K. H., Proc. Natl. Acad. Sci.
    USA 73: 221 (1976).
84. Hampl, H., and Nierhaus, K. H., manuscript in preparation.
85. Spillmann, S. and Nierhaus, K. H., manuscript submitted.
86. Geisser, M., Tischendorf, G. W., and Stöffler, G., Mol.
    Gen. Genet. 127: 129 (1973).

87. Geisser, M., Tischendorf, G. W., Stöffler, G., and Witt-mann, H. G., Mol. Gen. Genet. 127: 111 (1973).

88. Wool, I. G., and Stöffler, G., in "Ribosomes" (M. Nomura, P. Lengyel, and A. Tissières, ed.), Cold Spring Harbor Laboratory Press, New York, p. 417, 1974.

89. Higo, K. I., and Loertscher, J., J. Bact. 118: 180 (1974).

90. Yaguchi, M., Matheson, A. T., and Visentin, L. P., FEBS Lett. 46: 296 (1974).

91. Nomura, M., Traub, P., and Bechman, H., Nature (London) 219: 793 (1968).

92. Higo, K., Held, W., Kahan, L., and Nomura, M., Proc. Natl. Acad. Sci. USA 70: 944 (1973).

93. Visentin, L. P., Chow, C., Matheson, A. T., Yaguchi, M., and Rollin, F., Biochem. J. 130: 103 (1972).

94. Oda, G., Strøm, A. R., Visentin, L. P., and Yaguchi, M., FEBS Lett. 43: 127 (1974).

95. Wrede, P., and Erdmann, V. A., Proc. Natl. Acad. Sci. USA 74: 2706 (1977).

IMMUNOLOGICAL PROPERTIES OF PROTEIN CONJUGATES WITH NON-IMMUNOGENIC POLYMERS: STUDIES WITH RAGWEED POLLEN ALLERGEN, ANTIGEN E[*]

Te Piao King

The Rockefeller University
New York, New York

INTRODUCTION

A problem of considerable practical interest in immediate type of allergy is how to suppress effectively an established specific IgE antibody response. Experiences in the desensitization of allergic persons and in model experiments with animals have shown that specific IgE antibody levels can be lowered on repeated treatment with the offending allergen. The usual observation is that the larger the dose used for treatment the better is the suppression of IgE antibody. To overcome the problem of allergic shock in patients, people have suggested as early as the 1940s the use of chemically modified allergens as treatment reagents (1).

One possible way of chemical modification is to attach bulky side chains of large molecular weight to the protein molecule. This is shown schematically in Figure 1 for a protein antigen which is assumed to contain eight antigenic determinants. The

[*]This research is supported by grant AI-14422 of the National Institutes of Health.

FIGURE 1.   A schematic diagram of a globular
antigen coupled with large side chains to reduce
accessibility of antigenic determinants which are
represented by the shaded areas.

large bulky side chains reduce the accessibility of
the antigenic determinants; therefore, the number
and the affinity of antigenic determinants are decreased.   These two decreases will effect the allergenicity and immunogenicity of the protein antigen.   The allergenicity of an antigen depends on its
interaction with basophil or mast cell-bound specific IgE antibody.   Only the interaction of a multi-
valent antigen with cell-bound IgE antibody will
lead to the release of histamine and other chemical
mediators of immediate hypersensitivity, and uni-
valent fragments of antigens are inactive in this
regard (cf. 2).   Immunogenicity of a protein antigen for IgE and IgG antibody responses is also
known to require multivalency (cf. 3).   However,
the role of antigenic valency in the suppression of
specific antibody responses is not clear.
     In view of the above considerations, we have

prepared two conjugates of a model protein antigen with different bulky molecules and have studied their immunological properties in experimental animals.[‡] The model protein chosen is antigen E obtained from ragweed pollen. Ragweed pollen is the principal causative agent of autumn hay fever in man in the northeastern region of the United States, and it contains several protein allergens. Antigen E is the major allergen and it is an acidic protein of molecular weight of about 38,000 daltons (4).

The two bulky molecules which were used for coupling with antigen E are methoxypolyethylene glycol (PEG) and a random copolymer of D-glutamic acid and D-lysine (D-GL). These two molecules were chosen because of the reported findings of other workers (5, 6). PEG coupled bovine plasma albumin was reported to be non-immunogenic in rabbits and not to react with the antibodies specific for the native protein (5). Hapten conjugates of D-GL were reported to induce suppression of ongoing hapten specific IgE and IgG antibodies in experimental animals (6).

RESULTS

Preparation of Antigen E Conjugates

The procedure used for the preparation of PEG coupled antigen E is that reported previously (5). PEG is first coupled to cyanuric chloride, and the resulting 2,4-dichloro-6-PEG-s-triazine is then coupled with antigen E via the lysyl residues at pH 9 and r.t. These reactions are shown below.

[‡] Some of the results described in this paper have not appeared in print, and a full account of these studies will be published later.

$$CH_3(OCH_2CH_2)_nOH + \text{(cyanuric chloride)}$$

$$\longrightarrow CH_3(OCH_2CH_2)_nO\text{—(dichlorotriazine)}$$

$$AgE\text{-}NH_3^{\oplus} + CH_3(OCH_2CH_2)_nO\text{—(dichlorotriazine)} + H_2O$$

$$\longrightarrow \longrightarrow CH_3(OCH_2CH_2)_nO\text{—(hydroxytriazine)}\text{—}NH\text{-}AgE$$

The above coupling reaction with antigen E has been carried out with PEG preparations having average molecular weights of 2000 and 5000 daltons, corresponding to PEG with 45 and 114 ethylene oxide units, respectively. The maximal average number of PEG groups which can be introduced is 5-7 per molecule of antigen E. Antigen E contains 18 lysine residues per molecule, and 16 of them are accessible for modification with other reagents of small molecular size. The reduced degree of modification with PEG reflects steric hindrance by the bulky PEG groups (7).

The conjugates of antigen E with PEG are undoubtedly mixtures of derivatives varying in the number and the site of lysyl residues modified. When gel filtered on a column of Sepharose 6B, the conjugates gave slightly asymmetric peaks. For the antigen E PEG5000 conjugates, their elution position from Sepharose 6B was near that of a globular protein of about 160,000 daltons. They migrated as a broad diffuse zone on electrophoresis in 4% poly-

acrylamide gel containing SDS buffer, and their
mobility being in the range of dimeric and trimeric
forms of bovine plasma albumin corresponding to
molecular weights of 132,000 to 198,000 daltons,
respectively. The Sepharose gel filtration data
and the SDS gel electrophoretic data both suggest
the size of the conjugates to be greater than
100,000 daltons. The chemical data for a conjugate
of antigen E (molecular weight 38,000 daltons) and
PEG (molecular weight 5,000 daltons) in the molar
ratio of 1 to 6 will indicate a molecular weight of
about 68,000 daltons. This difference between the
chemical and the physical data is probably related
to shape differences of PEG and protein molecules.

Conjugates of antigen E with D-GL were prepared
via intermolecular disulfide bond formation, namely
antigen E containing thiol groups reacting with
D-GL containing 4-dithiopyridyl groups to yield a
conjugate with the release of 4-thiopyridone.
Thiol groups were introduced into antigen E upon
amidination of lysyl residues with 4-iminothiolane,
and 4-dithiopyridyl groups were introduced into
D-GL upon amidination with 4-iminothiolane in the
presence of 4,4'-dithiodipyridine. These reactions
are depicted below.

$$AgE\text{-}NH_3^{\oplus} + \text{(ring)}_{S}^{NH} \longrightarrow AgE\text{-}(NHCCH_2CH_2CH_2SH)_m$$

$$D\text{-}GL\text{-}NH_3^{\oplus} + \text{(ring)}_{S}^{NH} + N\text{(ring)}\text{-}SS\text{-}\text{(ring)}N$$

$$\longrightarrow D\text{-}GL\text{-}(NHCCH_2CH_2CH_2SS\text{-}\text{(ring)}N)_n + S=\text{(ring)}NH$$

$$\text{AgE} \cdots (\text{SH})_m + \text{D-GL} \cdots (\text{SS-}\langle \text{benzene} \rangle \text{N})_n$$

$$\longrightarrow \text{AgE} \cdots (\text{SS})_x \cdots \text{D-GL} + \text{S=}\langle \text{benzene} \rangle \text{NH}$$

By carrying out the amidinations at pH 8-9 and 25°, about 2-3 thiol or 4-dithiopyridyl groups were introduced into proteins. The disulfide bond formation reaction between the modified antigen E and D-GL took place rapidly at pH 6.6 and 25°. The above reactions are of general applicability for the preparation of other protein conjugates, as shown by us recently (8).

When separated on Sepharose CL6B, the antigen E·(D-GL) conjugates gave a broad zone which was eluted at a position well ahead of that of a globular protein of 160,000 daltons. The broad zone was divided into three equal fractions. Amino acid analysis of these fractions showed that they have similar compositions of antigen E and D-GL in the molar ratio of about 1 to 1.5. As no significant difference in the compositions of these cuts was found, a pool of the materials in these three fractions was used for the immunological studies given later.

It is not possible to estimate the molecular sizes of the conjugates of antigen E·(D-GL) from the elution positions from Sepharose CL6B, because of the shape difference between globular and non-globular solutes. This is indicated by the early elution position of D-GL as compared to that of immunoglobulin, even though their molecular weights were, respectively, 34,000 and 160,000 daltons.

Judging from other studies with conjugates of globular proteins prepared under similar conditions, e.g. AgE-horseradish peroxidase and AgE-sheep antibody (8), the conjugates of $[\text{AgE}_1\text{-}(\text{D-GL})_{1.5}]_n$ are probably in the size range of n = 1 to 3. In addition to this size heterogenity, the conjugates formed are also heterogenous with respect to their sites of disulfide bond formation since the chemical modifications of antigen E and D-GL occur randomly at different lysyl residues.

Immunological Studies of Antigen E Conjugates

Antigenic Activities. The two conjugates of antigen E with PEG or D-GL showed reduced antigenic activities in their reactions with antigen E-specific IgE and IgG antibodies. These results are summarized in Table I. The results are less extensive for the antigen E·(D-GL) conjugate than they are for the antigen E·PEG conjugate.

The antigen E·PEG conjugate did not form immune precipitate with rabbit antigen E-specific antibodies (test a). It inhibited the agglutination of antigen E coated red blood cells with specific antibodies (test b). It agglutinated red blood cells coated with antigen E-specific antibodies (test c). It gave positive passive cutaneous anaphylaxis test in rat sensitized with mouse antigen E-specific serum (test d) (9). It caused histamine release from ragweed sensitive human leukocytes (test e). In tests b through e, the concentration of conjugate required is 10-100 times greater than that for the native antigen, thus indicating that the conjugate has a reduced number of antigenic determinants and/or affinity for specific antibodies. Tests a, b and c measure primarily the interaction of IgG antibodies, as IgG is the major serum immunoglobulin. Tests d and e measure specifically the interaction of IgE antibodies because of its special biological property of binding to cells.

Immunogenicity. When rabbits were immunized with the antigen E·PEG2000 conjugate in the presence of complete Freund's adjuvant, they gave only one-tenth of the antibody response of the animals immunized with antigen E under the same conditions (7). When mice were immunized with the antigen E· PEG5000 conjugate with alum as adjuvant, no detectable IgE and IgG responses were found while those immunized with antigen E gave good responses under

Relative Antigenic Activities of Antigen E Conjugates

Tests	AgE	AgE·PEG$_{2K}$[1]	AgE·PEG$_{5K}$[2]	AgE·(D-GL)[2]
a. Precipitin reaction with anti-AgE ab	+	-	n.d.	n.d.
b. Inhibition of ag-glutination of AgE coated rbc with ab	1	0.01	0.05-0.10	0.05-0.10
c. Agglutination with ab coated rbc	1	n.d.	< 0.01	0.2
d. PCA in rat with mouse IgE ab	1	n.d.	0.01-0.10	0.1
e. Human leukocyte histamine release	1	0.02	n.d.	n.d.

1 taken from reference 7;
2 unpublished result;
Abbreviations used: + and - for presence and absence of immune precipitates; n.d. for not done; PCA for positive cutaneous anaphylaxis.

the same conditions.

Immunosuppression. It is known that treatment
of a previously immunized animal with large amounts
of the same immunizing antigen can lead to suppression of antibody response (10-12).

Studies were made to compare the effectiveness
of antigen E and its two conjugates in suppressing
ongoing antigen E-specific IgE and IgG antibody
responses in mice. These results are shown in
Figures 2, 3 and 4. In these figures PCA represents the serum titer of antigen E-specific IgE
antibodies as determined by passive cutaneous anaphylaxis in rats. HA represents the serum titer of
antigen E-specific IgG and IgM antibodies as determined by agglutination of antigen E coated red
blood cells; the contribution of IgM antibodies was
found to be negligible. Under the conditions used,
a PCA titer of 100 is estimated to be equivalent to
about 10 ng per ml of IgE antibody (2), and a HA
titer of 100 is equivalent to about 10 μg per ml of
IgG antibody (13).

Immune response to different antigens is known
to be under genetic control, and the (Balb/c x A/J)
F_1 strain of mice used in this work is known to be
a high responder to ragweed antigens (14). The results in Figures 2 to 4 were obtained with groups
of 3 mice. The mice were immunized intraperitone-
ally on day 0 with a mixture of antigen E (1 μg,
26 pmole) and alum (1 mg). Then they were treated
subcutaneously on days 21 and 24 with antigen E or
its conjugates (70 or 700 times the immunizing
dose). After a second immunization at week 4, all
treated groups showed continued decrease in their
IgE antibody level while the control group gave an
increased IgE antibody level. Following a third
and a fourth immunization at weeks 8 and 18, all
treated groups, as well as the untreated group,
showed transient increased IgE antibody levels.
Their IgE antibody levels peaked at about two weeks
after re-immunization, then they declined rapidly
for the treated groups and slowly for the untreated

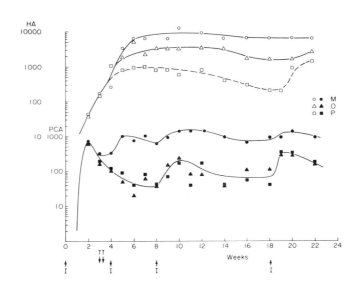

FIGURE 2. Immunosuppression of antigen E sensitized mice on treatment with antigen E·PEG conjugates. Mice were sensitized on day 0 with antigen E
(0.026 nmole), then treated with the conjugate 2
times, 1.8 and 18 nmoles, respectively, for groups O
and P on days 21 and 24. Group M is the untreated
control group. All groups were re-challenged with
antigen E on weeks 4, 8 and 18. HA (agglutination)
and PCA (passive cutaneous anaphylaxis) denote titers of antigen E-specific IgG and IgE antibodies,
respectively.

group. At their peak response and after decline
from their peak response, the IgE antibody levels of
all treated groups are, respectively, 3-6 and 4-16
times lower than that of the untreated group. The
results suggest that the antigen E·PEG conjugate is
slightly more effective than antigen E itself or its
D-GL conjugate in the suppression of IgE antibody
response.

The results in Figures 2 to 4 also show the
change in IgG antibody levels on treatment with antigen E or its conjugates. On repeated immunization
of all groups of mice, their IgG antibody levels did

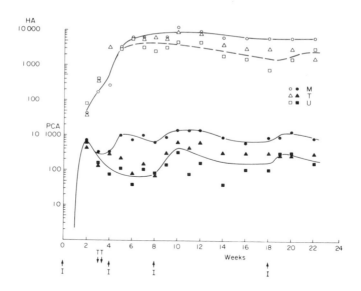

FIGURE 3. Immunosuppression of antigen E sensitized mice on treatment with antigen E·D-GL conjugates. Group M is the control group from Fig. 2 and
groups T and U were treated 2 times with 1.8 and 18
nmoles of conjugate, respectively. All other conditions are the same as those in Figure 2.

not rise and fall as readily as the IgE antibody
levels did. Treatment with antigen E·PEG conjugate
was clearly more effective than treatment with antigen E or its D-GL conjugate in the suppression of
IgG antibody response. Greater suppression of the
IgG antibody level was obtained at the higher than
at the lower treatment dose. For the group treated
with antigen E·PEG conjugate at the high and low
doses, the IgG antibody levels were, respectively,
16 and 2 times less than that of the untreated
groups at week 14.

Separate groups of mice were also treated with
antigen E or its conjugates on days 49 and 51 after
the initial sensitization with antigen E, while
keeping all subsequent challenges with antigen E at
the same schedule as those shown in Figures 2 to 4.

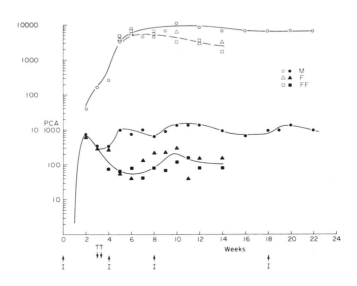

FIGURE 4. Immunosuppression of antigen E sensitized mice on treatment with antigen E. Group M is
the control group from Fig. 2 and groups F and FF
were treated 2 times with 1.8 and 18 nmoles of antigen E, respectively. All other conditions are the
same as those in Figure 2.

Under these conditions there was at best only a 2-
to 4-fold reduction in the IgE and IgG antibody
levels for the treated groups as compared to those
for the untreated group. This lack of immunosuppression may be related to the high IgG antibody
level at the time of treatment, which may effectively remove all the injected antigen E or its conjugates.
     Recently Lee and Sehon (15) reported similar
findings to those described above. They used a
mixture of PEG modified ragweed pollen proteins and
studied only its suppression of IgE antibodies in
mice.

DISCUSSION

The above results show that coupling of PEG or D-GL to antigen E both yielded modified antigens with reduced allergenic and immunogenic activities, yet retaining the immunosuppressive property of the native antigen. These immunological changes of the conjugates are interpreted as a consequence of steric hindrance by the coupled polymers which reduces the accessibility of antigenic determinants. Apparently the allergenicity and immunogenicity of a protein depends on the multi-valency of an antigen while its immunosuppressive property does not.

The size of an antigenic determinant of a protein reacting with IgE antibody is not known. It is probably the same as that of a determinant reacting with IgG antibody, in the size range of 4 to 8 amino acid residues. These amino acid residues are usually located on the exposed surface regions of the molecule (16, 17). For a protein in the size range of 40,000 daltons, each of its several antigenic determinants covers less than a few per cent of the surface region of the molecule. Therefore it may not be necessary to use polymers as large as those employed in the present work to reduce the accessibility of the determinants. Smaller polymers may even be more effective than the larger ones in this respect, since the increased density of the coupling will increase the number of determinants which can become inaccessible. In support of the preceding discussion are the similar immunological changes in the antigen E conjugates with PEG and D-GL of average molecular weights ranging from 2,000 to 34,000 daltons.

The observed immunosuppressive effect on treatment of antigen E sensitized mice with antigen E or its conjugates is of long duration and withstands at least three repeated challenges of antigen E. These results indicate strongly that the suppression is at the cellular level and that it can not be due to simple neutralization of the antibodies

by the antigens used for the treatment. The long
duration of suppression suggests that it is probably at the level of thymus derived lymphocytes, T
cells, rather than at the level of bone marrow de-
tived lymphocytes, B cells. This suggestion is
based on the report that the duration of unrespons-
iveness for B cells from a treated animal is much
shorter than that for T cells (10). If the suppression is at the T cell level, it can be due to the
activation of T-suppressor cells or to the inactivation of T-helper cells (11, 12, 18). Or it can be
due to impaired cell-cell interaction of macrophages,
T and B cells, which is believed to be of importance
in the initiation of antibody synthesis (11, 12, 18).

Several other chemical modifications of allergens or allergen extracts have been reported by different workers. Among them are formaldehyde or
glutaraldehyde treatment and urea denaturation.
Formaldehyde or glutaraldehyde treatment has been
applied to ragweed antigen E (19), and pollen extract (20), purified rye pollen allergen (21, 22),
and timothy pollen extract (23). The formaldehyde
or glutaraldehyde treated allergens showed reduced
allergenic activities about 10 to 1000-fold. They
all showed no change in their ability to induce allergen-specific IgG antibody production but variable
changes in their ability to induce specific IgE
antibody production were reported. These findings
are in contrast to the present findings of antigen
E·PEG conjugate. The difference may be related to
antigenic valency. The oligomers or polymers of
allergens formed on treatment with formaldehyde or
glutaraldehyde may not have the same extent of reduction of antigenic determinants as the antigen E·
PEG conjugate has.

Urea denatured antigen E showed greatly reduced
allergenic activity as it lacked the B cell-specif-
ic antigenic determinants of the native antigen.
It did not induce the formation of antigen E-specific IgE or IgG antibodies in mice. When antigen E
sensitized mice were treated with the denatured
antigen, their IgE antibody level was suppressed

about four-fold as compared to that of the untreated group, but their IgG antibody level was not affected. The immunosuppressive effect of the denatured antigen is believed to be at the T cell level (24). These findings differ from the present ones where treatment led to suppression of both IgE and IgG antibodies. This difference may be related to the loss of certain antigenic determinants on denaturation of antigen E. In the present study the conjugates are a mixture of modified antigens with a reduction in the density of antigenic determinants, but the mixture together retains all the determinants of the native antigen.

Glutaraldehyde modified antigen E (25) and timothy pollen extract (26), and formaldehyde modified ragweed pollen extract (27) have been tried for desensitization of allergic patients. In all three cases treatment caused significant increases in allergen-specific IgG antibodies; slight or no decrease in allergen-specific IgE antibodies was observed on treatment with glutaraldehyde modified antigen E or timothy pollen extract. On the basis of the immunosuppression data in experimental animals it appears likely that treatment of allergic patients with PEG or D-GL coupled antigen E will lead to suppression of both IgE and IgG antibodies. Therefore, the conjugates prepared in this work may have useful biological properties which are not present with other chemically modified antigens. Experimental verification of the preceding hypothesis will be forthcoming since we plan to study in the near future the immunosuppressive effect of antigen E·PEG conjugate in allergic man.

ACKNOWLEDGMENTS

I would like to thank Dr. Nicholas Chiorazzi of this University for teaching me some of the immunological methods used, Dr. Shu Man Fu, also of this University, for helpful discussions, and Dr. John J. Gavin of Miles Laboratories for a gift of D-GL.

REFERENCES

1. Stull, A.,, Cooke, R.A., Sherman, W.B., Hebald, S., and Hampton, S.F. (1940). J. Allergy 11:439.
2. Ishizaka, K. (1973). In "The Antigens" (M.Sela, ed.), Vol. 1, p. 479. Academic Press, New York.
3. Brawn, R.J. and Dandliker, W.B. (1977). In "Immunochemistry of Proteins" (M.Z. Atassi, ed.), Vol. 2, p. 45. Plenum Press, New York.
4. King, T.P. (1976). Adv. Immunol. 23:77.
5. Abuchowski, A., Van Es, T., Palczuk, N.C., and Davis, F.F. (1977). J. Biol. Chem. 252:3578.
6. Katz, D.H., Hamaoka, T., and Benacerraf, B. (1973). Proc. Natl. Acad. Sci. 70:2776.
7. King, T.P., Kochoumian, L., and Lichtenstein, L.M. (1977). Arch. Biochem. Biophys. 178:442.
8. King, T.P., Li, Y., and Kochoumian, L. (1978). Biochemistry 17: in press.
9. Mota, I. and Wong, D. (1969). Life Sci. 8:813.
10. Chiller, J.M. and Weigle, W.O. (1972). Contemp. Top. Immunobiol. 1:119.
11. Tada, T. (1975). Progr. Allergy 19:122.
12. Ishizaka, L. (1976). Adv. Immunol. 23:1.
13. King, T.P., Norman, P.S., and Tao, N. (1974). Immunochemistry 11:83.
14. Dorf, M.E., Newburger, P.E., Hamaoka, T., Katz, D.H., and Benacerraf, B. (1974). Eur. J. Immunol. 4:346.
15. Lee, W.Y. and Sehon, A.H. (1977). Nature 267:618.
16. Crumpton, M.J. (1974). In "The Antigens" (M. Sela, ed.), Vol. 2, p. 1. Academic Press, New York.
17. Goodman, J.W. (1975). In "The Antigens" (M. Sela, ed.), Vol. 3, p. 127. Academic Press, New York.
18. Katz, D.H. and Benacerraf, B. (1972). Adv. Immunol. 15:1.
19. Patterson, R. and Suszko, I.M. (1974). J. Immunol. 112:1855.

20. Patterson, R., Suszko, I.M., Zeiss, C.R., Pruzansky, J.J., and Bacal, E. (1978). J. Allergy Clin. Immunol. 61:28.
21. Marsh, D.G. (1971). Int. Arch. Allergy Appl. Immunol. 41:199.
22. Marsh, D.G., Lichtenstein, L.M., and Norman, P.S. (1972). Immunology 22:1013.
23. Moran, D.M., Wheeler, A.W., Overell, B.G., and Woroniecki, S.R. (1977). Int. Arch. Allergy Appl. Immunol. 54:315.
24. Ishizaka, K., Okudaira, H., and King, T.P. (1975). J. Immunol. 114:110.
25. Metzger, W.J., Patterson, R., Zeiss, C.R., Irons, J.S., Pruzansky, J.J., Suszko, I., and Levitz, D. (1976). New Engl. J. Med. 295:1160.
26. Johansson, S.G.O., Miller, A.C.M.L., Mullen, N., Overell, B.G., Tees, E.C., and Wheeler, A.W. (1974). Clin. Allergy 4:255.
27. Norman, P.S., Marsh, D.G., Lichtenstein, L.M., and Ishizaka, K. (1975). J. Allergy Clin. Immunol. 55:78.

β-LIPOTROPIN: PROHORMONE FOR β-ENDORPHIN
WITH POTENT ANALGESIC AND BEHAVIORAL ACTIVITIES

Choh Hao Li

Hormone Research Laboratory
University of California
San Francisco, California

β-Lipotropin (β-LPH) was first discovered and isolated from sheep pituitary glands (1,2). It has been detected in sheep serum (3) using rabbit antiserum to sheep β-lipotropin ($β_s$-LPH) and found to have a concentration of 10 ng/ml by radioimmuno- assay (4). Immunofluorescent and peroxidase con- jugated antibody reactions with sheep pituitary glands show that the hormone is located in baso- phils of both the anterior lobe and intermediate lobe cells (5). It is stored and released in the same vesicles as corticotropin (6).

Using specific antisera to human β-lipotropin ($β_h$-LPH), the hormone is found to be contained in the cytoplasm of cells and in the beaded axons of rat brain (7). Areas of greatest β-LPH content are hypothalamus, periventricular nucleus of the thalamus, ansa lenticularis, zona compacta of the substantia nigra, medial amygdaloid nucleus, zona incerta, periaqueductal central gray area, locus ceruleus, and a few fibers in the reticular forma- tion (7). The hormone has also been detected in extracts of bovine brain by radioimmunoassay (8).

β-LPH has also been obtained in highly purified form from bovine (9), porcine (10-12), rat (13) and human (14-16) pituitaries. The complete amino acid

Presented at the International Symposium on Proteins, March 6-9, 1978, Taipei. Supported in part by NIMH grant MH-30245 and the Hormone Research Foundation.

sequence of ovine (17-18), porcine (19-23), bovine (24) and human (16) hormones have been proposed. It consists of 91 amino acids with defined sequence as shown in Figure 1. From a comparison of the structures of various β-lipotropins, it is evident that the sequence of the COOH-terminal 36 residues is surprisingly homologous, whereas the amino acid sequence at the NH$_2$-terminal exhibits considerable variability (see Figure 1).

```
                                  5                    10
Human:     H-Glu-Leu-Thr-Gly-Gln-Arg-Leu-Arg-Gln-Gly-
Ovine:     H-Glu-Leu-Thr-Gly-Glu-Arg-Leu-Glu-Gln-Ala-
Porcine:   H-Glu-Leu-Ala-Gly-Ala-Pro-Pro-Glu-Pro-Ala-


                                 15                    20
           Asp-Gly-Pro-Asn-Ala-Gly-Ala-Asn-Asp-Gly-
           Arg-Gly-Pro-Glu-Ala-Gln-Ala-Glu-Ser-Ala-
           Arg-Asp-Pro-Glu-Ala-Pro-Ala-Glu-Gly-Ala-


                                 25                    30
           Glu-Gly-Pro-Asn-Ala-Leu-Glu-His-Ser-Leu-
           Ala-Ala-Arg-Ala-Glu-Leu-Glu-Tyr-Gly-Leu-
           Ala-Ala-Arg-Ala-Glu-Leu-Glu-His-Gly-Leu-


                                 35                    40
           Leu-Ala-Asp-Leu-Val-Ala-Ala-Glu-Lys-Lys-
           Val-Ala-Glu-Ala-Glu-Ala-Ala-Glu-Lys-Lys-
           Val-Ala-Glu-Ala-Gln-Ala-Ala-Glu-Lys-Lys-


                                 45                    50
           Asp-Glu-Gly-Pro-Tyr-Arg-Met-Glu-His-Phe-
           Asp-Ser-Gly-Pro-Tyr-Lys-Met-Glu-His-Phe-
           Asp-Glu-Gly-Pro-Tyr-Lys-Met-Glu-His-Phe-


                                 55                    60
           Arg-Trp-Gly-Ser-Pro-Pro-Lys-Asp-Lys-Arg-
           Arg-Trp-Gly-Ser-Pro-Pro-Lys-Asp-Lys-Arg-
           Arg-Trp-Gly-Ser-Pro-Pro-Lys-Asp-Lys-Arg-


                                 65                    70
           Tyr-Gly-Gly-Phe-Met-Thr-Ser-Glu-Lys-Ser-
           Tyr-Gly-Gly-Phe-Met-Thr-Ser-Glu-Lys-Ser-
           Tyr-Gly-Gly-Phe-Met-Thr-Ser-Glu-Lys-Ser-


                                 75                    80
           Gln-Thr-Pro-Leu-Val-Thr-Leu-Phe-Lys-Asn-
           Gln-Thr-Pro-Leu-Val-Thr-Leu-Phe-Lys-Asn-
           Gln-Thr-Pro-Leu-Val-Thr-Leu-Phe-Lys-Asn-


                                 85                          91
           Ala-Ile-Ile-Lys-Asn-Ala-Tyr-Lys-Lys-Gly-Glu-OH
           Ala-Ile-Ile-Lys-Asn-Ala-His-Lys-Lys-Gly-Gln-OH
           Ala-Ile-Val-Lys-Asn-Ala-His-Lys-Lys-Gly-Gln-OH
```

FIGURE 1.   Amino acid sequences of human, ovine (bovine) and porcine β-lipotropin.

A new form of β_s-LPH has been isolated by partition chromatography in agarose gel (25). All evidence is consistent with the conclusion that its structure is identical to that of β_s-LPH with the exception that the glutamic acid residue in position 1 was replaced by a pyroglutamic acid residue. The lipolytic activity of this new analog was about one-half that of β_s-LPH in isolated rabbit fat cells.

Another lipotropin (designated γ-LPH) with a structure corresponding to the 58 NH_2-terminal residues of β-LPH from sheep (26) and pig (27) glands has been isolated and characterized. It may be noted that the amino acid sequence of bovine β-melanotropin (28) is located in residues 41-58 of the β-LPH structure (see Figure 1).

Radioimmunoassay of β_h-LPH. Earlier immunological studies of β_s-LPH showed that the hormone is a poor antigen in rabbits (3). A sensitive radioimmunoassay (RIA) for β_s-LPH has subsequently been developed (4). Antiserum to the human hormone has recently been raised in rabbits and characterized by gel double diffusion, quantitative precipitin, microcomplement fixation and biological neutralization tests (29). Using this antiserum, an RIA system for β_h-LPH has been developed and shown the antiserum to be species specific (30,31). Human plasma concentrations of β_h-LPH were estimated by this RIA system to be 47.9 ± 5.7 pg/ml in 5 normal subjects (30). It appears that the plasma level of β_h-LPH in normal female subjects is higher than that of male subjects (31). It is of interest to note that plasma ACTH and β_h-LPH rose in parallel in response to insulin-induced hypoglycemia (30).

The half-life of β_h-LPH in adult male rats after a single intravenous injection of the hormone was found to be 4.2 ± 0.75 min as estimated by RIA (32). In the same study (32), the rate of disappearance of β_h-endorphin is slower ($t_{\frac{1}{2}}$ = 9.2 min) than that of β_h-LPH.

Biological properties of β-LPH. As a lipolytic agent, β-LPH is most active in rabbit adipose tissue, only weakly active in rat and appears to be inactive in mouse adipose tissue (3). Similar species specificity was observed on the levels of

serum free fatty acids when injected with β-LPH (33).
It stimulates lipolysis by activating the adenylate
cyclase system (34). In fasting mice, β-LPH exhibits fat mobilizing activity (3).

In addition to the lipolytic activity, β-LPH
has melanotropic activity of a potency similar to
ACTH (3). In the rabbit, it lowers plasma calcium
(35,36) and raises plasma phosphate (36). β-LPH
stimulates sebaceous gland activity in ovari-
ectomized-hypophysectomized rats (37).

Structure-Activity Relationship of β-LPH. The
lipolytic activity of γ_s-LPH is equal to that of
β_s-LPH (26). A fragment with a sequence of residues 1-65 (see Figure 1) obtained by CNBr cleavage
of the ovine hormone also exhibits comparable
activity (3). On the other hand, a fragment with
a sequence of residues 47-91 (see Figure 1)
possesses only one-tenth of the activity of β_s-LPH.
A synthetic peptide consisting of 50 amino acid
residues, β_s-LPH-(42-91) has about six times that of
β_s-LPH on a weight basis in isolated rabbit fat
cells (38).

Recently, Lemaire et al. (39) reported the
synthesis of β_s-LPH-(4$\overline{1}$-9$\overline{1}$), which spans the
complete amino acid sequences of both β-MSH and
β-endorphin, in good yield. The synthetic β_s-LPH-
(41-91) was homogenous as evidenced by partition
chromatography on agarose, thin-layer chromatography
in two solvent systems, paper electrophoresis at
pH 3.7 and 6.9, polyacrylamide gel electrophoresis
and amino acid analysis of acid and total enzymic
hydrolysates. The lipolytic activity of the
synthetic product in isolated rabbit fat cells is
5.4 times more active than the natural hormone on
a weight basis and it possesses approximately the
same potency of β_s-LPH-(42-91) as previously
reported (38). Table 1 summarizes the lipolytic
activity of various fragments of β_s-LPH. Apparently,
an extension of the hexa- or heptapeptide to the
NH_2-terminus of β-LPH-(47-91) greatly increases the
lipolytic activity. However, further extension
of the NH_2-terminal residues to Glu-1 lowers the
activity. The optimal number of residues that
could be added or subtracted from the NH_2-terminus
of β_s-LPH-(42-91) to give rise to maximal lipolytic
activity remains to be investigated.

TABLE 1. Lipolytic Potency of Various
Fragments of β_s-Lipotropin

Residue	Rabbit Adipose Tissue		Relative Potency[a]
1-91	pad/cells		1.0
1-47	pad		nil
1-58	pad		1.0
1-65	pad		1.1
66-91	pad		nil
41-91		cells	5.4
41-58	pad		$\simeq 10$
42-91		cells	6.0
48-65		cells	nil
47-91	pad		0.1
61-91		cells	nil

[a]On weight basis.

Total Synthesis of β_s-LPH. The total synthesis of β_s-LPH has now been achieved by Yamashiro and Li (40) employing the improved procedures of the solid-phase method. The anchoring linkage of the COOH-terminal glutamine was to brominated-styrene-1%-divinylbenzene polymer. This linkage when used in conjunction with N^α-t-butyloxycarbonyl protection has been shown to be cleaved only 0.03% per cycle and is approximately 15 to 50 times as stable as the standard linkage (41). Coupling was performed by a symmetrical anhydride technique which has been used to advantage for the β-LPH sequence (38). Trifluoroethanol was employed in the second stage of anhydride couplings to enhance the efficiency of this reaction (42). The side-chain of tryptophan was protected with the formyl group (43) in view of its susceptibility to destruction under conditions of repeated acidolysis of Boc groups. For histidine, the benzyloxycarbonyl group was used since it has been employed in synthesis of peptide analogs of the carboxyl terminal plasmin fragment of human somatotropin (44). For threonine, serine, and glutamic acid, the very stable p-halobenzyl protecting groups were used through residue 66

and benzyl protection thereafter. Selection of
the other protecting groups has been discussed
previously (45-49). Since methionine was not protected, the last Boc group was removed with TFA to
reduce t-butylation that occurs in HF (50).

The first purification step was accomplished
on CMC chromatography. A peak was detected close
to the position previously reported for the natural
hormone (18). It is evident from the recovery of
material that this procedure constituted the major
purification step in the scheme. The slower-moving
materials which represent the major side-products
of synthesis apparently consist of shorter sequences
as shown by both amino acid analysis and peptide
mapping. The formyl group on tryptophan was then
removed by brief treatment at pH 11.5 with NaOH
(51). Rechromatography on CMC gave a sharply defined peak as does natural β_S-LPH. The ultraviolet
spectrum of this material was identical with that
of the natural hormone in the region of 245-360 nm,
indicating that deformylation was complete.

Final purification was effected by partition
chromatography in a biphasic solvent system on
agarose. The R_f value of the major peak which is
a reflection of its distribution constant in the
solvent system was, within experimental error, indistinguishable from that of natural β_S-LPH. Under
these conditions, natural β_S-LPH gave R_f 0.285 and
< Glu^1-β_S-LPH gave R_f 0.37.

The final highly purified material was obtained after dialysis and CMC chromatography in
which a single sharp peak was obtained. The synthetic preparation was homogeneous and identical
to natural β_S-LPH on paper electrophoresis at two
pH values and on thin-layer chromatography. Amino
acid analysis of an HCl hydrolysate was in agreement with expected values. Amino acid analysis of
a total enzyme digest was in close agreement with
that previously reported for natural β_S-LPH (18).
The synthetic peptide showed only glutamic acid as
NH_2-terminal as does natural β_S-LPH. A peptide
map of a tryptic digest gave a pattern which was
close to that given by the natural hormone. Since
the behaviour of peptides on a map can vary from
one map to another, a map of a mixture of synthetic and natural materials was obtained. No
difference between the two could be discerned
including colorations of various peptides to
ninhydrin. The synthetic material showed behaviour

identical to that of natural β_S-LPH upon isoelectric focusing. It may be noted that natural
β_S-LPH and < Glu^1-β_S-LPH are not separable by
isoelectric focusing, presumably because no net
difference in charge exists.

The circular dichroism spectra of the synthetic
and natural hormones have been carried out. It was
found that the differences between the two spectra
are probably within the expected error as determined
by multiple runs of different preparations of a
single protein. The optical rotations of synthetic
hormone taken in a wavelength region of high sensi-
tivity were $[\alpha]_{300nm}^{27°}$ -604° and $[\alpha]_{250nm}^{27°}$ -1820°
at concentration 0.1% in 0.1 M Tris buffer of
pH 8.2. The corresponding values for natural
hormone were $[\alpha]_{300nm}^{27°}$ -590° and $[\alpha]_{250nm}^{27°}$ -1840°.

The biological activities of synthetic and
natural β_S-LPH were compared in isolated rabbit
fat cells as summarized in Table 2. The lipolytic
activities of the synthetic product is nearly identical to that for the natural hormone. In addition,
the immunoreactivity of the synthetic hormone is
the same as of the natural β_S-LPH as revealed by
both radioimmunoassay and complement fixation using
rabbit antiserum to the natural hormone.

TABLE 2. Lipolytic Activity of Synthetic β_S-LPH

β_S-LPH	Dose (µg/ml)	Glycerol production[a]
Natural	0.37	2.98 ± 0.07
	1.10	3.70 ± 0.13
Synthetic[b]	0.37	2.85 ± 0.05
	1.10	3.60 ± 0.11

[a] µmole/g of cells per hr; mean ± SE; determination
in triplicate.
[b] Relative potency to the natural hormone, 0.84
times, a confidence limit of 0.58-1.18 and λ=0.11.

Thus, the synthetic β_S-LPH has been found to
be indistinguishable from the natural hormone in
its R_f value on partition chromatography, mobility
in paper electrophoresis at two pH values, behaviour

on thin-layer chromatography, amino acid composition of both acid and enzymatic hydrolysates, NH_2-terminal residue, behaviour on peptide mapping, pI value in isoelectric focusing, circular dichroism spectra, optical rotation, lipolytic activity and immunoreactivity (Table 3).

TABLE 3. Properties of Synthetic β_s-LPH

Properties	Synthetic	Natural
Paper electrophoresis (R_f^{Lys})		
pH 3.7	0.56	0.56
pH 6.7	0.18	0.18
Thin Layer-chromatography[a] (R_f)	0.25	0.25
Partition chromatography[b]	0.275	0.285
NH_2-terminal residue	Glu	Glu
$[\alpha]_{589\ nm}^{27°}$	-110°	-105°
Circular dichroism[c] $[\theta]_\lambda$		
278 nm	-14,000	-11,700
220 nm	-25,000	-22,500
Isoelectric focusing[d], pI	6.33	6.33
Relative lipolysis potency[e]	0.84	1.0

[a]Solvent system: 1-BuOH/pyridine/HOAc/H_2O
   (5:5:1:4,v/v)
[b]Solvent system: 2-BuOH/H_2O/HOAc/10%
   Cl_3COOH/NaCl (150 nm/97 ml/4.3
   ml/7.5 ml/4.4 g)
[c]Solvent: 0.1 M HOAc
[d]On polyacrylamide gel
[e]Lipolytic activity in isolated rabbit fat
   cells

Isolation, Structure and Synthesis of β-Endorphin. During the course of isolation of melanotropins from camel pituitary glands (52), we were unable to find the existence of β-LPH, but obtained an untriakontapeptide (53) that has an amino acid sequence identical to the COOH-terminal 31-residues of ovine lipotropin [β_s-LPH-(61-91)] (See Figure 1). The peptide possesses very low

lipotropic activity but significant opiate
activity (53), as displayed in a preparation of
guinea pig ileum and in the opiate receptor binding
assay (54). This untriakontapeptide was designated
β-endorphin (β-EP) (53). A similar untriakontapeptide with opiate activity was obtained from
porcine (55,56), ovine (57), bovine (58) and
rat (59) pituitary glands.

Isolation of human β-EP has also been reported
(57,60). From 1,000 frozen human glands, only
3 mg of the peptide were obtained (60). Structural
analysis (60) established that the amino acid
sequence of the human peptide is identical to the
COOH-terminal 31-amino acid fragment of β_h-LPH
(see Figure 1). Figure 2 presents the amino acid
sequence of β-EP from various species. The only
variations occur in residues positions 23, 27
and 31.

Both camel (61) and human (62) β-EP have been
synthesized by the improved procedures of the
solid-phase method. The brominated-styrene-1%-
divinylbenzene polymer was used for the synthesis
of β_h-EP. Figure 3 presented the synthetic scheme
for the human β-EP. A yield of 32% was achieved
based on starting resin. The opiate activity of
the synthetic product was comparable to that of
the natural peptide by the guinea pig ileum assay.
It was found that the opiate activity of the synthetic camel and human β-EP is identical even
though their amino acid sequences are slightly
different (see Figure 2).

Synthetic Analogs of Human β-Endorphin with
Various Amino Acid Residues in Positions 1, 2, 4
and 5. Tables 4 and 5 summarized the in vivo
opiate activities of synthetic analogs with
various amino acid residues in positions 1, 2, 4
and 5 (63,64). It may be noted that substitutions
of Tyr[1] with D-isomer, residue in position 2 with
Sar, Ala, D-Leu and D-Lys, position 4 with D-Phe
and position 5 with D-Met, Pro, Leu and D-Leu
decrease the activity. Substitution of Gly[2] with
D-Ala retains full analgesic activity. The synthetic analog containing Leu-enkephalin segment has
only 17% analgesic activity of the parent molecule
and replacement Leu with D-isomer lowers the
activity to only 0.2%. Apparently, residue 5 in
the β-EP structure is more important for its
analgesic activity.

HUMAM:

H-Tyr-Gly-Gly-Phe-Met-Thr-Ser-Glu-Lys-Ser-
     5   10

Gln-Thr-Pro-Leu-Val-Thr-Leu-Phe-Lys-Asn-
      15   20

Ala-Ile-Ile-Lys-Asn-Ala-Tyr-Lys-Lys-Gly-Glu-OH
       25   31

PORCINE: Val   His   Gln-OH

CAMEL,OVINE,BOVINE: Ile   His   Gln-OH

AMINO ACID SEQUENCE OF HUMAN, PORCINE, CAMEL, OVINE, AND

BOVINE β-ENDORPHINS

FIGURE 2. The Amino acid sequence of β-endorphin from various species.

N^{α}-Boc-(Bzl)-Glu-Resin

> 1. TFA-CH$_2$Cl$_2$, 15 min.
> 2. N-methylmorpholine
> 3. preformed symmetrical anhydride
>   of Boc-Gly (CF$_3$CH$_2$OH to 20%)

Boc-Gly-(Bzl)-Glu-Resin

> 1. TFA-CH$_2$Cl$_2$, 15 min.
> 2. N-methylmorpholine
> 3. preformed symmetrical anhydride
>   of Boc-AA (CF$_3$CH$_2$OH to 20%)

Fully protected β_h-endorphin

> 1. TFA-CH$_2$Cl$_2$, 15 min.
> 2. N-methylmorpholine

Protected β_h-endorphin with free α-NH$_2$ group

> 1. HF, anisole, 0°, 75 min.

Crude β_h-endorphin

> 1. Sephadex G-10, 0.5 N HOAc
> 2. CMC chromatography
> 3. partition chromatography on G-50

$\underline{\beta_h}$-Endorphin (32% yield)

Resin, brominated styrene resin (1% divinylbenzene); side-chain
protecting groups: Lys, Z(o-Br); His, Boc; Asp, Bzl; Thr, Bzl;
Ser, Bzl; Glu, Bzl and Tyr (o-Br). Boc, t-butyloxycarbonyl;
Bzl, benzyl; TFA, trifluoroacetic acid; AA, amino acid.

FIGURE 3. The synthetic scheme of human
β-endorphin.

TABLE 4.   Opiate Activity of Synthetic
β_C-Endorphin with D-Amino Acids in
Positions 1, 2, 4 and 5

Synthetic peptides	Relative potency	
	g. pig ileum assay	ICV in mice
β_C-EP	100	100
[D-Tyr^1]-β_C-EP	<1	0.5
[D-Ala^2]-β_C-EP	43	100
[D-Phe^4]-β_C-EP	4	<0.3
[D-Met^5]-β_C-EP	4	1

TABLE 5.   Opiate Activity of Synthetic
β-Endorphin with Substitutions in
Positions 2 and 5

Synthetic peptides	Relative potency	
	g. pig ileum assay	ICV in mice
β_h-EP	100	100
[Sar^2]-β_C-EP	13	<0.3
[Ala^2]-β_C-EP	6	12
[D-Leu^2]-β_C-EP	12	48
[D-Lys^2]-β_C-EP	8	15
[Pro^5]-β_h-EP	<0.01	0.6
[Leu^5]-β_h-EP	18	17
[D-Leu^5]-β_h-EP	<0.01	0.2
[D-Ala^2, D-Leu^5]-β_h-EP	3	8

<u>Biological Properties of β-Endorphin</u>. In addition to the analgesic effect of β-EP, it exhibits a marked and prolonged state of catatonia (65-69) in rats when injected directly into the brain. β-EP has been shown to induce excess grooming behaviour in rats (70). Male sexual behaviour in rats was also found to be influenced by the peptide (71).

An intracerebroventricular dose of 94 μg β-EP in rats caused a profound hypothermia (69). The body temperature decreased at 30 min after injection and reached a maximum (4° lower than preinjection). A salivation effect of β-EP was also observed (72). Injections of β-EP into the anterior horn of the lateral ventricle in rats produced a profound increase in salivation along with a behavioral pattern described as sialogogic seizure.

In addition to its potent analgesic and behavioral activities, β-EP inhibits striatal dopamine release <u>in vitro</u> (73). It stimulates somatotropin (74,75) and prolactin (75,76) release when injected intracerebroventricularly in rats. β-Endorphin exhibits antidiuretic activity by intravenous injections in rats (77). This may be related to the observation that β-EP stimulates the release of vasopressin as measured by RIA (78).

<u>Synthesis of a Methionine Enkephalin with Potent Activity by Oral Administration</u>. Late in 1975, Hughes <u>et al.</u> (79) synthesized two pentapeptides in pig brain: H-Tyr-Gly-Gly-Phe-Met-OH (methionine enkephalin) and H-Tyr-Gly-Gly-Phe-Leu-OH (leucine enkephalin). Both peptides acted like morphine in the guinea pig ileum and mouse vas deferens preparations. Met-enkephalin is more potent than morphine in the receptor assay while leu-enkephalin is less potent than the methionine analog. Figure 1 shows that met-enkephalin is a peptide fragment of β-LPH corresponding to residues 61-65.

Comparison of the analgesic activity of met-enkephalin with leu-enkephalin shows the former to have measurable but fleeting activity while the latter shows none (81). A number of met-enkephalin analogs have been synthesized with an increase of its analgesic activity (81-88). Among these, [D-Thr^2,Thz^5]-enkephalinamide is most potent by oral administration (88).

TABLE 6. Median Antinociceptive Dose (AD_{50}) of [D-Thr^2,Thz^5]-Enkephalinamide, β$_h$-Endorphin and Morphine Sulfate After Various Routes of Administration.

Compound	I.C.V.[a] nmole/mouse	I.V.[a] μmole/kg	S.C.[b] μmole/kg	Oral[b] μmole/kg
[D-Thr^2,Thz^5]-Enkephalinamide	0.04	2.40	9.04(3.35-24.41)	9.11(5.69-22.42)
β$_h$-Endorphin	0.026	3.31	>47	—
Morphine sulfate	1.11	11.40	13.71(11.41-16.45)	15.81(9.83-25.46)

[a]Taken from (87)
[b]Taken from (88); 95% confidence limit in parentheses

The synthesis of [D-Thr^2,Thz^5]-enkephalinamide (87) was carried out by the solid-phase method. The synthetic product was homogenous on thin-layer chromatography, paper electrophoresis at pH 6.7 and pH 3.7, and amino acid analysis. When the analog was injected intravenously in mice, it was 4.8 times more potent than morphine. It is remarkable that [D-Thr^2,Thz^5]-enkephalinamide is 1.5-1.7 times more potent than morphine on a molar basis by oral administration (88). These data are summarized in Table 6. Römer et al. (86) recently reported that [D-Ala^2-Mephe^4-Met(0)^5]-enkephalin was active after enteric administration. The AD$_{50}$ of this analog was 102 mg/kg whereas [D-Thr^2,Thz^5]-enkephalinamide was less than 10 mg/kg. Thus, [D-Thr^2,Thz^5]-enkephalinamide is the most potent met-enkephalin analog ever reported by oral administration.

Summary and Concluding Remarks. In this paper, recent studies on β-lipotropin and β-endorphin are briefly reviewed. In addition, the most potent enkephalin analog, [D-Thr^2,Thz^5]-enkephalinamide, by oral administration, is presented. Among various fragments of β-LPH having opiate activity, only β-EP exhibits analgesic activity by intravenous injections. In addition, it is the most active peptide when administered directly into the brain. Recent clinical studies (89-92) indicate that synthetic β$_h$-EP (41) is active in patients with severe pain, narcotic abstinence, schizophrenic behaviour and deep depression by intravenous injections. Future structure-activity investigations of β$_h$-EP may lead to the development of a peptide with long-acting and/or oral active properties for extensive clinical studies.

ACKNOWLEDGMENTS

I thank Drs. J. Blake, W. C. Chang, D. Chung, B. A. Doneen, S. Lemaire, H. H. Loh, A. J. Rao, L-F. Tseng, and D. Yamashiro for their collaborations in these investigations.

REFERENCES

1. Li, C. H., Nature 201, 924 (1964).
2. Birk, Y., and Li, C. H., J. Biol. Chem. 239, 1048-1052 (1964).
3. Lohmar, P., and Li, C. H., Endocrinology 82, 898-904 (1967).
4. Desranleau, R., Gilardeau, C., and Chrétien, M., Endocrinology 91, 1004-1010 (1972).
5. Moon, H. D., Li, C. H., and Jenning, B. M., Anat. Rec. 175, 529-538 (1973).
6. Pelletier, G., Leclerc, R., Labrie, F., Cote, J., Chrétien, M., and Lis, M., Endocrinology 100, 770-776 (1977).
7. Watson, S. J., Barchas, J. D., and Li, C. H., Proc. Natl. Acad. Sci. USA 74, 5155-5158 (1977).
8. LaBella, F., Queen, G., Senyshyn, J., Lis, M., and Chrétien, M., Biochem. Biophys. Res. Commun. 75, 350-357 (1977).
9. Lohmar, P., and Li, C. H., Biochim. Biophys. Acta 148, 381-383 (1967).
10. Gilardeau, C., and Chrétien, M., Canad. J. Biochem. 48, 1017-1021 (1970).
11. Yudeau, M. A., and Pankov, Y. A., Probl. Endocrinol. 16, 49-51 (1970).
12. Cseh, G., Gráf, L., and Both, F., FEBS Lett. 2, 42-44 (1968).
13. Rubinstein, M., Stein, S., Gerber, L., and Udenfriend, S., Proc. Natl. Acad. Sci. USA 74, 3052-3055 (1977).
14. Scott, A. P., and Lowry, P. J., Biochem. J. 139, 593-602 (1974).
15. Chrétien, M., Gilardeau, C., Seidah, N., and Lis, M., Can. J. Biochem. 54, 778-782 (1976).
16. Li, C. H., and Chung, D., Nature 260, 622-624 (1976).
17. Li, C. H., Barnafi, L., Chrétien, M., and Chung, D., Nature 208, 1093-1094 (1965).
18. Li, C. H., Barnafi, L., Chrétien, M., and Chung, D., Excerpta Medica Int. Congr. Ser. 112, 334-364 (1966).
19. Chrétien, M., Gilardeau, C., and Li, C. H., Int. J. Pept. Prot. Res. 4, 263-265 (1972).
20. Gráf, L., and Li, C. H., Biochem. Biophys. Res. Commun. 53, 1304-1309 (1973).
21. Gráf, L., Barat, E., Cseh, G., and Sajco, M., Biochim. Biophys. Acta 229, 276-278 (1971).

22. Gilardeau, C., and Chrétien, M., in "Chemistry and Biology of Peptides" (J. Meienhofer, ed.) pp. 609-611. Ann Arbor Science Publ., 1972.

23. Pankov, Y. A., and Yudeau, M. A., Biokimia 37, 991-1031 (1972).

24. Li, C. H., Tan, L., and Chung, D., Biochem. Biophys. Res. Commun. 77, 1088-1093 (1977).

25. Yamashiro, D., and Li, C. H., Biochim. Biophys. Acta 451, 124-132 (1976).

26. Chrétien, M., and Li, C. H., Canad. J. Biochem. 45, 1153-1174 (1967).

27. Gráf, L., Cseh, G., and Medzihradszky-Schweiger, H., Biochim. Biophys. Acta 175, 444-447 (1969).

28. Geschwind, I. I., Li, C. H., and Barnafi, L., J. Am. Chem. Soc. 79, 6394-6401 (1957).

29. Rao, A. J., and Li, C. H., Int. J. Pept. Prot. Res. 10, 167-171 (1977).

30. Krieger, D. T., Liotta, A., and Li, C. H., Life Sciences 21, 1771-1778 (1977).

31. Wiedemann, E., Saito, T., Linfoot, J. A., and Li, C. H., J. Clin. Endocrinol. Metab. 45, 1108-1111 (1977).

32. Chang, W. C., Rao, A. J., and Li, C. H., Int. J. Pept. Prot. Res. 11, 93-94 (1978).

33. Tamasi, G., Cseh, G., and Gráf, L., Experientia 25, 360-361 (1969).

34. Lis, M., Gilardeau, C., and Chrétien, M., Proc. Soc. Exp. Biol. Med. 139, 680-683 (1972).

35. Lis, M., Gilardeau, C., and Chrétien, M., Acta Endocrinol. 59, 507-516 (1972).

36. Gildersleeve, D. L., Pearson, T. A., Baghdiantz, A., and Foster, G. V., Endocrinology 97, 533-534 (1975).

37. Thody, A. J., and Shuster, S., J. Endocrinol. 50, 533-534 (1971).

38. Yamashiro, D., and Li, C. H., Proc. Natl. Acad. Sci. USA 71, 4945-4949 (1974).

39. Lemaire, S., Yamashiro, D., and Li, C. H., Int. J. Pept. Prot. Res. 11, 193-199 (1978).

40. Yamashiro, D., and Li, C. H., J. Am. Chem. Soc., in press.

41. Li, C. H., Yamashiro, D., Tseng, L-F., and Loh, H. H., J. Med. Chem. 20, 325-328 (1977).

42. Yamashiro, D., Blake, J., and Li, C. H., Tetrahedron Lett. 18, 1469-1472 (1976).

43. Yamashiro, D., and Li, C. H., J. Org. Chem. 38, 2594-2597 (1973).

44. Blake, J., and Li, C. H., Int. J. Pept. Prot. Res. 11, in press (1978).
45. Yamashiro, D., Noble, R. L., Li, C. H., J. Org. Chem. 38, 3561-3565 (1973).
46. Yamashiro, D., J. Org. Chem. 42, 523-525 (1977).
47. Yamashiro, D., Noble, R. L., and Li, C. H., in "Chemistry and Biology of Peptides" (J. Meienhofer, ed.) pp. 197-202. Ann Arbor Science Publ., 1972.
48. Yamashiro, D., Blake, J., and Li, C. H., J. Am. Chem. Soc. 94, 2855-2859 (1972).
49. Yamashiro, D., and Li, C. H., J. Org. Chem. 38, 591-592 (1973).
50. Noble, R. L., Yamashiro, D., and Li, C. H., J. Am. Chem. Soc. 98, 2324-2328 (1976).
51. Lemaire, S., Yamashiro, D., and Li, C. H., J. Med. Chem. 19, 373-376 (1976).
52. Li, C. H., Danho, W. O., Chung, D., and Rao, A. J., Biochemistry 14, 947-952 (1975).
53. Li, C. H., and Chung, D., Proc. Natl. Acad. Sci. USA 73, 1145-1148 (1976).
54. Cox, B. M., Goldstein, A., and Li, C. H., Proc. Natl. Acad. Sci. USA 73, 1821-1823 (1976).
55. Bradbury, A. F., Smyth, D. G., and Snell, C. R., Biochem. Biophys. Res. Commun. 69, 950-956 (1976).
56. Gráf, L., Barat, E., and Patthy, A., Acta Biochim. Biophys. Acad. Sci. Hung. 11 (2-3), 121-122 (1976).
57. Chrétien, M., Benjannet, S., Dragon, N., Seidah, N. G., and Lis, M., Biochem. Biophys. Res. Commun. 72, 472-478 (1976b).
58. Li, C. H., Tan, L., and Chung, D., Biochem. Biophys. Res. Commun. 77, 1088-1093 (1977).
59. Rubinstein, M., Stein, S., and Udenfriend, S., Proc. Natl. Acad. Sci. USA 74, 4969-4972 (1977).
60. Li, C. H., Chung, D., and Doneen, B. A., Biochem. Biophys. Res. Commun. 72, 1542-1547 (1976).
61. Li, C. H., Lemaire, S., Yamashiro, D., and Doneen, B. A., Biochem. Biophys. Res. Commun. 71, 19-25 (1976).
62. Li, C. H., Yamashiro, D., Tseng, L-F., and Loh, H. H., J. Med. Chem. 20, 325-328 (1977a).

63. Yamashiro, D., Tseng, L-F., Doneen, B. A., Loh, H. H. and Li, C. H., Int. J. Pept. Prot. Res. $\underline{10}$, 159-166 (1977).

64. Yamashiro, D., Li, C. H., Tseng, L-F., and Loh, H. H., Int. J. Pept. Prot. Res. $\underline{11}$, in press (1978).

65. Jacquet, Y. F., Marks, N., and Li, C. H., in "Opiates and Endogenous Peptides" (H. Koster-litz, ed.), pp. 411-414. Elsevier/North Holland Biomedical Press, Amsterdam, 1976.

66. Jacquet, Y. F., and Marks, N., Science $\underline{194}$, 632-635 (1976).

67. Bloom, F., Segal, D., Ling, N., and Guillemin, R., Science $\underline{194}$, 630-632 (1976).

68. Motomatsu, T., Lis, M., Seidah, N., and Chrétien, M., Canad. J. Neurol. Sci. $\underline{4}$, 49-52 (1977).

69. Tseng, L-F., Loh, H. H., and Li, C. H., Biochem. Biophys. Res. Commun. $\underline{74}$, 390-396 (1977).

70. Gispen, V. M., Wiegand, V. M., Bradbury, A. F., Hume, E. C., Smyth, D. G., Snell, C. R., and DeWied, D., Nature $\underline{264}$, 794 (1976).

71. Meyerson, B. J., and Terenius, L., J. Pharmacology $\underline{42}$, 191-192 (1977).

72. Holaday, J., Loh, H. H., and Li, C. H., Life Sci., in press.

73. Loh, H. H., Brase, D. A., Sampath-Khanna, S., Mar, J. B., Way, E. L., and Li, C. H., Nature $\underline{264}$, 567-568 (1975).

74. Dupont, A., Cusan, L., Garon, M., Labrie, F., and Li, Proc. Natl. Acad. Sci. USA $\underline{74}$, 758-759 (1976).

75. Rivier, C., Vale, W., Ling, N., Brown, M., and Guillemin, R., Endocrinology $\underline{100}$, 238-241 (1977).

76. Dupont, A., Cusan, Labrie, F., Coy, D. A., and Li, C. H., Biochem. Biophys. Res. Commun. $\underline{75}$. 76-82 (1977).

77. Tseng-L-F., Wei, E. T., Loh, H. H., and Li, C. H., Science, in press.

78. Weitzmann, R. E., Fisher, D. A., Minick, S., Ling, N., and Guillemin, R., Endocrinology $\underline{101}$, 1643-1646 (1977).

79. Hughes, J., Smith. T. W., Kosterlitz, H. W., Fothergill, L. A., Morgan, B. A., and Morris, H. R, Nature $\underline{258}$, 577-579 (1975).

80. Chang, J. K., Fong, B. T. W., Pert, A., and Pert, C. B., Life Sci. 18, 1473-1482 (1976).

81. Hambrook, J. M., Morgan, B. A., Rance, M. J., and Smith, C. F. C., Nature 262, 782-783 (1976).

82. Pert, C. B., Pert, A., Chang, J. K, and Fong, B. T. W., Science 194, 330-332 (1976).

83. Bajusz, S., Ronai, A. Z., Székely, J. I., Dunai-Kóvacs, Zs., Berzetei, I., and Gráf, L., Acta Biochem. Biophys. Acad. Sci. Hung. 11, 305-309 (1976).

84. Bajusz, S., Ronai, A. Z., Székely, J. I., Gráf, L., Dunai-Kóvacs, Zs., and Burzetei, I, FEBS Letters 76, 91-92 (1977).

85. Coy, D. H., Kastin, A. J., Schally, A. V., Morin, O., Caron, N. G., Labrie, F., Walker, J. M., Fertel, R., Bertson, G. G., and Sandman, C. A., Biochem. Biophys. Res. Commun. 73, 632-638 (1976).

86. Römer, D., Büscher, H. H., Hill, R. C., Pless, J., Bauer, W., Cardinaux, F., Closse, A., Hauser, D., and Hugenin, R., Nature 268, 547-549 (1977).

87. Yamashiro, D., Tseng L-F., and Li, C. H., Biochem. Biophys. Res. Commun. 78, 1124-1129 (1977).

88. Tseng, L-F., Loh, H. H., and Li, C. H., manuscript in preparation.

89. Kline, N. S., Li, C. H., Lehmann, H. E., Lajtha, A., Laski, E., and Cooper, T., Arch. Gen. Psychiatry 34, 1111-1113 (1977).

90. Catlin, D. H., Hui, K. K., Loh, H. H., Li, C. H., Commun, Psychopharmacology 1, 493-500 (1977).

91. Su, C. Y., Lin, S. H., Wang, Y. T., Li, C. H., Loh, H. H., Chen, C. S., Hung, L. H., Lin, C. S., and Lin, B. C., J. Formosan Med. Assoc., in press.

92. Hosobuchi, Y., and Li, C. H., Commun. Psychopharmacology 2, 33-37 (1978).

MOLECULAR DYNAMICS AND REGULATION OF THE COMPLEMENT SYSTEM[1]

Hans J. Müller-Eberhard[2]

Department of Molecular Immunology
Research Institute of Scripps Clinic
La Jolla, California

INTRODUCTION

Complement (C) constitutes an integral part of the immune system. It consists of a set of proteins that occurs in plasma in inactive, but activatable form. These proteins have the potential to generate biological activity by entering into complex protein-protein interactions with each other. In the course of these interactions indigenous C enzymes are assembled and activated, and fission and fusion products of C proteins are formed. Although C can function in cell-free solution, it has the unusual ability to transfer itself from solution to the surface of biological particles and to function as a solid phase enzyme system. The capacity of C to mark and prepare particles for ingestion by phagocytic cells and its potential to attack and lyse cells is based on this ability to transfer from fluid phase to solid phase. The firm

---

[1]This is publication number 1495 from the Research Institute of Scripps Clinic. This work was supported by United States Public Health Service Grants AI 07007 and HL 16411.
[2]Dr. Müller-Eberhard is the Cecil H. and Ida M. Green Investigator in Medical Research, Research Institute of Scripps Clinic.

binding of C molecules to the surface of cells and
other biological particles is afforded by enzymatic
generation of transient binding sites on several C
proteins. The cytolytic activity of C is due to
the strong interaction of its membrane attack
complex with cell membrane lipids.

Fragments arising from enzymatic cleavage of C
molecules express a plethora of biological activities. Thus, C is increasingly emerging as a
humoral system capable of generating molecular
effectors of cellular functions. Cells responding
to the stimuli of C reaction products include poly-
morphonuclear leukocytes, monocytes, lymphocytes,
macrophages, mast cells, smooth muscle cells and
platelets. *In vivo*, C participates together with
antibodies and various cellular elements in host
defense against infections. This physiological
function of C has become apparent through the
observation and clinical study of genetic C defi-
ciencies in man. C also participates actively in
inflammation and immunologic tissue injury, as it
occurs in autoimmune diseases.

The regulation of the C system is as complex as
the system itself. The enzymes are either labile
or susceptible to inhibition or disassembly by
regulatory proteins of serum. The transiently
revealed binding regions of certain C molecules
decay spontaneously, or, as in the case of the
membrane attack complex, can be blocked by specific
serum inhibitors. At least four biologically
active complement fragments are under the control
of serum exo- or endopeptidases . Regulation in
favor of the system is exercised by properdin which
increases the effectiveness of two C enzymes.
Background information on the molecular biology and
biochemistry of complement may be obtained from
recent reviews (1-8).

THE PROTEINS

To date, twenty distinct proteins have been
recognized as constituents of the C system: four-
teen are considered components proper and six are
regulatory proteins (Table I). The components
include five proenzymes, Clr, Cls, C2, Factor B
and Factor D. These are serine proteases inasmuch
as their activated forms are inhibitable by DFP.

TABLE I. The 20 Proteins of the Complement System

Components

     Zymogens (5):   C1r, C1s, C2, B, D

     Non-Enzymes (9):   C1q, C3, C4, C5, C6, C7, C8, C9, P

Regulators

     Enzymes (2):   C3bINA, SCPB(AT-INA)

     Non-Enzymes (4):   β1H, C1 INH, S-protein, C1qINH

The nine non-proteolytic enzyme components include C1q, the recognition protein of the classical pathway, C3, the most versatile component of the C system, C4, the modulator of C2, C5,6,7,8 and 9, the precursors of the membrane attack complex, and properdin (P), the modulator of the alternative C3/C5 convertase.

Among the regulators two are enzymes. The endopeptidase C3b inactivator (C3bINA) cleaves and inactivates C3b and C4b. It is greatly enhanced in this function by β1H, its cofactor (9). The exopeptidase, serum carboxypeptidase B (SCPB), also called anaphylatoxin inactivator (AT-INA), removes the COOH-terminal arginine residue from the two anaphylatoxins and thereby abrogates their activities (10). This enzyme also inactivates bradykinin. The C1q inhibitor (C1qINH) is a little characterized protein that is capable of physically binding to C1q and thereby interfering with its function (11). The C1 inhibitor (C1 INH) blocks the activity of C1r̄* and C1s̄ by firmly combining with these enzymes (2). The S-protein is synonymous with the membrane attack complex (MAC)

---

* The bar denotes the activated form of a complement enzyme.

inhibitor (MAC-INH) and blocks the membrane binding
region of the forming MAC (12). β1H, in addition
to being a cofactor of C3bINA, by itself causes
active disassembly of the alternative C3/C5
convertase (13).

Some of the properties of the components and
the regulatory proteins are listed in Tables II-V.
Clq has the largest molecular size and is the most
complex. It is a collagen-like protein that is
composed of eighteen polypeptide chains. Three
similar, but distinct chains of 20,000 to 22,000
dalton occur six times in the molecule, forming
six structural and functional subunits (5,14).
The amino acid sequence of the chains as well as
the ultrastructure and chemical subunit composition
of the protein have been elucidated (15). Clr
consists of two apparently identical, non-
covalently linked chains (2). Cls is a single
chain protein that dimerizes in presence of
Ca^{++} (2). C2 has two free sulfhydryl groups that
upon oxidation with iodine form an intramolecular
disulfide bond. As a result, C2 activity in the
C reaction increases up to twenty-fold (1). C2
is also the precursor of a kinin-like peptide (16).
C3 is the precursor of several biologically active
fragments (17). It supplies one of the two anaphylatoxins (C3a), an essential subunit (C3b)

TABLE II. Proteins of the Classical Human
Complement System

Protein	Molecular weight	Number of chains	Electroph. mobility	Conc. (μg/ml)
Clq	400,000	6x3	γ_2	65
Clr	190,000	2	β	50
Cls	88,000	1	α	40
C2	117,000	1	β_1	25
C3	180,000	2	β_2	1600
C4	206,000	3	β_1	640

TABLE III. Proteins of the Classical Human Complement System

Protein	Molecular weight	Number of chains	Electroph. mobility	Conc. (μg/ml)
C5	180,000	2	β_1	80
C6	128,000	1	β_2	75
C7	121,000	1	β_2	55
C8	154,000	3	γ_1	55
C9	79,000	1	α	60

of three C enzymes, the major opsonin (C3b) and a leukocytosis factor (C3e) (18). C5 supplies the second anaphylatoxin which is also a potent chemotactic factor (C5a), as well as the nucleus for MAC assembly (C5b) (8,17). Factor B is the precursor of two fragments, Ba, a chemotactic factor (19), and Bb, a subunit of the alternative C3/C5 convertase, and, by itself, a macrophage spreading factor (20). Properdin is composed of four apparently identical subunits which are held together by non-covalent forces (21,22).

Several pairs of C proteins may have arisen from a common ancestral protein. Clr and Cls are very similar in amino acid composition and overall structure and function. C2 and Factor B are both linked to the HLA system and their respective genes are located on chromosome six in man (23,24). The substrate and bond specificity of C2 and Factor B are identical. C3 and C5 have been shown to be homologous in primary structure, at least as far as their activation peptides, C3 and C5a are concerned (25). C6 and C7 are similar in function, molecular size and amino acid composition.

The sites of biosynthesis of C proteins are actively being explored. Macrophages synthesize Clq, Clr, Cls, C2, C3, C4, Factors B, D and properdin (26,27). Fibroblasts, intestinal

TABLE IV.   Proteins of the Human Properdin System

Protein	Symbol	Mol.wt.	Number of chains	Electroph. mobility	Conc. (μg/ml)
C3	C3	180,000	2	β	1600
Proactivator	B	93,000	1	β	200
Proactivator Convertase	D	24,000	1	α	2
Properdin	P	184,000	4	γ	20
C3b Inactivator	C3bINA	88,000	2	β	34
β1H	β1H	150,000	1	β	500

TABLE V. The Regulatory Proteins of the Complement System

Protein	Symbol	Mol.wt.	Number of chains	Electroph. mobility	Conc. (μg/ml)
C$\bar{1}$ inhibitor	C$\bar{1}$ INH	105,000	1	α	240
C1q inhibitor	C1qINH	85,000	1	β	20
Membrane attack complex inhibitor	MAC-INH, S	71,000	1	α	600
Antithrombin III	AT III	64,000	1	α	230
Anaphylatoxin inhibitor	AT-INH, SCPB	310,000	Multiple	α	50
C3b inactivator	C3bINA	80,000	2	β	34
β1H	β1H	150,000	1	β	530

epithelium, peripheral mono- and lymphocytes also
synthesize some of the C proteins.

Properties of the regulatory proteins are
listed in Table V. The MAC inhibitor is a relatively new protein (12). It is only weakly immunogenic and therefore antisera prepared to whole
human serum do not contain anti-MAC-INH, although
the protein occurs in human serum at a concentration of 600 µg/ml. Antithrombin III (AT III)
shares with MAC-INH the ability to block the
membrane binding site of MAC (29). The anaphylatoxin inactivator is a high molecular weight
carboxypeptidase B (10).

All of the proteins described in Tables II-V
can be obtained in highly purified form and are
available in this laboratory for the experimental
elucidation of the molecular mechanisms and
biological effects of the C system.

THE PATHWAYS

The C proteins organize themselves to form two
pathways of activation, the classical and the
alternative, which merge into the common, terminal
pathway of membrane attack. The two pathways of
activation have a similar molecular organization
(Fig. 1). An initial enzyme catalyzes the formation of the target bound C3 convertase which in
turn catalyzes the formation of the target bound
C5 convertase. The C5 convertase of either pathway, by cleaving C5, can set in motion the self-
assembly of the membrane attack complex, C5b-9.

The classical pathway (CP) is activated by
immune complexes of the IgG and IgM type. The
alternative pathway (AP) is activated by bacterial
and fungal polysaccharides and lipopolysaccharides
in particulate form (6). It is also activated by
certain immunoglobulin aggregates, i.e., guinea
pig γ_1 (30), human IgA and IgG in conjunction with
cells infected with measles, influenza or herpes
simplex virus (31). Experimentally convenient
activators of the human AP are particulate inulin,
zymosan (6) or rabbit erythrocytes (32).

Several biologically important entities arise
as byproducts of the C reaction. Both pathways
generate C3a and C5a, the two anaphylatoxins (17),
and C3b, the opsonin of the C system. The CP
also elaborates a kinin-like peptide from C2.

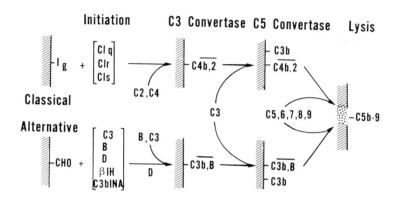

FIGURE I.   Complement pathways.

The AP liberates Ba, the activation fragment of
Factor B, which has chemotactic activity for poly-
morphonuclear leukocytes (19), and Bb, which is
the decay product of the AP enzymes and a macrophage spreading factor (20).

THE METASTABLE BINDING SITE

Transfer of C proteins from solution to the
surface of target particles is accomplished
through activation of metastable binding sites (1).
These binding sites are transiently revealed by
the respective activating enzymes. Apparently
cleavage of a critical peptide bond leads to dissociation or dislocation of the activation fragment
of a given component and to exposure of structures
that are concealed in the native molecule. Owing
to the revealed site, a molecule can bind to a
suitable acceptor and establish a firm association
with it. Failing collision with the acceptor
within an instant after activation, the site decays
and the molecule remains unbound in the fluid
phase. It cannot be activated again. The reaction
may be represented as follows:

$$C \xrightarrow{\quad E \quad} C^* \xrightarrow{\qquad} C\text{-}A$$

$$C^* \xrightarrow{\qquad\qquad} C_i$$

where C denotes the native molecule, E the activating enzyme, C^* the activated component, C-A the acceptor bound molecule and C_i the decayed and unbound form. Through this mechanism, activated C3 and C4 can bind directly to biological membranes, activated C2 and Factor B to their acceptors C4b and C3b, respectively, and activated C5 to C6. An additional metastable binding site is generated by a non-enzymatic mechanism. Collision of the bimolecular complex C5b,6 with C7 results in formation of the trimolecular C5b,6,7* complex which for a few milliseconds has the ability to bind to biological membranes (see below).

THE CLASSICAL PATHWAY (CP)

Activation and regulation of Cl, the first component of the CP, may be described as shown in Fig. 2. Cl is a Ca^{++} dependent complex of Clq, Clr and Cls. The subcomponents are present in equimolar amounts. Through Clq, which has six or more stable binding sites for immunoglobulin (33), Cl can attach to immune complexes or to antibody molecules on the surface of a target cell. Cl binding is reversible. It leads to activation of Clr within the Cl complex. Both chains of Clr are cleaved in the process into 55,000 and 35,000 dalton fragments which are linked by S-S bonds. Since Clq is not known to be an enzyme, the mechanism of the enzymatic activation of Clr is not understood (34). Clr̄ activates Cls by cleavage of both chains of the Cls dimer into fragments that are similar in size to the Clr̄ fragments and are likewise linked by S-S bonds. Thus, one molecule of enzyme can act on only one molecule of substrate.

Activation results in labilization of the Cl complex and to its becoming susceptible to the action of Cl INH. The inhibitor efficiently binds to both Clr̄ and Cls̄, thereby totally abrogating their enzymatic activity.

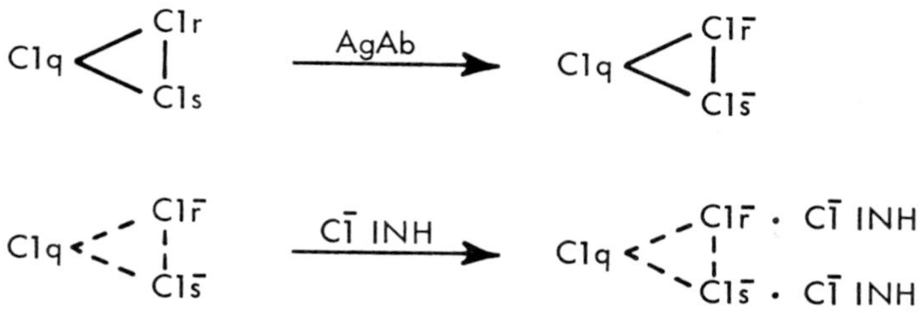

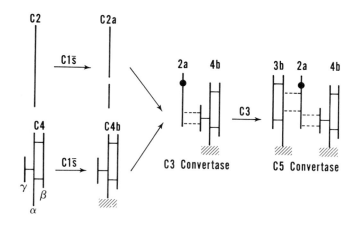

FIGURE 2.   Activation and control of Cl.

FIGURE 3.   Formation of the C3 and C5
convertases of the classical pathway.

The assembly of the classical C3 and C5 convertases is schematically depicted in Fig. 3.
Attack of C4 by Cl$\overline{s}$ in the Cl complex activates
the metastable binding site in C4 and allows C4b
to attach to the target cell surface.   Cleavage of
C2 by Cl$\overline{s}$ liberates the C2a fragment (85,000
dalton) and allows it to bind firmly to C4b.   The
γ-chain (33,000 dalton) of the three-chain C4
molecule is primarily involved in C2a binding.
The four-chain 280,000 dalton complex has C3
cleaving activity and is called C3 convertase.   The
enzymatic site is located in the C2a subunit (2).
The enzyme cleaves the bond between residue 77
and 78 of the α-chain of C3.   Dissociation of C3a

(9000 dalton) allows C3b to attach to acceptors
on the surface of the target cell. A C3b molecule
bound in the immediate micro-environment of a
C4b,2a complex confers on this enzyme the ability
to cleave and activate C5. Cr convertase thus
consists of the six-chain 450,000 dalton C4b,2a,3b
complex.

Formation and function of the enzymes are
rigidly controlled. Three regulatory mechanisms
are apparent. First, the metastable binding site
of activated C4 imposes a spatial constraint on
the location of the forming enzyme. Second, decay-
dissociation of C2a from both the C3 and the C5
convertase limits their function in time. Third,
C3b and C4b can be irreversibly inactivated by the
enzyme C3bINA so that they can no longer serve as
enzyme subunit. This reaction is thirty-fold
enhanced by the presence of β1H (35).

THE C3b-DEPENDENT FEEDBACK AND ITS REGULATION

Basic to the understanding of the alternative
pathway is the C3b initiated positive feedback
mechanism (36) (Fig. 4). A molecule of C3b and a
molecule of Factor B form, in presence of Mg^{++},
the loose bimolecular complex C3b,B. In complex
with C3b, Factor B is presented such that Factor D
is able to cleave Factor B, which results in formation of the C3 convertase of the alternative pathway, C3b,Bb. In acting upon C3, the enzyme
supplies in a short period of time many molecules
of C3b, each of which initiates the formation
of a molecule of C3 convertase provided the supply
of Factor B is not limiting. Since Factor D is
not incorporated into the enzyme complex, it can
activate many C3b,B complexes (37). In its
uncontrolled form, the process resembles a chain
reaction.

Regulation is provided by C3bINA and by β1H
(Fig. 5). β1H binds to C3b and that binding is
competitive with that of Bb. As a result, β1H
disassembles C3b,Bb, dissociating Bb in inactive
form (38,13). In complex with β1h, C3b is readily
cleaved and inactivated by C3bINA (9). With the
formation of inactive C3b(C3b$_i$), β1H is released
and both control proteins are free to attack the
next molecule of enzyme or of C3b.

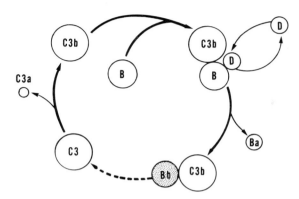

FIGURE 4.   C3b-dependent positive feedback.

(1)   C3b + β1H ⟶ C3b,β1H

(2) $\overline{C3b,B}$ + β1H ⟶ C3b,β1H + B$_i$

(3)   C3b,β1H + C3bINA ⟶ C3b$_i$ + β1H + C3bINA

FIGURE 5.   Control of the alternative pathway
by β1H and the C3b inactivator.

THE ALTERNATIVE PATHWAY (AP)

The AP proceeds essentially in three steps
(Fig. 1). First, at or near the surface of an
activating particle the initial enzyme is assembled,
which is a C3 convertase. The assembly of the initial enzyme requires native C3, Factors B, D and
Mg^{++}, but not properdin. It is the apparent
function of the enzyme to cleave C3 and to deposit
C3b on the surface of the particle under attack.
Second, at the site of C3b deposition, the solid
phase C3 convertase is assembled. Factor B first
forms a Mg^{++} dependent inactive complex with the
bound C3b which enables Factor D to cleave Factor
B into the 63,000 dalton Bb fragment and the
30,000 dalton Ba fragment. The latter is

dissociated and the active complex C3b,Bb is formed.
The assembled C3 convertase is a three-chain
233,000 dalton enzyme with its active site being
located in the Bb subunit (Fig. 6). Third, by
increasing the multiplicity of target bound C3b,
C5 convertase activity as well as immune adherence
reactivity are generated. Through the action of
the C5 convertase on C5, complement dependent
cytolysis is initiated. The last, and apparently
unessential, event of the sequence is the recruit-
ment of properdin (P). Upon collision of native P
with the solid phase C3/C5 convertase, P is
adsorbed to the enzyme in a non-enzymatic reaction.
The function of P is to lend physical stability to
the enzyme, increasing its half-life at 37° for
1.5 min. to 8-10 min. The described sequence of
molecular events was reported in 1976 (39-41) and
has since been further elaborated upon with regard
to control and initiation.
     The ability of the AP to proceed on the surface
of activating particles depends entirely on the
initial deposition of C3b, which is accomplished
by the initial C3 convertase. How this enzyme
is generated is still uncertain. Several hypotheses have invoked additional factors, such as a
special class of antibody, a Clq analogue, or even
properdin. These models of initiation are predi-
cated on the assumption that without an activator,
there is no initial enzyme formation. Another

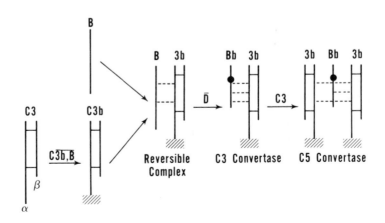

FIGURE 6. Formation of the C3 and C5
convertases of the alternative pathway.

model invokes spontaneous C3 turnover with random
attachment of C3b to nearby particles as the
initial events (42). This model lacks an entity
that allows to discriminate between an activator
and a non-activator surface. If, however, the
surface of an activator is capable of restricting
the control by the regulatory proteins, then
focusing of C3b deposition on activators becomes
possible by default of control. Such deregulation
of AP control on the surface of activators has been
shown for rabbit erythrocytes, *E. coli* and yeast
cell walls (43,44). This restriction of control
could be explained entirely by reduced accessibility
to β1H of C3b bound to the surface of activators (45).
Regulation of the AP is exerted by multiple
mechanisms: 1) spontaneous decay of C3/C5 convertase, 2) stabilization of the enzyme by properdin,
3) disassembly of the enzyme by β1H, 4) inactivation of C3b, C3bINA and β1H, and 5) restriction of
β1H control on the surface of certain AP activators.

THE TERMINAL PATHWAY OF MEMBRANE ATTACK

Complement dependent cytolysis is entirely
a function of the C5b-9 complex, also designated
membrane attack complex (MAC) (46,47). Cleavage
of C5 by either the classical or the alternative
C5 convertase constitutes the biochemical signal
for the self-assembly of MAC. After dissociation
of C5a (anaphylatoxin and chemotoxin), nascent C5b
forms a stable bimolecular complex with C6. The
C5b,6 complex in binding C7 becomes a trimolecular
complex which for an instant has the ability to
bind to the surface of biological membranes. Subsequently, C8 and C9 are bound to C5b-7 (Fig. 7).
As the assembly of the complex progresses, the
ability of the complex to bind detergent or phospholipid increases (48). Present evidence suggests
that the nascent C5b-7 complex overcomes the
charge barrier of a biological membrane and that
the fully assembled MAC interacts with the hydrophobic interior of such membranes (49,50). It
is proposed that as the forming MAC establishes
contact with the target membrane, it firmly binds
and thereby rearranges the phospholipid molecules

in its immediate microenvironment. As a result,
ultrastructural and functional membrane impairment
ensues.

The forming MAC is controlled by the rapid
decay of its membrane binding potential and by the
ability of the MAC-INH to block this binding
site (12).

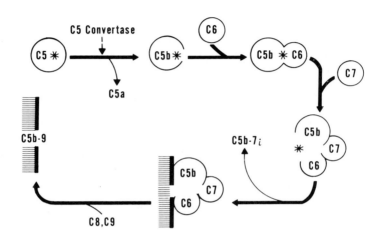

FIGURE 7.   Assembly of the membrane attack
complex.

REFERENCES

1.   Müller-Eberhard, H. J., Ann. Rev. Biochem.
     44:697 (1975).
2.   Cooper, N. R., and Ziccardi, R. J., in
     "Proteolysis and Physiological Regulations,"
     Miami Winter Symposia 11 (D. W. Ribbons and
     K. Breed, eds.), p. 167. Academic Press,
     New York, 1976.
3.   Müller-Eberhard, H. J., in "Textbook of
     Immunopathology," Second Edition (P. A.
     Miescher and H. J. Müller-Eberhard, eds.),
     p. 45. Grune and Stratton, New York, 1976.
4.   Osler, A. G., in "Foundations of Immunology
     Series." Prentice-Hall, Inc., Englewood
     Cliffs, New Jersey, 1976.

5. Reid, K.B.M., and Porter, R. R., in "Contemporary Topics in Molecular Immunology," Vol. 4 (F. P. Inman and W. J. Mandy, eds.), p. 1. Plenum Press, New York, 1975.

6. Götze, O., and Müller-Eberhard, H. J., Adv. Immunol. 24:1 (1976).

7. "Comprehensive Immunology," Vol. 2, "Biological Amplification Systems in Immunology" (N. K. Day and R. A. Good, eds.). Plenum Publishing Co., New York, 1977.

8. Hugli, T. E., and Müller-Eberhard, H. J., Adv. Immunol., in press (1978).

9. Pangburn, M. K., Schreiber, R. D., and Müller-Eberhard, H. J., J. Exp. Med. 146:257 (1977).

10. Bokisch, V. A., and Müller-Eberhard, H. J., J. Clin. Invest. 49:2427 (1970).

11. Ghebrehiwet, B., and Müller-Eberhard, H. J., J. Immunol. 120:27 (1978).

12. Podack, E. R., Kolb, W. P., and Müller-Eberhard H. J., Fed. Proc. 36:1209 (1977).

13. Whaley, K., and Ruddy, S., J. Exp. Med. 144:1147 (1976).

14. Calcott, M. A., and Müller-Eberhard, H. J., Biochemistry 11:3443 (1972).

15. Porter, R. R., Fed. Proc. 36:2191 (1977).

16. Donaldson, V. H., and Rosen, F. S., Personal Communication.

17. Müller-Eberhard, H. J., in "Molecular and Biological Aspects of the Acute Allergic Reactions," Nobel Symposium 33, p. 339. Plenum Publishing Co., New York, 1976.

18. Ghebrehiwet, B., and Müller-Eberhard, H. J., J. Exp. Med., in press (1978).

19. Hadding, U., Hamuro, J., and Bitter-Suermann, D., Abstract, Seventh Intl. Complement Workshop, J. Immunol., in press (1978).

20. Götze, O., Bianco, C., and Cohn, Z. A., Abstract, Seventh Intl. Complement Workshop, J. Immunol., in press (1978).

21. Götze, O., Medicus, R. G., and Müller-Eberhard, H. J., J. Immunol. 118:525 (1977).

22. Minta, J. O., and Lepow, I. H., Immunochemistry 11:361 (1974).

23. Fu, S. M., Kunkel, H. G., Brusman, H. P., Allen, F. H., Jr., and Fotino, M., J. Exp. Med. 140:1108 (1974).

24. Allen, F. H., Jr., Vox Sang. 27:382 (1974).

25. Fernandez, H. N., and Hugli, T. E., J. Biol. Chem. 252:1826 (1977).
26. Colten, H. R., Adv. Immunol. 22:67 (1976).
27. Loos, M., Müller, W., and Storz, R., Abstract, Seventh Intl. Complement Workshop, J. Immunol., in press (1978).
28. Brade, V., Fries, W., and Bentley, C., Abstract, Seventh Intl. Complement Workshop, J. Immunol., in press (1978).
29. Curd, J. G., Podack, E. R., Sundsmo, J. S., Griffin, J. H., and Müller-Eberhard, H. J., J. Clin. Invest., in press (1978).
30. Sandberg, A. L., Oliveira, B., and Osler, A.G., J. Immunol. 106:282 (1971).
31. Joseph, B. S., Cooper, N. R., and Oldstone, M.B.A., J. Exp. Med. 141:774 (1975).
32. Platts-Mills, T.A.E., and Ishizaka, K., J. Immunol. 113:348 (1974).
33. Schumaker, V. N., Calcott, M. A., Spiegelberg, H. L., and Müller-Eberhard, H. J., Biochemistry 15:5175 (1976).
34. Ziccardi, R. J., and Cooper, N. R., J. Immunol. 116:504 (1976).
35. Pangburn, M. K., and Müller-Eberhard, H. J., Abstract, Seventh Intl. Complement Workshop, J. Immunol., in press (1978).
36. Müller-Eberhard, H. J., and Götze, O., J. Exp. Med. 135:1003 (1972).
37. Lesavre, P., and Müller-Eberhard, H. J., Fed. Proc., in press (1978).
38. Weiler, J. M., Daha, M. R., Austen, K. F., and Fearon, D. T., Proc. Natl. Acad. Sci. USA 73:3268 (1976).
39. Medicus, R. G., Schreiber, R. D., Götze, O., and Müller-Eberhard, H. J., Proc. Natl. Acad. Sci. USA 73:612 (1976).
40. Schreiber, R. D., Götze, O., and Müller-Eberhard, H. J., J. Exp. Med. 144:1062 (1976).
41. Medicus, R. G., Götze, O. and Müller-Eberhard, H. J., J. Exp. Med. 144:1076 (1976).
42. Lachmann, P. J., and Halbwachs, L., Clin. Exp. Immunol. 21:109 (1975).
43. Fearon, D. T., and Austen, K. F., Proc. Natl. Acad. Sci. USA 74:1683 (1977).
44. Fearon, D. T., and Austen, K. F., J. Exp. Med. 146:22 (1977).

45. Pangburn, M. K., and Müller-Eberhard, H. J.,
    Proc. Natl. Acad. Sci. USA, in press (1978).
46. Kolb, W. P., and Müller-Eberhard, H. J.,
    J. Exp. Med. 138:438 (1973).
47. Mayer, M. M., Scientific American 54:229
    (1973).
48. Podack, E. R., Halverson, C., Esser, A. F.,
    Kolb, W. P., and Müller-Eberhard, H. J.,
    Abstract, Seventh Intl. Complement Workshop,
    J. Immunol., in press (1978).
49. Esser, A. F., and Müller-Eberhard, H. J.,
    Abstract, VIIth Intl. Conference on Magnetic
    Resonance in Biological Systems, Quebec,
    Canada (1976).
50. Michaels, D. W., Abramowitz, A. S., Hammer,
    C. H., and Mayer, M. M., Proc. Natl. Acad.
    Sci. USA 73:2852 (1976).

FIBRINOGEN: A HIGHLY EVOLVED REGULATORY

AGENT FOR MAINTAINING THE INTEGRITY

OF THE VERTEBRATE CIRCULATORY SYSTEM[1]

R. F. Doolittle

K. W. K. Watt

B. A. Cottrell

T. Takagi[2]

Department of Chemistry
University of California, San Diego
La Jolla, California

I. INTRODUCTION

The blood of vertebrate animals serves indispensable transport and communication functions, but, because of its very fluidity, it is constantly in danger of being lost should a defect occur in the containment system. It is not surprising, then, that regulatory processes evolved which are aimed at preventing blood loss under various traumatic circumstances. These protective measures include muscle contractions which throttle down the blood supply to injured tissues, the plugging of small holes or breaks with sticky cells or cell fragments (platelets), and finally the formation of a gelatinous plug, made of the material called fibrin. It is this latter aspect with which we will deal in this article. Our intention is to show how a simple protein

---

[1]Supported by NIH Grants HL 18,576 and GM 17,702.

[2]Present address: Biological Institute, Tohoku Univ., Sendai, Japan.

molecule, fibrinogen, can regulate the fluidity of the blood
merely by exhibiting preferential susceptibility to a pair of
highly specific proteases--thrombin on the one hand, and plasmin on the other. Thus, the snipping of two-four bonds by
thrombin can transform this nicely soluble molecule into a
form which spontaneously polymerizes into the insoluble fibrin
gel. Ultimately, the gel itself will be dissolved by further
proteolysis inflicted by plasmin. Moreover, the resulting
fibrin degradation products can slow or prevent the polymerization of more fibrin, and/or--depending on the circumstances--
stimulate the biosynthesis of more fibrinogen.

   This overall scheme of fibrin formation and dissolution
has been generally accepted for the better part of a generation. The details of the process have until recently remained
mysterious, however, primarily because the structure of the
fundamental fibrin(ogen) unit was not known, or at least not
with any certainty or precision. During the past few years a
plethora of primary and secondary structure data have been assembled in several laboratories. When cast with earlier notions about the three-dimensional structure based on physico-
chemical measurements, certain electron microscope observations, and some insightful interpretations of various enzymatic and chemical fragmentation patterns, these data allow a
quite detailed model of the molecule to be developed which
offers remarkably simple and satisfying interpretations of
polymerization and fibrinolysis. The model is also completely
consistent with--indeed, was partly built on--observations
involving clot stabilization, a phenomenon whereby covalent
bonds are introduced between fibrin units by the transglutaminase designated factor XIII.

II. AMINO ACID SEQUENCE STUDIES

   Vertebrate fibrinogen molecules are composed of three
pairs of non-identical chains $(\alpha_2\beta_2\gamma_2)$, all of which are inter-
connected by disulfide bonds. In all the species examined so
far, γ-chains have molecular weights of about 47,000, β-chains
are somewhat larger (ca. 55,000) and α-chains are the largest,
ranging upwards from 60,000 depending on the organism (1).
The overall molecular weight of the native molecule for most
species is about 340,000.

   In recent years major efforts have been under way in several laboratories to determine the primary structure of the
three chains as they occur in the human molecule. The bulk
of these studies has been conducted in three laboratories:
Blombäck's in Stockholm, Henschen's in Munich, and ours in

La Jolla. All three groups have published numerous reports
dealing with various portions of the structure, and no attempt
will be made to review these comprehensively here. Suffice it
to say that at this point complete sequences have been reported for the γ-chain (2) and the β-chain (3). About three-
quarters of the α-chain sequence is now known (4).

The most significant feature to emerge from the primary
structure studies so far, to our way of thinking, has been
the extensive homology observed among the three non-identical
chains (5). There is no doubt that all three chains have descended from a common ancestor. This being the case, we can
expect residual three-dimensional similarities and structure-
function relationships which are equivalent for all three
chains. On the other hand, each of the three chains have certain distinct characteristics, and these also ought to be reflected in their amino acid sequences. In the paragraphs
which follow we attempt to show how the similarities and differences in the three sequences bear out the overall structural features of the fibrinogen molecule.

A.   Distribution of Cysteines

The human fibrinogen molecule contains 58 cysteine residues, all of which are bound into disulfide bonds (Fig. 1).
They are distributed such that there are 8 in each α-chain,
11 in each β-chain, and 10 in each γ-chain. As in many other
proteins, nowhere is the common evolutionary heritage of the
non-identical chains better revealed than with regard to the
distribution of cysteine residues, the majority of which occur

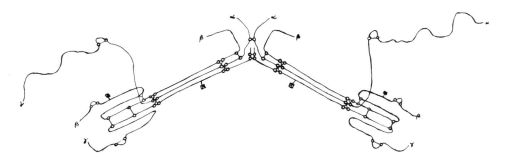

Fig. 1. Proposed arrangement of 29 disulfide bonds in
human fibrinogen. Circles - cysteine residues; squares -
carbohydrates.

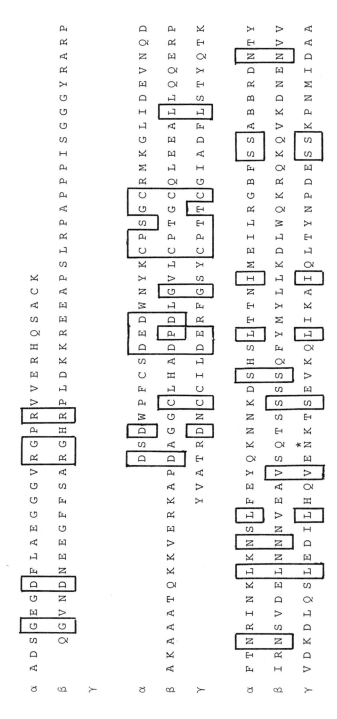

Fig. 2. Comparison of amino acid sequences of α-, β-, and γ-chains of human fibrinogen (aligned to maximize homologous segments; identical residues are boxed). Data are from: α-chain, Doolittle and Cottrell (1978); β-chain; Watt et al (1978); γ-chain, Henschen et al (1977).
* = carbohydrate attachment sites. The code used is: A, Ala; B, Asx; C, Cys; D, Asp; E, Glu; F, Phe; G, Gly; H, His; I, Ile; K, Lys; L, Leu; M, Met; N, Asn; P, Pro; Q, Gln; R, Arg; S, Ser; T, Thr; V, Val; W, Trp; Z, Glx (24).

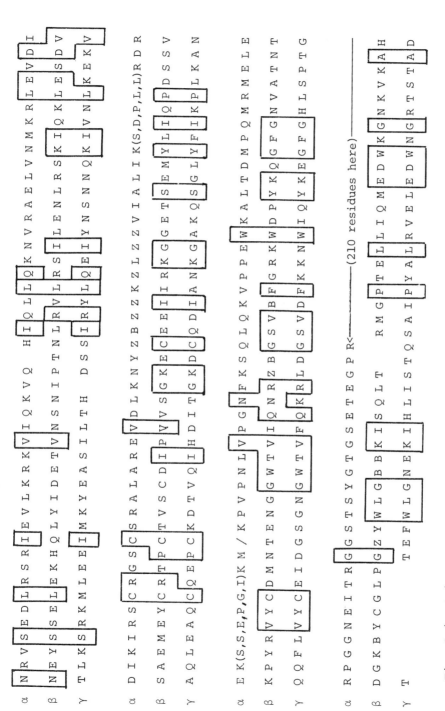

Fig. 2 (cont.)

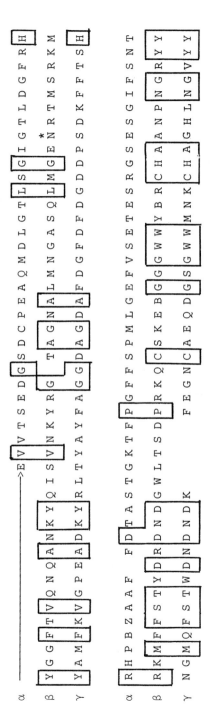

Fig. 2 (cont.)

at exactly the same locations in two or three of the chains
(Fig. 2). In the cases of the γ- and β-chains, 9 of their
10 and 11 cysteines can be matched.

B. Divergence of the Chains

Several interesting features are revealed by a comparison
of the sequences when the three chains are optimally aligned
by placing their equivalent cysteines in register (Fig. 2).
First, it immediately becomes apparent why there is not a
fibrinopeptide C, since a terminal deletion (measuring 57
residues in the case of the human molecule) has excised the
thrombin-sensitive portion of the γ-chain. Similarly, the
non-involvement of β-chains in the covalent fibrin stabilization process is more understandable when it is observed that
a carboxy-terminal deletion has removed the residues which
correspond to those which are cross-linked in the γ-chain.

A quantitative assessment of the numbers of identical residues in the chains suggests that β- and γ-chains have evolved
more recently from a common ancestor than has either from an
α-chain progenitor. This approach may be somewhat misleading,
however, because when only the amino-terminal "halves" of the
molecules are compared, all three chains appear to be equally
similar. In contrast, with the exception of some faint homology near the very end of the chain, the number of observed
identities in the carboxy-terminal half is virtually at a random level when the α-chain is compared with either of the
other two chains.

		α	β	γ	
	α	-	17	13	amino-terminal "half"
carboxy-terminal "half"	β	6	-	15	
	γ	5	35	-	

Percent Identical Residues

It is as though some radical intrusion has taken place during
α-chain evolution which has changed the entire character of
its carboxy-terminal half. That this may indeed be the case
is suggested by the fact that the carboxy-terminal half of
the α-chain has a very polar amino acid composition, with almost half of the residues being glycine or serine throughout
one long stretch. This unusual situation could conceivably

have come about by a frame shift mutation or some sort of gene-splicing. Although the absence of such a distinctive region from β- and γ-chains could readily be explained by a deletion in advance of the duplicative event which led to their divergence, the fact that the prototype fibrinogen chain appears to have been built up to its full length by a series of tandem duplications argues against such a course, since in none of the other segments does there appear to be any history for the existence of such an unusual amino acid composition. Whichever the case, the structural consequences are significant.

III. THREE-DIMENSIONAL ASPECTS

If our reasoning is correct, then the observed amino acid sequences would indicate that all three chains ought to have some common or equivalent three-dimensional aspects in their amino-terminal halves, whereas only β- and γ-chains ought to have equivalent structural features in their carboxy-terminal halves. Indeed, there is a good deal of independent evidence that this is the case. In this section we will attempt to show how the amino acid sequence data bear out previous suggestions when such a route is followed.

To begin with, it must be noted that the proposed disulfide bond arrangement (Fig. 1) is completely consistent with early electron microscopy studies which suggested that fibrinogen has a triglobular structure in which two terminal domains are connected to a central domain by thread-like strands (7). The depiction is also compatible with established physicochemical properties of fibrinogen, which indicate that the molecule is elongated. Finally, the arrangement fits--indeed is based upon--long-noted enzymatic and chemical fragmentation patterns (6, 8).

If we now use the amino acid sequence data to build upon the scaffolding available both from the longstanding observations noted above and the proposed arrangement of disulfide bonds, then the following structural picture emerges. First, all six amino-terminals of the α-, β- and γ-chains are girdled together in a central domain. This central domain, which includes the thrombin-sensitive bonds which lead to the release of the fibrinopeptides, is dimeric, and it is connected to the two terminal domains by two sets of three-stranded ropes. These three-stranded connectors, as we have argued elsewhere (9), are in actuality coiled α-helices. They are demarcated by two sets of unusual disulfide arrangements which lead to three-strand symmetrical junctions we call "disulfide rings." The distance between these delineating disulfide rings is 111 residues in the cases of the α- and γ-chains (10) and 112

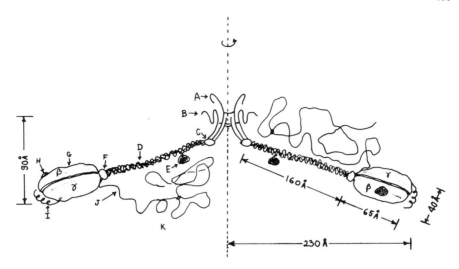

Fig. 3. Detailed model of fibrinogen indicating key features: A, fibrinopeptide A; B, fibrinopeptide B; C, first disulfide ring; D, coiled-coil connection between domains; E, carbohydrate attached to γ-chain; F, second disulfide ring; G, terminal domain; H, carbohydrate on β-chain; I, γ-chain cross-linking site; J, protease-sensitive region; K, polar carboxy-terminal half of α-chain. Note pseudosymmetry of the halves of the terminal domain contributed by β- and γ-chains respectively.

residues in the case of the β-chain (3). If we presume a translation distance of 1.5A consistent with the existence of α-helices, then the interdomainal connections are approximately 160 A long. The terminal domains are comprised primarily of the carboxy-two-thirds of the β- and γ-chains respectively, and whereas a three-fold symmetry was assumed throughout the interdomainal connectors and in the design of the "disulfide rings," a two-fold pseudosymmetry can be predicted for this domain because of the large degree of homology which exists between the β- and γ-chains in these regions. Finally, the aberration of the α-chain, as detected by a sequence which can no longer be aligned with the other chains, begins shortly after the cysteine residues which hold that chain together with the others in the "second disulfide ring." Coincidentally, it is in this very region that a number of proteases, including plasmin, are wont to attack, releasing this large polar appendage which is the carboxy-terminal two-thirds of the chain (11). As shown above, all of these features can be incorporated into a detailed model which we feel approximates the real structure of the human fibrinogen molecule (Fig. 3).

IV. FIBRIN FORMATION

    Given a fibrinogen molecule of the design shown in Fig. 3,
what would be a realistic mechanism for bringing about its
polymerization? To start with, it must be presumed that it is
the presence of the fibrinopeptides in the central domain
which keeps these molecules soluble. This may be accomplished by the significant polar constitution which typifies
these peptides, or by the direct masking of key polymerization sites. The inexact nature of this blocking and/or fend-
ing action is attested to by the great variability of these
peptides which has occurred during evolution.
    Thrombin, exhibiting an exquisite preferential proteoly-
sis, is able to release these peptides, an event which triggers the spontaneous polymerization of the parent protein.
In all the species examined to date (upwards of 50) and in
the cases of both the fibrinopeptides A and B, the bond
cleaved by thrombin is arginyl-glycine (Table I). Interestingly enough, six other pairs of arginyl-glycine bonds occur
in the sequence of human fibrinogen (Fig. 2), including five
in the α-chain and one in the β-chain, and none of these appears to be cleaved by thrombin, even though at least four
are in exposed regions which are vulnerable to attack by
plasmin and other proteases. It is also noteworthy that although the fibrinopeptides--which are on the amino-terminal
side of the junctions split by thrombin--are among the most
changeable peptide structures known--the sequences exposed
on the carboxy-terminal side of the attacked bonds are highly conserved (Table I). These exposed regions must interact
with sites on the terminal domains of other fibrin(ogen)
molecules which have also been subject to thrombin attack
such that a reciprocal pairing can take place (Fig. 4). The
half-molecule staggered overlap which results is consistent
not only with early ideas about how polymerization ought to
proceed (12) but also with the observed periodicity of
fibrin (13).
    After the initial dimerization, polymerization may pro-
ceed in either direction, subsequent to the appropriate removal of fibrinopeptides. It should be noted that a second
set of contacts comes into effect with the addition of the
third unit in the system (Fig. 4). After the formation of
polymers of intermediate length, a lateral involvement ordinarily occurs whereby the growing protofibril can increase
in breadth as well as length (Fig. 5). This latter aspect
is quite sensitive to the solution environment, being dis-
couraged by high pH and/or ionic strength (14).

TABLE I

Amino Acid Sequences Around the Thrombin-Sensitive
Bonds in Fibrinogen from Various Species

α-Chain   Thrombin

Human[a]   ..Glu-Gly-Gly-Gly-Val-Arg Gly-Pro-Arg-Val-Val-Glu..

Bovine[b,c]   ..Glu-Gly-Gly-Gly-Val-Arg Gly-Pro-Arg-Leu-Val-Glu..

Dog[d]   ..Glu-Gly-Gly-Gly-Val-Arg Gly-Pro-Arg-Ile-Val-Glu..

Chicken[e,f]   ..Glu-Gly-Gly-Gly-Gly-Arg Gly-Pro-Arg-Ile-Leu-Glu..

Lamprey[c]   Asp-Asp-Ile-Ser-Leu-Arg   Gly-Pro-Arg-Leu-Xxx[g]-Glx..

β-Chain

Human[a]   ..Gly-Phe-Phe-Ser-Ala-Arg Gly-His-Arg-Pro-Leu-Asp..

Bovine[b,c]   ..Val-Gly-Leu-Gly-Ala-Arg Gly-His-Arg-Pro-Tyr-Asx..

Dog[d]   ..Ser-Thr-Val-Asp-Ala-Arg Gly-His-Arg-Pro-Leu-Asp..

Chicken[f]   ..(still undetermined)Arg Gly-His-Arg-Pro-Leu-Asx..

Lamprey[c]   ..Ala-Ala-Leu-Asp-Val-Arg Gly-Val-Arg-Pro-Leu-Pro..

---

[a] Blombäck & Blombäck (6)

[b] Dayhoff (24)

[c] Cottrell & Doolittle (25)

[d] Birken et al (26)

[e] Takagi et al (27)

[f] Murano et al (28)

[g] Xxx is probably Ser or Thr

1. Fibrin Stabilization. Under physiological circumstances, the developing fibrin polymer is reinforced by the
introduction of a number of intermolecular covalent bonds
involving certain glutaminyl and lysyl sidechains. The formation of these cross-links is brought about through the action
of the calcium-dependent plasma transglutaminase known as
factor XIII, an enzyme which is itself activated by the
thrombin-catalyzed cleavage of an arginyl-glycine bond (15).
The first cross-links to be formed in the polymerizing system
couple adjacent units in the polymer by a set of reciprocal
bonds formed near the carboxy-terminals of γ-chains (16).
These bonds are introduced rapidly and can even occur during
the formation of intermediate polymers in advance of gela-
tion (1). Subsequently, another set of cross-links is incorporated which involves α-chains (17). The sites in this
case have been localized to certain peptides in the polar

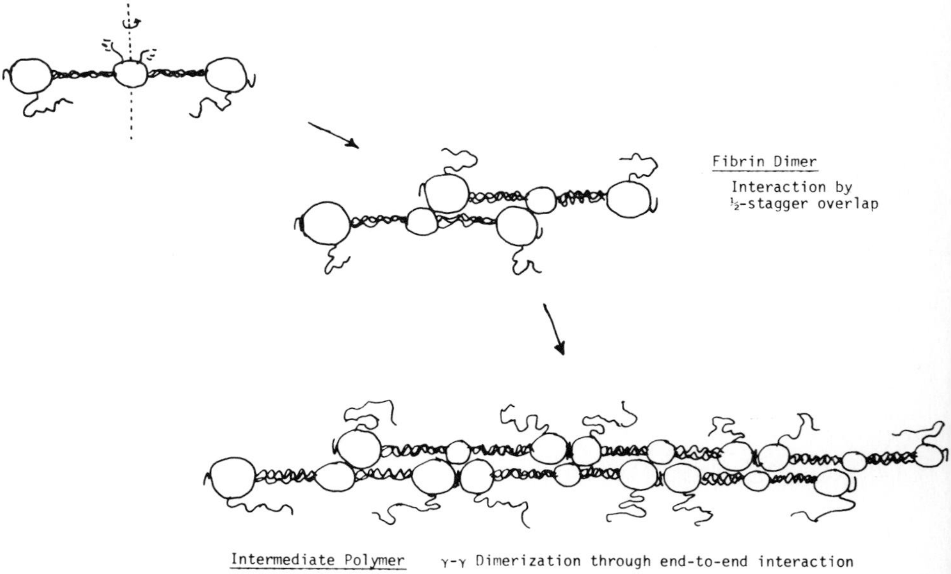

Fibrin Dimer

Interaction by
½-stagger overlap

Intermediate Polymer   γ-γ Dimerization through end-to-end interaction

Fig. 4. Schematic outline of initial events involved in
polymerization of fibrin. Release of fibrinopeptide(s) from
central domain allows reciprocal interaction with terminal
domain of another fibrinogen molecule which has also lost fi-
brinopeptide(s). A second set of interactions comes into play
when a third molecule is added, at which time γ-chains in
neighboring molecules come into contact and can be covalently
bonded by factor XIII (redrawn from Ref. 1).

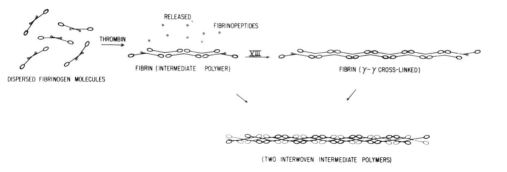

RELEASED FIBRINOPEPTIDES

THROMBIN

DISPERSED FIBRINOGEN MOLECULES

FIBRIN (INTERMEDIATE POLYMER)

XIII

FIBRIN (γ-γ CROSS-LINKED)

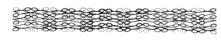

(TWO INTERWOVEN INTERMEDIATE POLYMERS)

FULLY FORMED FIBRIN

Fig. 5. Subsequent events in polymerization which lead to fully developed fibrin strands. The intermediate polymers are interwoven by a different set of molecular contacts than are involved in initial contacts (Ref. 29).

appendage which is the carboxy-terminal half of the α-chain (18). The formation of cross-links involving α-chains is sensitive to pH and ionic strength in the same way as is fiber breadth, suggesting that the attachements are in the lateral dimension (1).

V. FIBRINOLYSIS

Although the formation of blood clots is of vital importance, some mechanism must also be provided for the removal of these materials when they are no longer needed or for dissolving any fibrin which might break loose into the general circulation. Indeed, the maintenance of intravascular fluidity while still having the capability of sealing leaks in the system involves a very delicate biochemical balance. In this regard, the complicated system for generating thrombin has its counterpart in a system geared to the generation of plasmin (fibrinolysin), and the two systems are likely coupled in some subtle way.

Plasmin is not nearly as narrow in its specificity as thrombin when it comes to attacking fibrinogen or fibrin. In fact, the destruction of fibrin can be viewed as being largely dependent on the latent vulnerability of the fibrin(ogen) molecule, more than as a phenomenon involving a highly devel-

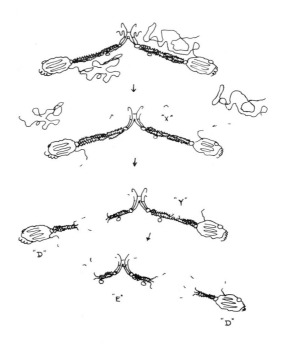

Fig. 6. Pictorial description of fibrinogenolysis by
plasmin. Note early release of polar carboxy-terminal portions of α-chains and polar amino-terminal "elbows" of β-
chains, then concerted attack on interdomainal connectors.
Fibrinolytic pathway follows a similar course.

oped and specific attacking agent (19). Thus, the initial
stages of the digestion of fibrinogen or fibrin by plasmin appear not to differ significantly from that effected by trypsin
(19). On the other hand, plasmin is more efficient in its
fibrinolytic action than trypsin and can bring about the actual dissolution of clots by cleaving many fewer bonds than
trypsin under the same circumstances (20).

The plasmin-catalyzed degradation of fibrinogen and fibrin
follow very similar courses, similar--although not absolutely
identical--sets of core fragments resulting. Many regulatory
functions have been claimed for these digestion products over
the years, the most significant of which being inhibition they
can effect with regard to further polymerization (21). They
may also play a role in regulating the biosynthesis of fibrinogen (22).

Long before any significant amount of sequence data was
available for fibrinogen, Marder (23) showed how the plasmin-
digestion course of fibrinogen was in perfect harmony with the
triglobular model of fibrinogen proposed by Hall and Slayter
(7). What we would like to stress in this section is how the
amino acid sequence data have completely borne out that contention, especially now that we know many of the exact sites
of plasmin cleavage (Table II). The general course of fibrinogenolysis is depicted in Fig. 6. First, the polar appen-
dages which are extensions of the α-chains are cleared away.
At the same time, or shortly thereafter, an amino-terminal
segment of the β-chain is also excised. The usual interpretation of these events is that both of those entities must be
highly exposed protrusions. The most important aspect of
fibrinolysis, however, is an interdomainal attack in the
central region of the three-stranded rope (Fig. 7). The mode
of fibrin polymerization proposed (Fig. 5) is such that
these connectors remain fully exposed to solvent and ought
to be accessible to degradative attack. In the same sense
that a chain is no stronger than its weakest link, breakage
of a significant number of the interdomainal connectors ought
to reduce fibrin gels to the intermediate polymer stage, at
which point they are soluble. Further plasmin attack results
in the characteristic core fragments.

The importance of interdomainal attack for fibrinolysis
may be attested to in the observed sequences in this region.
Although we showed that the sequences of these portions of the
three chains are noteworthy in that they are mostly consistent

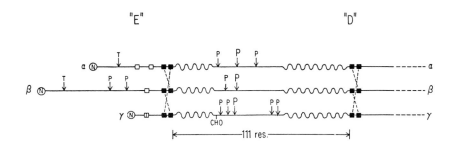

Fig. 7. Schematic depiction of the connections between
fragments D and E and the disulfide rings (---). Thrombin
attack points for the release of fibrinopeptides A and B are
designated by T. Primary plasmin attack points are indicated
by P (Ref.10).

TABLE II

Some Identified Plasmin Cleavage Points Which
Occur During Fibrinogenolysis

Stage	α-Chain	β-Chain	γ-Chain
	Lys–Met 600 601	Arg–Ala 42 43	
	Lys–Ala 234 235		
Early	Arg–Met 243 244		
	Arg–Gly 256 257	Lys–Asp 133 134	
	Lys–Ser 223 224		Lys–Ala 62 63
Intermediate	Lys–Met 210 211	Lys–Asp 122 123	
	Arg–Asp 104 105	Lys–Ala 58 59	Lys–Ser 85 86
	Lys–Asn 78 79		Lys–Gln 58 59
	Arg–Val 110 111		Lys–Thr 53 54
Late	Arg–Gly 16 17		Lys–Met 88 89
	Arg–Val 19 20		Lys–Gln 405 406

Arbitrarily taken as events occurring during a 2
hour time course of digestion. Naturally there is no
sharp demarcation between stages. Moreover, the order
of cleavage is not entirely obligatory. Numbering re-
fers to sequence of human fibrinogen (Fig. 2).

with the existence of the α-helices, there are some exceptions to this generalization, and the exceptions mainly occur
in the central region of where the plasmin attack occurs.
Since a perfectly coiled-coil might be resistant to proteolytic attack, it may be that these non-helical portions have
evolved as better potential targets. In other words, fibrinogen appears to have a built-in vulnerability which facili-
tates subsequent fibrinolysis.

VI. SUMMARY

What we have tried to do in this brief article is show how
studies on the primary structure of human fibrinogen can be
used to bolster some longstanding ideas about fibrinogen
structure and behavior. In particular, we have used the data
to flesh out a model in which two terminal domains are connected to a central focus by sets of three-stranded ropes.
We have invoked a general principle that common ancestry for
polypeptides ought to be reflected in equivalent three-dimensional structures, and following that rule have tried to show
how a three-fold symmetry in the amino-terminal halves of the
peptides can give rise to each half of the central domain and
the three-chain connectors which run to the terminal domains.
In contrast, the aberrational sequence observed in the carboxy-terminal half of the α-chain exempts that structure from
any equivalence with the carboxy-terminal halves of the β-
and γ-chains. Instead, the terminal domains ought to exhibit
a twofold symmetry. The detailed model we have developed
offers perfectly simple solutions for fibrin formation and
stablization, as well as for fibrinolysis.

REFERENCES

1. Doolittle, R. F. Adv. Prot. Chem., 27, 1-109 (1973)

2. Lottspeich, R. and Henschen, A. Hoppe-Seyler's Z.
    Physiol. Chem., 358, 935-938, (1977).

3. Watt, K. W. K., Takagi, T. and Doolittle, R. F. Proc.
    Natl. Acad. Sci., U.S., In Press.

4. Doolittle, R. F. and Cottrell, B. A. 17th Cong. Intern.
    Soc. Hemat. Abstracts (Paris) (1978)

5. Doolittle, R. F. Federation Proc., 35, 2145-2149 (1976).

6. Blombäck, B. and Blombäck, M. Annals N.Y. Acad. Sci.,
    202, 77-97 (1972).

7. Hall. C. E. and Slayter, H. S.   J. Biophys. Biochem.
    Cytol., 5, 11-16 (1959)

8. Marder, V. J.   Scand. J. Haematol., supp. 13, 21-36
    (1971).

9. Doolittle, R. F., Goldbaum, D. M. and Doolittle, L. R.
    J. Mol. Biol., In Press.

10. Doolittle, R. F., Cassman, K. G., Cottrell, B. A.,
    Freizner, S. J. and Takagi, T.   Biochemistry, 16, 1710-
    1715 (1977).

11. Takagi, T. and Doolittle, R. F.   Biochemistry, 14, 5149-
    5156 (1975).

12. Ferry, J. D., Katz, S. and Tinoco, I., Jr.   J. Polymer
    Sci., 12, 509-516 (1954).

13. Stryer, L., Cohen, C. and Langridge, R.   Nature (London),
    197, 793-794 (1963).

14. Ferry, J. D. and Morrison, P. R.   J. Am. Chem. Soc., 69,
    388-400 (1947).

15. Takagi, T. and Doolittle, R. F.   Biochemistry, 13, 750-
    756 (1974).

16. Chen, R. and Doolittle, R. F.   Proc. Natl. Acad. Sci.,
    U.S., 66, 472-479 (1970).

17. McKee, P. A., Mattock, P. and Hill, R. L.   Proc. Natl.
    Acad. Sci., U.S., 66, 738-744 (1970).

18. Doolittle, R. F., Cassman, K. G., Cottrell, B. A. and
    Freizner, S. J.   Biochemistry, 16, 1715-1719 (1977).

19. Mihalyi, E.   Thromb. Diath. Haem., supp. 39, 43-61 (1970).

20. Weinstein, M. J. and Doolittle, R. F.   Biochim. Biophys.
    Acta, 258, 577-590 (1972).

21. Latallo, Z. S., Budzynski, A. Z., and Kowalski, E.
    Nature (London), 203, 1184 (1964).

22. Barnhart, M. I., Cress, D. C., Noonan, S. M. and Walsh,
    R. T.   Thromb. Diath. Haem., supp. 39, 143-159 (1970).

23. Marder, V. J.   Thromb. Diath. Haem., supp. 39, 187-195
    (1970).

24. Dayhoff, M. O.   Atlas Protein Sequence and Structure,
    Natl. Biomedical Res. Found., Silver Spring, MD. (1972).

25. Cottrell, B. A. and Doolittle, R. F.   Biochim. Biophys.
    Acta, 453, 426-438 (1976).

26. Birken, S., Wilner, G. D. and Canfield, R. E. Thromb. Res., 7, 599-610 (1975).

27. Takagi, T., Finlayson, J. S. and Iwanaga, S. Biochim. Biophys. Acta, In Press.

28. Murano, G., Walz, D., Williams, L., Pindyck, J. and Mossesson, M. W. Thromb. Res., 11, 1-10 (1977).

29. Doolittle, R. F. Horizons in Biochemistry, 3, 164-191 (1977).

STUDIES OF PRESYNAPTICALLY NEUROTOXIC
AND MYOTOXIC PHOSPHOLIPASES A$_2$

David Eaker[1]

Institute of Biochemistry
University of Uppsala
Box 576, S-751 23 Uppsala
Sweden

I. INTRODUCTION

Although the neurotoxic nature of certain snake venoms
has been recognized for centuries, snake venoms and various
fractions derived therefrom were until little more than a
decade ago regarded as exotic substances by the international
scientific community at large and received serious attention
only in a very few laboratories scattered around the world.
Owing in no small part to the unerring pioneering work done
here in Taiwan by C.C. Chang, C.Y. Lee, and C.C. Yang on the
physiological and chemical definition of elapid venom fractions,
snake venoms have ascended to a leading position as specific
tools for the study of transmission across the neuromuscular
junction and other cholinergic synapses.

The α-bungarotoxin described in the trailblazing paper by
Chang and Lee on the neurotoxins of the many-banded krait,
Bungarus multicinctus (1), and many other homologous *postsynap-*
tic neurotoxins of the curarimimetic type isolated subsequently
from other elapid venoms are now used routinely in laboratories
around the world as specific reagents for the isolation and
study of nicotinic acetylcholine receptors from brain, muscle,
and electric organs (2). More than 50 postsynaptic neurotoxins
of this type have been sequenced (3), and crystallographic
structures have been reported from two laboratories (4-6).

---

[1]Supported by NFR grant 2859-011.

In the same paper (1) Chang and Lee reported the presence
in krait venom of *presynaptically* active neurotoxins, designated β- and γ-bungarotoxins, which interfered with the release
of acetylcholine from motor nerve terminals. The characterization of these and related pre-synaptically active neurotoxins
has been considerably more difficult than was the case with
the postsynaptic curarimimetic neurotoxins for three reasons.

Firstly, the detection of presynaptic activity in the
crude venoms can be technically difficult owing to the presence
of curarimimetic toxins. The situation is further complicated
by the fact that most of the presynaptically active neurotoxins
from snake venoms are not exclusively presynaptic, but also
have some degree of musculotropic action. Postsynaptic effects
can thus nearly obscure the presynaptic activity unless transmitter release is monitored directly.

Secondly, the presynaptically active neurotoxins are larger
and more complicated than the postsynaptic ones and often consist of two or more subunits. If these are inadvertently separated from each other prematurely in the venom fractionation
scheme the activity can drop so far that the relevant fractions
become screened out in the simple lethality assays that are usually used to monitor neurotoxin isolation.

Thirdly, until it became evident very recently that the
presynaptically active neurotoxins are highly specialized phospholipases A2 there was probably a tendency to attribute the
lethality of venom fractions exhibiting phospholipase activity
to relatively non-specific destruction of membranes, thereby
exempting these fractions from further study.

Already in 1938 Slotta and Fraenkel-Conrat isolated in
crystalline form from the venom of the Brazilian rattle-snake,
Crotalus durrissus terrificus, a neurotoxic protein, designated
crotoxin, which also exhibited indirect hemolytic activity and
they suggested then that the neurotoxicity might depend on the
evident ability of the substance to split phospholipids (7).
Although crotoxin was thus isolated more than two decades before any other toxin of the same type its *presynaptic* action
was not demonstrated until 1971 (8). Furthermore, the significance of the phospholipase activity was not fully appreciated
until we reported in 1974 that the presynaptic neurotoxin,
notexin, which we had isolated earlier from the venom of the
Australian tiger snake, *Notechis scutatus scutatus,* was simply
an enzymatically active structural homolog of porcine pancreatic phospholipase A2 (9-11). We suggested then that crotoxin,
β-bungarotoxin and other presynaptically active snake venom
toxins such as *taipoxin* had evolved from an ancestral phospholipase structure, but it did not seem necessary to implicate
the phospholipase activity in the neurotoxic activity since β-
bungarotoxin and taipoxin seemed at that time to lack catalytic
activity.

We now know that all of these presynaptic neurotoxins are indeed homologs of mammalian pancreatic phospholipases A2 and that all of them manifest phospholipase activity under certain conditions. It is also clear that various snakes, especially of the elapid family, have evolved phospholipases with varying degrees of specificity toward nerve terminals, muscles, lungs, kidneys and perhaps other tissues as well.

I shall here review briefly what we know about the chemistry of these "phospholipase" type toxins and shall describe some recent efforts to define at the molecular level their mode(s) of action and relationships of structure to function.

II. STRUCTURAL CHEMISTRY

A. Structural Properties of Individual Toxins

The seven most well characterized presynaptic neurotoxins of the phospholipase type are listed below approximately in the order of decreasing presynaptic specificity. β-Bungarotoxin appears to be completely specific for nerve terminals, but the other toxins all exhibit some degree of myotoxic action which even overshades the presynaptic action in the case of the *Enhydrina schistosa* myotoxin. The *N. nigricollis* phospholipase appears to preferentially attack lung and kidney upon intravenous injection into mice.

The four different structural types represented by these six toxins can be illustrated schematically as shown in Fig. 1. The phospholipase-type subunits indicated by the presence of the "catalytic cleft" are in all cases single peptide chains of 118-140 amino acid residues cross-linked by 6-8 disulfide bridges.

1. β-Bungarotoxin. The first chemical composition data for this toxin from the venom of the many-banded krait, *Bungarus multicinctus,* were reported by Lee *et al.* in 1972 (12). Kelly and Brown subsequently found that the toxin consisted of two subunits which were apparently linked together by at least one disulfide bridge (13). The recently completed amino acid sequence analyses of both subunits (14) show that the larger chain contains 120 amino acid residues and is homologous in sequence to other vertebrate phospholipases A2, whereas the smaller subunit contains 60 residues and bears no obvious homology with any other known venom protein. Although the phospholipase activity of the intact toxin must reside in the larger "phospholipase" subunit this has not yet been demonstrated directly because the conditions required for reduction of the inter-chain disulfide bridge lead to cleavage of intrachain

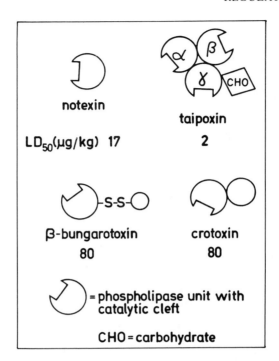

Fig. 1. The four known structural types of presynaptically active phospholipase neurotoxins. Notechis II-5, the *Enhydrina schistosa* myotoxin and the *N. nigricollis* phospholipase
are single-chain toxins like notexin. The LD_{50} doses apply for
intravenous injection into mice.

disulfides as well.

   2. *Taipoxin*. The isolation and characterization of this
toxin from the venom of the Australian taipan, *Oxyuranus scutellatus scutellatus* was described in 1976 (15). Taipoxin is a
non-covalent ternary complex of three subunits designated α,
β, and γ, all of which are homologous in sequence to the pork
panreatic phospholipase A2. The α-subunit contains 119 amino
acid residues and 7 disulfide bridges and is very basic. The
β-subunit (which is represented in the complex by at least two
iso-forms) contains 120 residues and 7 disulfide bridges and
is neutral. The γ-subunit contains 133 residues and 8 disulfide bridges and is very acidic, owing in part to the presence
of a large carbohydrate moiety containing four sialic acid residues. The complete amino acid sequence of the γ-chain and
the site of attachment of the carbohydrate have been published
(16).

When the separation of the subunits is done without exposure of the γ-subunit to low pH the subunits can be recombined to form the native complex. Of the three subunits, only the basic α-subunit shows neurotoxic activity on its own, and this subunit also seems to be responsible for the phospholipase activity of the complex. Upon recombination of the α-subunit with the other two the lethality increases 500-fold on a molar basis.

3. Crotoxin. The properties of this toxin from the venom of the Brazilian rattlesnake, *Crotalus durrissus terrificus,* have been studied for many years in the laboratories of Fraenkel-Conrat in Berkeley and Habermann in Germany. A review of the research on crotoxin has been published recently by Habermann and Breithaupt (17). Crotoxin is a non-covalent complex of two subunits, both of which must be present for the manifestation of high neurotoxic activity. Although the compositions of the two subunits vary depending on the source of the venom used the data reported by Breithaupt *et al.* (18) were obtained with highly purified components and serve to illustrate the nature of the complex. According to these authors the larger subunit, which is called basic crotalus phospholipase A by the Habermann group and crotoxin B by the Fraenkel-Conrat group, is a single chain of 140 amino acid residues cross-linked by eight disulfide bridges. Partial sequence data indicate that this component is homologous to the other phospholipase mentioned herein. The smaller component called crotapotin (Habermann) or crotoxin A (Fraenkel-Conrat) is very acidic and consists of three peptide chains of 40, 34 and 14 residues held together by disulfide bridges.

The phospholipase activity of the larger subunit is suppressed in some assay systems by the presence of the acidic crotapotin, which seems to have no enzymatic or toxic activity alone.

4. Notexin. This presynaptically neurotoxic and myotoxic protein from the venom of the Australian tiger snake, *Notechis scutatus scutatus,* was the first phospholipase toxin to be characterized in terms of amino acid sequence (19). The molecule is a single peptide chain of 119 residues cross-linked by 7 disulfide bridges. Notexin crystallizes easily and the crystals diffract well to at least a resolution of 1.8 Å (20), which should permit complete solution of the three-dimensional structure.

5. Notechis II-5. This toxin was isolated from the same venom and differs in sequence from notexin at only seven positions (21). It is one-third as lethal and about 5-10 fold less myotoxic as notexin.

6. E. schistosa Myotoxin VI:5. This strongly myotoxic phospholipase from the venom of the common sea snake, *Enhydrina schistosa,* consists of a single, very basic peptide chain of 120 residues cross-linked by seven disulfide bridges (22). The sequence analysis now in progress shows that this toxin is homologous to the other phospholipases already described.

7. Basic N. nigricollis Phospholipase. This phospholipase from the venom of the black-necked spitting cobra, *Naja nigricollis,* is a single peptide chain of 118 residues cross-linked by seven disulfide bridges (23). This phospholipase differs in sequence from the phospholipase CM-III of *Naja mossambica mossambica* venom (24) by a single Phe/Leu substitution at position 70 (homology numbering in Fig. 2).

B. Charge Properties of the Toxins and Their Subunits

In Table I the various neurotoxins and their subunits or chains are compared in terms of their net charge at neutral pH as calculated from sequence data. Histidine was for this purpose assigned a charge value of $+1/2$.

TABLE I. Charge Properties of the Toxins

Toxin	Net charge at pH 7	Isoelectric point
β-Bungarotoxin	+ 5.5	
phospholipase chain	+ 0.5	
smaller chain	+ 5	
Taipoxin		pH 5 (15)
α-subunit		>pH 10
β-subunit		pH 7
γ-subunit	-19	pH 2.5
Crotoxin		pH 5
crotapotin		pH 3.4 (18)
phospholipase A		pH 9.7
Notexin	+ 3	
Notechis II-5	+ 4	
E. schistosa myotoxin VI:5		>pH 9
N. nigricollis phospholipase	+ 9.5	>pH 10

Each of these toxins either is a basic protein or contains one very basic subunit. In the cases of taipoxin and crotoxin the basic subunits are the only ones which show either phospholipase activity or toxicity in the absence of the other subunit(s). In β-bungarotoxin, on the other hand, the "phospholipase" chain is neutral and the basicity resides in the smaller "moderating" subunit. This is interesting in view of the facts that β-bungarotoxin is the only one of these toxins which is exclusively presynaptic and is also the only one in which the subunits are covalently bound.

C. Sequence Homology

In Fig. 2 the phospholipase chains of the snake venom neurotoxins for which sequence data are available are compared with the prophospholipase A2 from pig pancreas. Although only six disulfide bridges were located in the original structural work on the latter protein (25,26), the presence of a seventh disulfide bridge was revealed and a few additional corrections in the sequence were made in a recent reinvestigation (27). A three-dimensional structure based on 3.8 Å crystallographic data has also been reported (28). In the model the invariant residues Tyr-28, His-48 and Asp-49 (numbering in Fig. 2) and the calcium ion that is required for catalytic activity by the pancreatic enzyme and all known homologues are located in a classical "active site" cleft and were suggested to play direct roles in the enzymatic action. An important role for His-48 was proposed earlier on the basis chemical modification experiments done with the pork enzyme (29) and notexin (30).

In an elegant series of papers (31-32) the Utrecht group has shown that the catalytic site of the pancreatic enzyme is intact and functional in the zymogen and is not disturbed by removal of Ala-1 in addition to the usual "activation" heptapeptide of the zymogen, because both the zymogen and the des-Ala-1 form of the enzyme can digest monomeric substrates nearly as well as does the enzyme proper. However, neither the zymogen nor the des-Ala-1 form of the enzyme can attack the *micellar* phospholipid which is the proper substrate for the intact activated enzyme. These results were interpreted to indicate that the immediate N-terminal sequence of the enzyme is part of an interface recognition site which is required for specific interaction between the enzyme and organized lipid-water interfaces, and that it is this recognition site rather than the catalytic site which is "masked" by the N-terminal heptapeptide of the zymogen.

The γ-subunit of taipoxin has an N-terminal sequence analogous to that of the pancreatic zymogen, but the replacement of Arg by Ile at position -1 obviously prevents "activation" by

```
           -5                                  1                   5                                        10
Ser-GLU-Leu-Pro-Gln-Pro-SER-Ile-Asp-Phe-Glu-GLN-PHE-Ser-Asn-MET-ILE-Gln-CYS-Thr-Ile-PRO-
----GLU-Glu-Gly-Ile-Ser-SER-Arg-Ala-LEU-Trp-GLN-      -Arg-Ser-MET-      -Lys-CYS-Ala-Ile-PRO-
                         ASN-LEU-Tyr-GLN-      -Lys-Asn-MET-      -His-CYS-Thr-Val-PRO-
                         ASN-LEU-Val-GLN-      -Ser-Tyr-Leu-      -Gln-CYS-Ala-Asn-His-
                         ASN-LEU-Val-GLN-      -Ser-Tyr-Leu-      -Gln-CYS-Ala-Asn-His-
                         ASN-LEU-Ile-Asn-      -Met-Glu-MET-      -Arg-Tyr-Thr-Ile-PRO-

    15                      20                  25                      30              35
CYS(Gly,Ser)Glu-CYS-Leu-Ala-Tyr-Met-ASP-TYR-GLY-CYS-TYR-CYS-GLY-Pro-GLY-GLY-SER-GLY-THR-
Gly-Ser-His-PRO-Leu-Met-Asp-Phe-Asn-Asn-      -      -      -Leu-      -      -SER-      -THR-
----Ser-Arg-PRO-Trp-Trp-His-Phe-Ala-ASP-      -      -      -Arg-      -      -Lys-      -THR-
Gly-Arg-Arg-PRO-Thr-Arg-His-Tyr-Met-ASP-      -      -      -Trp-      -      -SER-      -THR-
Gly-Lys-Arg-PRO-Thr-His-Tyr-Met-ASP-      -      -      -Ala-      -      -SER-      -THR-
CYS-Glu-Lys-Thr-Trp-Gly-Tyr-Ala-ASP-      -      -      -Ala-      -      -SER-      -Arg-

    40                      45                  50                      55
PRO-Ile-ASP-LEU-ASP-ARG-CYS-CYS-Lys-Thr-HIS-ASP-Glu-CYS-TYR-Ala-Glu-ALA-Gly-LYS-Leu-
-VAL-      -Glu-      -Glu-Thr-      -Asn-      -Arg-Asp-      -Lys-Asn-Leu-
-VAL-      -Asp-      -Gln-Val-      -Asn-      -Glu-Lys-      -Gly-LYS-Met-
-VAL-      -Glu-      -Lys-Ile-      -Asp-      -Ser-Asp-      -Glu-LYS-Lys-
-VAL-      -Glu-      -Lys-Ile-      -Asp-      -Asp-Glu-      -Gly-LYS-Lys-
-Ile-      -Ala-      -Tyr-Val-      -Asn-      -Gly-Asp-      -Glu-LYS-Lys-

    60                      65                       75             80
Ser-Ala-CYS-LYS-Ser-Val-Leu-Ser-Glu-PRO-Asn-Asp-Thr-TYR-Ser-TYR-Glu-CYS-----Asn-Glu-
Asp-Ser-      -LYS-Phe-Leu-Val-Asp-Asn-      -Tyr-Thr-Glu-Ser-      -Ser-CYS-----Ser-Asn-
-Gly-            -Trp-          -Tyr-Leu-Thr-Leu-      -Lys-      -Lys-CYS-Ser-Gln-Gly-
-Gly-            -Ser-          -Lys-Met-Ser-Ala-      -Asp-      -Tyr-CYS-Gly-Glu-Asn-
-Gly-            -Phe-          -Lys-Met-Ser-Ala-      -Asp-      -Tyr-CYS-Gly-Glu-Asn-
His-Lys-         -Asn-          -Lys-Thr-Ser-Gln-      -Ser-      -Lys-Leu-Thr-Lys-Arg-
```

CHO

```
                85                    90                   95                  100
Gly-Gln-Leu-Thr-CYS----Asn-Asp-Asp-Asn-Asp-Glu-CYS-Lys-Ala-PHE-Ile-CYS-Asn-CYS-ASP-Arg-
Thr-Glu-Ile-Thr------Asn-Ser-Lys-Asn-Asn-Ala-  -Glu-Ala-PHE-Ile-  -Asn-    -Arg-
Lys-Leu-Thr------    -----Ser-Gly-Gly-Asn-Ser-Lys-  -Gly-Ala-Ala-VAL-  -Asn-    -Leu-
Gly-Pro-Tyr------    -----Arg-Asn-Ile-Lys-Lys-Lys-  -Leu-Arg-PHE-VAL-  -Asp-    -Val-
Gly-Pro-Tyr------    -----Arg-Asn-Ile-Lys-Lys-Lys-  -Leu-Arg-PHE-VAL-  -Asp-    -Val-
Thr-Ile-Ile------    -----Tyr-Gly-Ala-Ala-Gly-Gly-Thr-  ----Arg-Ile-VAL-  -Asp-    -Arg-
                                              Val

               105                   110                  115                 120
Thr-ALA-Val-Thr-CYS-PHE-Ala-Gly-ALA-PRO-TYR-Asn-Asp-Asp-Leu-Tyr-ASN-Ile-Gly-Met-Ile-Glu-
Asn-    -ALA-Ile-    -    -Ser-Lys-ALA-PRO-  -ASN-Lys-Glu-His-Lys-  -Leu-Asp-THR-LYS-LYS-
Val-    -ALA-Asn-    -    -Ala-Gly-ALA-Arg-  -Ile-Asp-Ala-Asn-Tyr-  -ILE-Asn-Phe-LYS-LYS-
Glu-    -ALA-Phe-    -    -Ala-Lys-ALA-PRO-  -ASN-Asn-Ala-Asn-Trp-  -ILE-Asp-THR-LYS-LYS-
Glu-    -ALA-Phe-    -    -Ala-Lys-ALA-PRO-  -ASN-Asn-Ala-Asn-Trp-  -ILE-Asp-THR-LYS-LYS-
Thr-    -ALA-Leu-    -    -Gly-Gln-Ser-Asp-  -Ile-Glu-Glu-His-Lys-  -ILE-Asp-THR-Ala-Arg-

               125
----CYS-His-Lys
Tyr-    -------
Arg-    -GLN-----
Arg-    -GLN-----
Arg-    -GLN-----
Phe-    -GLN-----
```

γ-subunit of taipoxin (16)
prophospholipase A2 from porcine pancreas (25,27)
basic phospholipase from Naja nigricollis venom (24)
notechis II-5 (21)
notexin (19)
phospholipase chain of β-bungarotoxin (14)

Fig. 2. Alignment showing the homology between 5 chains of different "phospholipase" neurotoxins and the prophospholipase A2 from porcine pancreas. Highly conserved or invariant residues are written in capital letters, as are the odd Cys-15 in the β-bungarotoxin chain and the extra Cys-15 and Cys-19 in the γ-subunit of taipoxin. Blanks (—) denote identity with the top sequence and --- denotes a deletion. The γ-subunit of taipoxin has a large carbohydrate moiety (CHO) attached to Asn-70.

421

trypsin. The presence of this N-terminal peptide might explain the nearly negligible catalytic activity of the γ-subunit. Another similarity between the γ-subunit and the pancreatic enzyme is the presence of the pentapeptide stretch 62-66 which is deleted in the other venom proteins shown in Fig. 2.

Unfortunately, comparison of the sequences of the neurotoxic phospholipases with the sequences of pancreatic and other non-neurotoxic phospholipases of snake venom origin have not yet provided any clues to the basis for the toxicity.

III. THE NATURE AND BASIS OF THE TOXIC ACTIVITY

All of the toxins listed in section II.A. can upon intravenous administration at sufficiently high doses cause rapid death by asphyxiation by presynaptic blockade of transmission across the neuromuscular junctions of the breathing muscles. Other peripheral cholinergic junctions are of course affected as well. As was already mentioned above, β-bungarotoxin is the only one of these toxins which has no known toxic effect other than its blocking action on nerve terminals. All of the other toxins listed show some degree of direct local myotoxic action upon intramuscular injection, notexin being the most potent myotoxin by this mode of administration. However, the presynaptic activity of taipoxin, notexin, Notechis II-5 and crotoxin is so potent that intravenous injection of lethal doses into mice kills the animal before other symptoms appear. Although the *Enhydrina schistosa* myotoxin VI:5 is less myotoxic than notexin upon direct intramuscular injection it is also sufficiently less potent presynaptically to allow the development of muscle necrosis and myoglobinuria when injected intravenously at dose levels near the LD_{50} (22). In the case of the *Naja nigricollis* phospholipase the most conspicuous effects of low doses administered intravenously are damage to the lungs and kidneys.

A. Physiology and Pharmacology of Intoxication

The presynaptic effects of all these toxins are qualitatively similar and are characterized by gradual reduction to complete stop of acetylcholine release from the poisoned nerve terminals (8,34-38). Irregular spontaneous release of transmitter producing short "bursts" of miniature end plate potentials - sometimes of giant amplitude - are usually observed at some stage prior to the total shutdown. Ultrastructural examination of nerve terminals at an early stage of poisoning by taipoxin

or notexin revealed a reduction in the content of synaptic
vesicles and the presence of vesicles of abnormally large size
(39). These latter were suggested to account for the giant
miniature end plate potentials. At late stages of intoxication
the nerve terminals were shrunken and devoid of vesicles and
the axolemma showed numerous omega-shaped indentations which
were suggested to represent arrested vesicle reformation. The
blocking action of the toxins was thus attributed impaired recycling of vesicles rather than to direct interference with
the release mechanism.

The pathology of the myotoxic action of notexin has been
studied in detail by Harris *et al.* (40). A basic phospholipase
very similar to the *Naja nigricollis* enzyme described above
was reported by Dumarey *et al.* (41) to cause congestion and
hemorrhage of the lungs leading to hemoptis and damage to the
kidneys leading to hematuria. We have made similar observations
with our enzyme. Although the LD_{50} of the *N. nigricollis* enzyme
(300 μg/kg) is much higher than those of the other toxins mentioned herein the effects on the animal are devastatingly cruel.

B. Biochemical Studies of the Toxins

Attempts to find biochemical correlates to the effects on
transmitter release have been made by studying the effects of
the various neurotoxins on the uptake and retention of transmitter or transmitter precursors by synaptosomes (also called
nerve terminal sacs or T-sacs) from brain and electric tissue.
β-Bungarotoxin was found to inhibit the high-affinity uptake
of γ-aminobutyric acid (GABA) by synaptosomes prepared from
mouse (42) and rat brain (43). In the former study the toxin
also caused release of previously accumulated transmitter, but
this occurred without lysis of the synaptosomes. In a more recent investigation Dowdall *et al.* (44) compared the effects of
notexin, notechis II-5, taipoxin, β-bungarotoxin and the *E.
schistosa* VI:5 on the high affinity uptake of choline by T-sacs
prepared from the electric tissue of the stingray, *Torpedo mar-
morata*. Several other basic proteins, such as protamine and
the *post*synaptic neurotoxins siamensis 3 and α-bungarotoxin
were used as controls. Notexin, notechis II-5, taipoxin and *E.
schistosa* VI:5 all inhibited the uptake completely and irreversibly, notexin being the most potent. β-Bungarotoxin gave maxim-
ally about 50 % inhibition. Under the conditions used, taipoxin
could totally block uptake without causing any release of previously accumulated choline, indicating that the toxin did not
cause gross damage to the membrane. Bee venom phospholipase A2,
which is not a structural homolog of the toxins, lysed the T-
sacs and released all accumulated choline. Of the control proteins, only the very basic protamine caused any inhibition, but

the effect was reversible.

C. The Role of the Phospholipase Activity

As was mentioned above, all of these presynaptically active neurotoxins show calcium-dependent phospholipase A activity, and attempts to establish whether and to what extent this
is involved in the toxic action have been made in several laboratories (30,35,42,44-51).

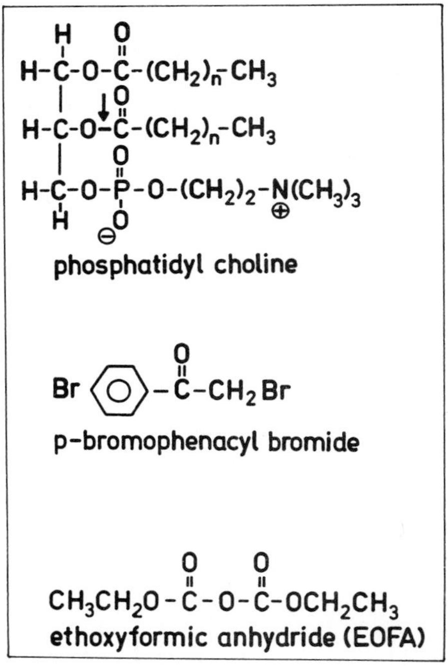

$$H-\underset{\underset{H}{|}}{\overset{\overset{H}{|}}{C}}-O-\overset{\overset{O}{\|}}{C}-(CH_2)_n-CH_3$$

phosphatidyl choline

p-bromophenacyl bromide

ethoxyformic anhydride (EOFA)

Fig. 3. Chemical formulas of the generic substrate and
two reagents which have been used to specifically inactivate
phospholipases A2.

Although all of these "phospholipase" neurotoxins can and
do liberate fatty acids from synaptosomes (T-sacs) and even
other cell membranes under "physiological" conditions it is by
no means easy to prove that this is an *essential* feature of
their presynaptic blocking action instead of a secondary consequence of their binding to membranes that happen to contain
phospholipids which are susceptible to attack. However, very
strong evidence that the catalytic activity, or at least the

structural integrity of the catalytic site, is essential for
the toxic action has been obtained both by manipulating the
ionic environment and by direct chemical modification of the
toxins.

The former line of experimentation is based on the fact
that both the release of transmitter from stimulated nerves
and the high-affinity uptake of transmitter or transmitter-
precursor by T-sacs can proceed almost normally in media containing strontium instead of calcium, whereas, with the possible exception of taipoxin (35), the phospholipase activity
of the toxins is absolutely calcium-dependent and strontium is
a competitive inhibitor. In strontium media β-bungarotoxin (35,
48-50), and crotoxin (35) do not inhibit transmitter release
from stimulated nerves. Although taipoxin showed about 30 % of
its normal blocking potency in Sr^{2+}-Tyrode, it also showed appreciable phospholipase activity in media containing Sr^{2+} instead of Ca^{2+}, was not inhibited by Sr^{2+} in the presence of
Ca^{2+}, and furthermore, was activated by Mg^{2+}, which is abundant
in the axoplasm (35). However, neither taipoxin, notexin, notechis II-5, *E. schistosa* VI:5 nor β-bungarotoxin inhibited the
high-affinity uptake of choline by *Torpedo* T-sacs in Sr^{2+}-
Ringer (44).

D. Chemical Modification Studies

Many homologous phospholipases A2 can be inactivated with
surprisingly high specificity by chemical modification with the
two reagents shown in Fig. 3, neither of which shows any re-
semblance to the substrate. The inactivation of the pork panc-
reactic phospholipase A by specific alkylation of the invariant
His-48 with p-bromophenacyl bromide was first demonstrated by
Volwerk *et al.* (29). This reaction, which is competitively inhibited by Ca^{2+}, seems to be an extremely specific "active site
probe" for phospholipases homologous to the pancreatic enzyme.
This reaction was first applied to a presynaptically neurotoxic
phospholipase by Halpert *et al.* (30), who showed that the
neurotoxicity, myotoxicity and phospholipase activity of notexin were reduced at least 100-fold by alkylation of His-48.
The reaction goes equally well with notechis II-5 and with the
same results.

Treatment of intact taipoxin with p-bromophenacyl bromide
results in the alkylation of single histidine residues (presumably His-48) in the α- and β-subunits (46). The γ-subunit does
not react. The (PBP)$_2$-taipoxin thus formed (Fig. 4) has no significant enzymatic or toxic activity. When the separated subunits were individually treated with the reagent alkylation
again occurred only with the α- and β-subunits. Although the
γ-subunit contains the invariant His-48 and, like the α-subunit,

binds Ca^{2+} and exhibits phospholipase activity, it does not
react with the reagent. The β-subunit, on the other hand, is
enzymatically inactive, apparently owing to complete lack of
affinity for Ca^{2+}, but is nevertheless specifically alkylated.
Reactivity of His-48 with p-bromophenacyl bromide is thus evidently not an infallible indicator of a functional catalytic
site.

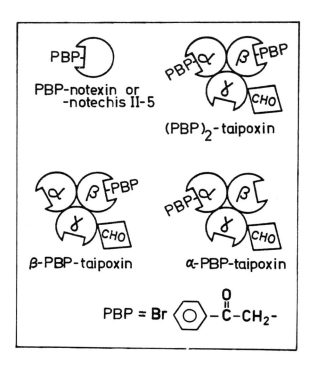

Fig. 4. Schematic representation of chemically modified
forms of notexin, notechis II-5 and taipoxin prepared by specific alkylation with p-bromophenacyl bromide. The PBP-group is
attached to the invariant His-48 in the active-site cleft.

Since the alkylation of the α- and β-subunits does not
interfere with reformation of the taipoxin complex it was easy
to prepare reconstituted taipoxin alkylated on either the α-
or β-subunit. The β-PBP-taipoxin was nearly as toxic as the
native toxin, but the α-PBP-taipoxin (Fig. 4) was inactive (46).
These results show that the strongly basic and enzymatically
active α-subunit is the essential toxic component of the taipoxin complex, the other subunits having some important "helper" roles.

Chemical modification experiments with ethoxyformic anhydride (EOFA) have been done with β-bungarotoxin (47) and notechis II-5 (45). Although EOFA can react with histidine, tyrosine, and lysine and its use for the modification of these toxins is thus not as straight-forward as the alkylation with p-bromophenacyl bromide, selectivity can nevertheless be achieved by manipulation of the reaction conditions and by selective removal of EOF-groups from tyrosine and histidine with hydroxylamine.

In the case of notechis II-5, a single residue of tyrosine (perhaps Tyr-28), one residue of lysine and two residues of histidine were readily ethoxyformylated at pH 6 (45). Under these conditions, the ethoxyformylation of the faster-reacting histidine caused a parallel decrease in the lethality and the phospholipase activity toward egg yolk. The slower-reacting histidine could be selectively de-blocked with low concentrations of hydroxylamine without restoration of lethality or catalytic activity toward egg yolk, indicating that this residue was not important for either activity. Unfortunately, the faster-reacting histidine could not be deblocked without also deblocking the ethoxyformylated tyrosine, so it was not possible to determine whether ethoxyformylation of the tyrosine might also affect these activities. The most interesting observation was that the ethoxyformylation of the fast-reacting histidine did not diminish the enzyme activity toward purified egg yolk lecithin, indicating that the catalytic site per se was not affected. Apparently the fast-reacting histidine is part of a site that allows the enzyme to interact with more complicated arrangements of the substrate such as occur in micelles of egg yolk lipoprotein or the nerve terminal membrane. Since two histidine residues were also reactive toward ethoxyformic anhydride in the PBP-His-48 derivative of the toxin, the remaining two histidine residues His-14 and His-21 must be the sites of the ethoxyformylation.

Halpert (45) also showed that more extensive treatment with ethoxyformic anhydride led to an inactivation which could not be reversed with hydroxylamine and which was therefore ascribed to ethoxyformylation of a lysine reside. This irreversible modification decreased the lethality and the enzymatic activity toward both egg yolk and purified lecithin to the same extent, indicating that a lysine amino group might be somehow involved in the catalytic site itself. On the other hand, Howard and Truog reported that ethoxyformylation of amino groups in the presence of Ca^{++} and substrate abolished the neurotoxicity of β-bungarotoxin without affecting the catalytic activity (47).

E. The Roles of Helper Subunits

The observations reviewed above suggest that a highly bas-

ic surface might be required for the specific interaction between these toxins and their target(s) in the membranes. However, excessive basicity must apparently be shielded to prevent
less specific binding and absorption of these relatively small
molecules by many types of cells, which would reduce their
specific toxicity by a dilution effect. This kind of "chaperone"
function might be the role of the acidic crotapotin of crotoxin
and the acidic γ-subunit of taipoxin. In the case of notexin a
good balance of charge has apparently been achieved without the
aid of any helper. Notechis II-5 is more basic and has higher
phospholipase activity but is less neurotoxic and myotoxic than
notexin.

The helper subunit(s) when present, might also have an important function at the target level. This is almost certainly
the case with β-bungarotoxin, where the helper subunit seems to
provide, rather than shield basicity and is, moreover, covalently bound to the neutral phospholipase subunit. The observation
of Hendon and Fraenkel-Conrat (52) that crotoxin A (crotapotin)
strongly enhanced the neurotoxicity of crotoxin B (the basic
phospholipase) even when the two were injected separately suggests that the non-toxic helper subunit might have an affinity
for and function at the target.

VI. CONCLUSIONS AND SPECULATIONS

Although the phospholipase activity of these toxins seems
to be an essential feature of their neuro- and/or myotoxic actions it seems unlikely that the primary target to which they
bind should be the substrate itself. Oberg and Kelly (53) have
reported saturable, high-affinity binding of labelled β-bungarotoxin to cell membranes. The sites were characterized by dissociation constants in the nanomolar range, which is compatible
with the concentration of toxin required to block transmitter
release. This high affinity suggests that the binding site
might include protein in addition to phospholipid. This hypothesis is further supported by the observation of Howard and
Truog (47) that the release of fatty acids from purified bacterial membranes by β-bungarotoxin was inhibited in the presence of trypsin or Pronase, and since neither of these proteases inactivated the toxin itself the inhibition was attributed
to alteration of the membranes.

Strong and Kelly showed that membranes undergoing phase
transitions are preferentially hydrolyzed by β-bungarotoxin,
and suggested that the toxin might therefore attack lipids at
a phase boundary between a protein recognition site imbedded
in the membrane and the surrounding lipid bilayer (51). Phase
transitions might also occur in the nerve-terminal membrane
during the exocytotic release of transmitter and the endocytot-

ic retrieval of vesicles. This would account for the fact that the blockade of transmitter release by the toxins is hastened and aggravated by nerve stimulation.

Labelled β-bungarotoxin did not appear to cross synaptosomal membranes (53). Further evidence that the toxic action does not require entrance of the toxin into the terminal is the observation of Howard and Wu that β-bungarotoxin immobilized on beads of Sepharose 2B still inhibited the uptake of GABA by brain synaptosomes (54). This should open the way to the use of immobilized toxins as affinity supports for the isolation of the target protein, as has been done with emminent success with the nicotinic acetylcholine receptor by the use of immobilized post-synaptic neurotoxins.

One would also like to know whether any particular protein(s) might be specifically released from the affected membranes by the action of these toxins. If so, these toxins might be elegant tools for microsurgical dissection of the nerve terminal membrane.

ACKNOWLEDGMENTS

I am deeply grateful to my colleagues Jan Fohlman, Jim Halpert, Peter Lind, and Evert Karlsson for most of the experimental work on which our contributions to the literature of these toxins are based.

REFERENCES

1. Chang, C.C., and Lee, C.Y., Arch. Int. Pharmacodyn. 144:241 (1963).
2. Lester, H., Sci. Am. 236:107 (1977).
3. Karlsson, E., in "Handbook of Pharmacology", Vol. 52 (C.Y. Lee, ed.) Springer, Berlin, in press.
4. Tsernoglou, D., and Petsko, G.A., FEBS Lett. 68:1 (1977).
5. Tsernoglou, D., and Petsko, G.A., Proc. Natl. Acad. Sci. USA 74:971 (1977).
6. Low, B.W., Preston, A.S., Sato, A., Rosen, L.S., Searl, J. E., Rudko, A.D., and Richardson, J.S., Proc. Natl. Acad. Sci, USA 73:2991 (1977).
7. Slotta, K.H., and Fraenkel-Conrat, H., Ber. Dtsch. Chem. Ges. 71:1076 (1938).
8. Brazil, O.V., and Excell, B.J., J. Physiol. 212:34P (1971).
9. Eaker, D., Toxicon 13:90 (1975).
10. Eaker, D., Halpert, J., Fohlman, J., and Karlsson, E., in "Animal, Plant, and Microbial Toxins", Vol. 2 (A. Ohsaka, K. Hayashi and Y. Sawai, eds.) p. 27. Plenum Publishing Co., New York, 1976.
11. Karlsson, E., Eaker, D., and Rydén, L., Toxicon 10:105 (1972).

12. Lee, C.Y., Chang, S.L., Kau, S.T., and Luh, S.H., J. Chromatogr. 72:71 (1972).
13. Kelly, R.B., and Brown, F.R., J. Neurobiol. 5:135 (1974).
14. Kondo, K., Narita., and Lee. C. Y., J. Biochem. 83:101 (1978).
15. Fohlman, J., Eaker, D., Karlsson, E., and Thesleff, S., Eur. J. Biochem. 68:457 (1976).
16. Fohlman, J., Lind, P., and Eaker, D., FEBS Lett. 84:367 (1977).
17. Habermann, E., and Breithaupt, H., Toxicon 16:19 (1978).
18. Breithaupt, H., Rübsamen, K., and Habermann, E., Eur. J. Biochem. 49:333 (1974).
19. Halpert, J., and Eaker, D., J. Biol. Chem. 250:6990 (1975).
20. Kannan, K.K., Lövgren, S., Cid-Dresdner, H., Petef, M., and Eaker, D., Toxicon 15:435 (1977).
21. Halpert, J., and Eaker, D., J. Biol. Chem. 251:7343 (1976).
22. Fohlman, J., and Eaker, D., Toxicon 15:385 (1977).
23. Obidairo, T.K., Tampitag, S., and Eaker, D., in preparation.
24. Joubert, F., Biochim. Biophys. Acta 493:216 (1977).
25. De Haas, G.H., Slotboom, A.J., Bonsen, P.P.M., van Deenen, L.L.M., Maroux, S., Puigserver, A., and Desnuelle, P., Biochim. Biophys. Acta 221:31 (1970).
26. De Haas, G.H., Slotboom, A.J., Bonsen, P.P.M., Niewinheizen, W., and van Deenen, L.L.M., Biochim. Biophys. Acta 221:54 (1970).
27. Puijk, W.C., Verheij, H.M., and De Haas, G.H., Biochim. Biophys. Acta 492:254 (1977).
28. Drenth, J., Enzing, C.M., Kalk, K.H., and Vessies, J.C.A., Nature 264:373 (1976).
29. Volwerk, J.J., Pieterson, W.A., and De Haas, G.H., Biochemistry 13:1446 (1974).
30. Halpert, J., Eaker, D., and Karlsson, E., FEBS Lett. 61:72 (1976).
31. Pieterson, W.A., Vidal, J.C., Volwerk, J.J., and De Haas, G.H., Biochemistry 13:1455 (1974).
32. van Dam-Mieras, M.C.E., Slotboom, A.J., Pieterson, W.A., and De Haas, G.H., Biochemistry 14:5387 (1975).
33. Slotboom, A.J., and De Haas, G.H., Biochemistry 14:5394 (1975).
34. Chang, C.C., Lee, J.D., Eaker, D., and Fohlman, J., Toxicon 15:571 (1977).
35. Chang, C.C., Su, M.J., Lee, J.D., and Eaker, D., Naunyn-Schmiedebergs Arch. Pharmacol. 299:155 (1977).
36. Chang, C.C., Chen, T.F., and Lee, C.Y., J. Pharmacol. Exp. Ther. 184:339 (1973).
37. Harris, J.B., Karlsson, E., and Thesleff, S., Br. J. Pharmac. 47:141 (1973).

38. Kamenskaya, M.A., and Thesleff, S., Acta Physiol. Scand. 90:716 (1974).

39. Cull-Candy, S.G., Fohlman, J., Gustavsson, D., Lüllmann-Rauch, R., and Thesleff, S., Neuroscience 1:175 (1976).

40. Harris, J.B., Johnson, M.A., and Karlsson, E., Clin. Exp. Pharmacol. Physiol. 2:383 (1975).

41. Dumarey, C., SMet, M.D., Joseph, D., and Boquet, P., C.R. Acad. Sci. Paris 280:D1633 (1975).

42. Wernicke, J.F., Vanker, A.D., and Howard, B.D., J. Neurochem. 25:483 (1975).

43. Sen, I., Grantham, P.A., and Cooper, J.R., Proc. Natl. Acad. Sci. USA 73:2664 (1976).

44. Dowdall, M.J., Fohlman, J.P., and Eaker, D., Nature 269: 700 (1977).

45. Halpert, J., to be published.

46. Fohlman, J., Dowdall, M.J., Eaker, D., Sjödin, T., Leander, S., and Lüllmann-Rauch, R., Eur. J. Biochem., in press.

47. Howard, B.D., and Truog, R., Biochemistry 16:122 (1977).

48. Strong, P.N., Goerke, J., Oberg, S.G., and Kelly, R.B., Proc. Natl. Acad. Sci. USA 73:178 (1976).

49. Kelly, R.B., Oberg, S.G., Strong, P.N., and Wagner, G.M., Cold Spring Harbor Symp. Quant. Biol. 40:117 (1975).

50. Strong, P.N., Heuser, J.E., and Kelly, R.B., in "Cellular Neurobiology", p. 227, Alan R. Liss, Inc., New York, 1977.

51. Strong, P.N., and Kelly, R.B., Biochim. Biophys. Acta 469: 231 (1977).

52. Hendon, R.A., and Fraenkel-Conrat, H., Toxicon 14:283 (1976).

53. Oberg, S.G., and Kelly, R.B., Biochim. Biophys. Acta 433: 662 (1976).

54. Howard, B.D., and Wu, W.C.S., Brain Research 103:190 (1976).

PHARMACOLOGY OF SNAKE TOXINS WITH PHOSPHOLIPASE A$_2$ ACTIVITY

C. Y. Lee
C. L. Ho

Pharmacological Institute
College of Medicine
National Taiwan University and
Institute of Biological Chemistry
Academia Sinica
Taipei, Taiwan, R. O. C.

INTRODUCTION

Phospholipase A$_2$ (PLA2) is widely distributed in all of the snake venoms so far investigated. It catalyzes the hydrolysis of the fatty acyl ester group of phospholipids at the 2-position. The enzyme that acts at the 1-position has not been detected in snake venoms.

Since phospholipids are one of the main constituents of biological membranes, this enzyme has been extensively used as an enzymatic probe in the study of the essentiality of phospholipids to the structure and function of biological membranes.

Opinions concerning the pharmacology, especially the toxicity of snake venom PLA have dramatically changed over the past two decades. In the literature up to 1950's, it was often claimed that PLA was the component responsible for most of the venom actions. Most of these studies were, however, carried out with either impure preparations or even heated crude venom. For example, Tobias claimed that PLA was responsible for the conduction block of motor nerve caused by cobra venom (1). He used acid-boiled cobra venom as the source of PLA. However, cardiotoxin as well as neurotoxin in cobra venom are also heat stable and we have demonstrated that cardiotoxin, but not PLA, is responsible for the conduction block of nerve caused by cobra venom (2).

433

Then, opinions in the 1950's and 1960's changed dramatically. Since toxic principles such as neurotoxin and cardiotoxin could be separated from PLA and moreover, a few purified preparations of PLA were found to be pharmacologically weak, generalizations were then made that all venom PLAs have little toxicity.

In contrast, recent studies on so-called presynaptic neurotoxins indicate that certain snake venoms do contain toxic PLA and that PLA may be responsible for the venom effects on the neuromuscular transmission. There are several snake toxins with high neuro-toxicity, which act presynaptically on the motor nerve terminals to affect acetylcholine release. These presynaptic neurotoxins, in most cases, contain a subunit which is a basic PLA (e.g. crotoxin, taipoxin, β-bungarotoxin). Crotoxin, isolated in a crystalline form from the venom of the South American rattle-snake *Crotalus durissus terrificus* in 1938 by Slotta and Fraenkel-Conrat (3), has been found to be a complex between a basic PLA_2, also called crotoxin B, and an acidic protein, crotoxin A or crotapotin. The basic PLA_2 is only moderately toxic but upon recombination with crotapotin, practically full lethality can be restored (4,5). Taipoxin from the Australian taipan *Oxyuranus scutellatus scutellatus* is a noncovalent ternary complex of α-, β-, and γ- subunits. The α-subunit is a basic PLA_2 with an LD_{50} of 300 μg/kg mouse. The β-subunit is a neutral PLA-analogue but practically non-

TABLE I. Lethality of Snake Toxins with
Phospholipase A_2 Activity

Toxin	LD_{50} (μg/kg mouse)		Reference
Taipoxin	2.1	(i.v.)	Fohlman et al (6)
β-Bungarotoxin	14	(i.p.)	Lee et al (8)
Notexin	17	(i.v.)	Karlsson et al (10)
Crotoxin	35	(i.p.)	Hendon & Fraenkel-Conrat (5)
N. nigricollis			
basic PLA_2	250	(i.v.)	Dumarey et al (12)
	1400	(i.p.)	Ho (unpublished)
neutral PLA_2	>5000	(i.p.)	" "
N. naja atra			
PLA_2 (acidic)	8000	(i.p.)	Chang et al (13)
N. melanoleuca			
PLA_2 (acidic)	>5000	(i.p.)	Ho (unpublished)

toxic. The γ-subunit is an acidic glycoprotein and also non-
toxic. When the three subunits are combined, the enzymatic activity of the α-subunit is suppressed and the lethality is increased about 500-fold on a molar basis (6). β-Bungarotoxin
isolated from the venom of *Bungarus multicinctus* (7,8) consists
of two dissimilar polypeptide chains cross-linked by disulfide
bonds. The amino acid sequence of the A-chain is similar to
that of notexin and PLA_2 from other snake venoms. The amino
acid sequence of the B-chain shows some homology with that of
protease inhibitors (9). Notexin from the Australian tiger
snake *Notechis scutatus scutatus* is a single-chain basic PLA_2
without having a subunit (10). Besides presynaptic effects,
notexin also displays a high myotoxicity (11).

As shown in Table I, there are large variations in lethality among PLAs from different sources. In general, basic PLA
preparations are more toxic than acidic or neutral PLAs. Although the pharmacological reasons for the differences in lethality are not quite clear, it is apparent that toxins acting
at the presynaptic nerve endings are extremely toxic.

In this presentation, various pharmacological properties
and enzyme activities of PLA_2 preparations isolated from snake
venoms of different species will be compared and the relationship between their pharmacological actions and enzyme activity
will be discussed.

MATERIALS AND METHODS

The PLA_2 preparations and toxins with PLA_2 activity used
in this study are listed in Table II.

Assay of enzyme activity of PLA_2

The method of Strong et al. (18) was slightly modified.
Egg lecithin (Sigma) was emulsified with equimolar sodium de-
oxycholate in the presence of 0.05 mM EDTA and 2.5 mM Ca^{2+}.
Enzyme activity was determined by measuring the amount of NaOH
(2mM) needed to titrate the fatty acid liberated from lecithin
by the action of PLA_2. The reaction was performed at $37^\circ C$ and
the titration end-point was set at pH 7.4.

Indirect hemolysis

Indirect hemolysis was assayed with the three-time washed
erythrocytes (RBC) of the guinea-pig, suspended in Tris buffer
in the presence of 2.5 mM Ca^{2+}. One μg of PLA_2 was added to

0.1 ml lecithin emulsion and incubated at 37°C for 15 min. Immediately after the incubation, 0.5 ml of RBC suspension was
added and incubated for another 60 min. Hemolytic reaction was
stopped by addition of 1 ml of cooled (4°C) Tris buffer solution. After centrifuging at 1500 rpm for 10 min the supernatant
was read for released hemoglobin at the wavelength of 540 nm.

TABLE II.   PLA2 Preparations and Toxins with PLA2 Activity
              Studied in the Present Investigation

PLA2 or Toxin	Abbreviation	Mothod of Purification
Acidic PLA2 from		
Naja naja atra	NNA	Lo & Chang (14)
Acidic PLA2 from		
N. melanoleuca	NM-4	Wu (15)
Neutral PLA2 from		
N. nigricollis	NG-2	Wu et al (16)
Basic PLA2 from		
N. nigricollis	NG-4	Wu et al (16)
β-Bungarotoxin	β-BuTX	Lee et al (8)
Notexin*	NoTX	Karlsson et al (10)
Ceruleotoxin**	CeTX	Bon & Changeux (17)

* A gift from Dr. E. Karlsson, Inst. of Bioch., Uppsala.
** A gift from Dr. C. Bon, Institut Pasteur, Paris.

Direct hemolysis

Half milliliter of the washed guinea-pig RBC suspension
(in Tris buffer with 2.5 mM Ca^{2+}) was incubated with 0.05 ml
of 1 mg/ml PLA2 preparation at 37°C for 60 min. Measurement of
hemolysis was the same as for the indirect hemolysis.

Biventer cervicis nerve-muscle preparation of the chick

The biventer cervicis nerve-muscle preparation isolated
from the baby chick according to the method of Ginsborg and
Warriner (19) was suspended in 10 ml of Krebs solution, aerated
with 95% O_2 + 5% CO_2 at 37°C. The preparation was stimulated
indirectly with supramaximal rectangular pulses of 0.5 msec at
a frequency of 0.2 Hz. Muscle contraction was recorded with a
Grass force-displacement transducer connected to a Grass 7P
polygraph.

Measurement of membrane potentials

The resting membrane potential was measured from the superficial muscle layer of the mouse diaphragm by the conventional microelectrode recording technique of Fatt and Katz (20)

Isolated guinea-pig ileum

A segment of about 2.5 cm of the guinea-pig ileum was suspended in an organ bath containing 10 ml Krebs solution, oxygenated with 95% O_2 + 5% CO_2 at 30°C. The longitudinal contraction was recorded isometrically with a Grass force-displacement transducer connected to a Grass 7P polygraph.

Blood pressure of the rat

The carotid arterial blood pressure of the rat, anesthetized with 30 mg/kg of pentobarbital sodium, was recorded with a Statham P_{23} AC pressure transducer. PLA_2 was injected intravenously into the femoral vein.

Myonecrotic effects of PLA_2

The PLA_2 preparations was injected subcutaneously into the right thigh of mice (NIH strain) weighing from 25 to 30 gm. After 24 hrs, the mice were killed and the treated thigh muscle was cut into small pieces and fixed with 3% glutaldehyde followed by osmication. The tissues were then dehydrated and embedded in Epon. Thin sections were stained with uranyl acetate followed by lead citrate and examined with a Hitachi HU 12A electron microscope.

RESULTS

Relationship between indirect hemolytic action and enzyme activity of PLA_2

Although indirect hemolysis has been widely used as a biological method for the assay of enzyme activity of PLA_2, the relative potencies of indirect hemolysis induced by PLA_2 preparations of different sources are only roughly correlated with their enzyme activities assayed by the chemical method (Fig. 1). The acidic PLA_2 preparations such as NNA and NM-4 showed the highest indirect hemolytic activity, while those

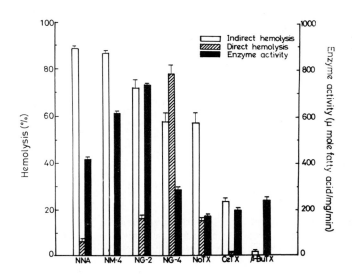

Fig. 1. Comparison of the indirect-and direct-hemolytic and enzyme activities of various PLA2s. Each PLA2 is indicated by the abbreviation (see Table II) at bottom of the histogram bars. Standard errors are shown at the top of histogram bars. Numbers of experiments are seven for indirect hemolysis, ten for direct hemolysis and five for enzyme activity.

PLA2 preparations comprising subunits, such as β–bungarotoxin and ceruleotoxin, exhibited the lowest hemolytic activity, the basic PLA2 preparations such as NG-4 and notexin being in between.

Direct hemolytic action of PLA2

Direct hemolysis induced by cobra venom has been attributed to DLF (direct lytic factor, cardiotoxin, cytotoxin)(21). Recently, we (22) have demonstrated that the basic PLA2 from *Naja nigricollis* venom possesses cardiotoxin-like activities including direct hemolytic action. In the present study, it was observed that both basic and neutral PLA2s from *Naja nigricollis* (NG-4 and NG-2) as well as notexin induced direct hemolysis on guinea-pig RBC, whereas acidic PLA2 (NNA and NM-4) and those toxins with subunits (CeTX and β-BuTX) showed negligible or no direct hemolytic activity (Fig. 1). Among those

PLA$_2$ preparations that induced direct hemolysis, the basic PLA$_2$ (NG-4) was the most active.

Effects of PLA$_2$ on the chick biventer cervicis nerve-muscle preparation

PLA$_2$ preparations from different snake venoms or even those from the same venom but with different isoelectric points show different potencies in causing blockade of neuromuscular transmission and direct depression of muscle contraction. As shown in Table III, acidic and neutral PLA$_2$ preparations (NNA, NM-4 and NG-2) are weak in these effects. At a concentration as high as 3×10^{-5} g/ml, all of them required more than 3 hrs to abolish the contraction of indirectly stimulated chick biventer cervicis muscle. The neutral PLA$_2$ appears to be more active than the acidic PLA$_2$s. After complete cessation of the contraction the muscle shows decreased responsiveness to acetylcholine (5×10^{-6} g/ml). The neuromuscular blockade by these PLA$_2$s is accompanied with some elevation in the baseline tension. The basic PLA$_2$ preparations such as notexin and NG-4 show stronger neuromuscular blocking and musculotropic actions.

TABLE III.   Effects of PLA$_2$ on the Chicken Biventer
            Cervicis Nerve Muscle Preparation

PLA$_2$ (g/ml)	Time for N-M block* (min, mean ± S.E.M.)	ACh response after N-M block	Contracture
NNA (3×10^{-5})	318.0±26.2 (n=5)**	decreased	+
NM-4 (3×10^{-5})	345.4±20.2 (n=5)	decreased	+
NG-2 (3×10^{-5})	232.0±16.3 (n=5)	decreased	+
NG-4 (2×10^{-5})	117.5± 7.0 (n=6)	greatly decreased	++
NoTx (1×10^{-5})	81.0± 3.8 (n=3)	greatly decreased	++
CeTx (1×10^{-5})	134.0± 8.7 (n=3)	greatly decreased	++
β-BuTx (2×10^{-5})	40.3± 5.5 (n=3)	no decrease	−

*   N-M block: Neuromuscular block.
** n: Number of experiments.

Notexin at a concentration of 10^{-5} g/ml and NG-4 at 2 x 10^{-5} g/ml produce marked contracture with a greatly reduced responsiveness to acetylcholine as reported previously (22,23). β-Bungarotoxin possesses a high selectivity to the motor nerve terminal (24). At a concentration as high as 2 x 10^{-5} g/ml, the acetylcholine response was not affected. Ceruleotoxin has been claimed to be an acidic protein capable of inhibiting the membrane ion permeability without binding to the acetylcholine receptor site (17). In our experiment, however, it behaves like a basic PLA_2 in inducing muscle contracture and causing reduced responsiveness to acetylcholine.

Effects of PLA_2 on the resting membrane potential of skeletal muscle

The resting membrane potential of the mouse diaphragmatic muscle was reduced most rapidly by the basic PLA_2 (NG-4). The neutral PLA_2 (NG-2) was much weaker while the acidic PLA_2 (NNA and (NM-4) were weakest among PLA_2 preparations tested in depolarizing the cell membrane (Fig. 2). Ceruleotoxin at a concentration of 2 x 10^{-5} g/ml depolarized the membrane to a grea-

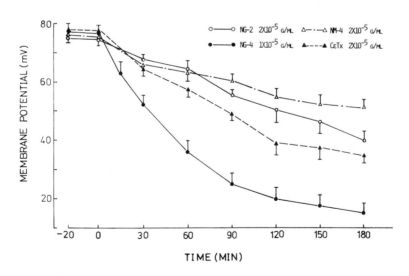

Fig. 2. Effects of various PLA_2s on the resting membrane potential of the isolated mouse diaphragm. Different PLA_2s are represented by different symbols as indicated at the upper right portion of the figure. Each point represents mean ± S.E.M. of 50 observations from 5 animals.

ter extent than did both acidic and neutral PLA$_2$ preparations. β-Bungarotoxin has no effect on the resting membrane potential.

Myonecrotic effects of PLA$_2$

Notexin has been shown to be a potent myonecrotic toxin causing necrosis of muscle fibers within 12 - 24 h (11). Our present studies reveal that not only the basic PLA$_2$ (NG-4) but also the neutral PLA$_2$ (NG-2) causes necrotic changes of muscle fibers of the injected leg, although the basic PLA$_2$ is far more potent than the neutral one (Fig. 3).

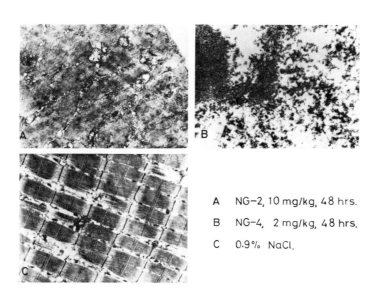

A   NG-2, 10 mg/kg, 48 hrs.

B   NG-4, 2 mg/kg, 48 hrs.

C   0.9% NaCl.

Fig. 3. Electron micrographs demonstrating necrotic actions of the neutral and basic PLA$_2$s from *Naja nigricollis* venom. Neutral (NG-2, 10 mg/kg) (A) and basic (NG-4, 2 mg/kg) (B) PLA$_2$s are injected subcutaneously into the right thigh of the mouse (X 10,000). Note the loss of striation of myofibrils and the swelling of mitochodria in (A), and the necrotic change of the entire muscle fibers, forming an amorphorus mass in (B).

Effects of PLA$_2$ on the isolated guinea-pig ileum

The isolated guinea-pig ileum is very sensitive to the acidic PLA$_2$ from cobra venom. As shown in Fig. 4, at a concen-

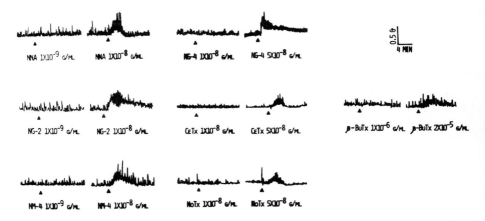

Fig. 4. Stimulant effects of PLA2s on the isolated guinea-pig ileum. Triangles indicate the time of application of
PLA2. The tone increased by PLA2 recovered spontaneously without washing. Tension and time calibrations are given at the
uper right corner.

tration as low as 10^{-8} g/ml, the acidic PLA2s (NNA and NM-4)
are capable to induce a contracture of the intestinal smooth
muscle. The basic PLA2s (NG-4 and notexin) as well as ceruleotoxin require five times the concentration to evoke a similar
contracture of the ileum. β-Bungarotoxin is the weakest in
stimulating the ileum. Its minimal effective concentration is
about 2000 times that of the acidic PLA2. Contracture induced
by PLA2 is easily blocked by pretreatment with indomethacin
(10^{-6} g/ml), suggesting possible involvement of prostaglandin
release.

Hypotensive effects of PLA2 in the anesthetized rat

PLA2, when given intravenously, causes a transient hypotension in rats, cats and rabbits. Rats are the most sensitive
species among these animals. As shown in Fig. 5, at a dose as
low as 0.01 mg/kg, both the acidic (NNA) and neutral PLA2
(NG-2) produce an appreciable fall in arterial blood pressure
in rats. A second larger dose of PLA2 given after recovery
from the initial fall fails to produce hypotensive response,
indicating development of tachyphylaxis. The basic PLA2s (NG-4
and notexin) are slightly weaker; they need five times higher
dose to induce a hypotension of similar extent. β-Bungarotoxin
is much weaker in this action. It requires a dose of 0.5 mg/kg
to produce a depressor action of slow onset in rats.

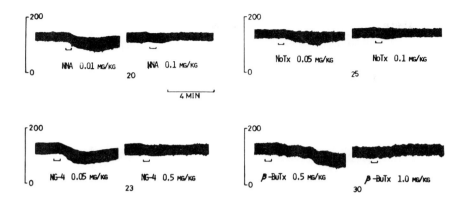

Fig. 5. Depressor action of PLA₂ in the anesthetized rat. The symbol ⌣ denotes the duration of i.v. injection of PLA₂. Pressure calibration in mmHg is given at the left side of each tracing. Time calibration is given in the middle of the figure. Note the development of tachyphylaxis after the first administration of each PLA₂. The number below the pressure tracing denotes the time in min after the first administration of each PLA₂.

DISCUSSION

The pharmacology of PLA₂ from snake venoms has recently been reviewed by Rosenberg (25). It is now well-established that PLA₂ is a toxic enzyme, possessing a great variety of pharmacological actions. PLA₂ preparations from various snake venoms, however, show marked differences not only in their enzyme activities but also in their toxicities and pharmacological properties. In general, acidic PLA₂ preparations are of low lethality, while basic PLA₂ preparations are of relatively high lethality (Table I). There are several snake toxins with extremely high neurotoxicity (e.g. β-bungarotoxin, taipoxin, notexin, crotoxin, etc) which act presynaptically on motor nerve terminals to affect acetylcholine release. These presynaptic toxins are either basic PLA₂ or contain a subunit which is a basic PLA₂. Although the pharmacological reasons for the differences in lethality are not quite clear and there is apparently no correlation between the lethality and their PLA₂ activity, the selectivity of toxins to act at the presynaptic site of motor nerve endings seems to be responsible for the high lethality of these snake toxins.

From the results of our present study, the PLA$_2$ preparations from elapid venoms may be classified into the following three categories:

1. Acidic or neutral PLA$_2$s, such as NNA, NM-4 and NG-2, possess relatively high catalytic and indirect-hemolytic but negligible or no direct-hemolytic activities. These preparations are also highly effective in causing hypotensive response and contracture of the guinea-pig ileum, both of which may result from the release of autopharmacologic substances such as histamine, serotonin, prostaglandins, etc. However, their effects on the nerve and skeletal muscle are rather weak, being consistent with their relatively low lethality.

2. Basic PLA$_2$s, such as NG-4 and notexin, have moderate catalytic and indirect-hemolytic activities, being less effective in producing hypotensive response and contracture of the intestinal smooth muscle but more potent in inducing direct hemolysis, neuromuscular blockade, contracture of skeletal muscle and myonecrosis. These effects are rather similar to those of cardiotoxin from cobra venom. As pointed out by Strysom (26) it is interesting to note that there is a number of correspondences in sequences of PLA$_2$s at the half portion from C-terminal and the full sequences of cardiotoxins.

Ceruleotoxin exhibits similar pharmacological actions to those of basic PLA$_2$s, although it has been reported to be an acidic protein (17). At present, this is the only exception from the above classification, and we have no explanation for such a deviation.

3. Presynaptic toxins, as exemplified by β-bungarotoxin, show relatively low catalytic and negligible or no hemolytic activities. These toxins, being more selective to act on the motor nerve endings, have less effects on the muscle itself. They are also weak to affect the smooth muscle and the circulatory system.

It has been argued whether or not the enzyme activity is responsible for the neurotoxicity of these toxins. Since treatments which inhibit or abolish the catalytic activity, such as replacement of Ca^{2+} by Sr^{2+} (18), or reaction with P-bromophenacyl bromide (27), also inhibit or abolish the neurotoxicity, it seems very likely that the PLA$_2$ activity plays a direct role in the presynaptic neurotoxicity. However, it remains to be elucidated which part of the molecules of these presynaptic toxins is responsible for the selectivity. It is interesting to note that those toxins containing PLA$_2$ as a subunit, such as β-bungarotoxin, taipoxin and crotoxin, are more selective in the presynaptic action than the single-chain basic PLA$_2$ such as notexin.

SUMMARY

Pharmacological actions of PLA$_2$ preparations from several elapid venoms including β-bungarotoxin, notexin and ceruleotoxin were compared along with their enzyme activities.

In general, acidic and neutral PLA$_2$s are of low lethality, despite their high catalytic and indirect-hemolytic activities. They are also highly effective in causing hypotensive response and contracture of the intestinal smooth muscle but are rather weak in their actions on the nerve and skeletal muscle.

By contrast, basic PLA$_2$s are of relatively high lethality and are more potent in inducing direct hemolysis, neuromuscular block, contracture of skeletal muscle and myonecrosis, although their catalytic and indirect-hemolytic activities are less potent than are acidic and neutral PLA$_2$s. They are also less effective in producing hypotensive response and contracture of the intestinal smooth muscle.

Ceruleotoxin behaves like a basic PLA$_2$ in a number of pharmacological tests, although it has been reported to be an acidic neurotoxin.

On the other hand, the presynaptic toxins which contain PLA$_2$ as a subunit are highly toxic despite their relatively low catalytic and negligible or no hemolytic activities. These toxins act selectively on the motor nerve endings and have less or no direct effects on the muscle itself. They are also weak to affect the circulatory system and to stimulate the intestinal smooth muscle.

ACKNOWLEDGMENTS

The authors wish to express their appreciation to Dr. T. B. Lo for the supply of purified phospholipase A from *Naja naja atra*, Dr. E. Karlsson for the supply of notexin and Dr. C. Bon for the supply of ceruleotoxin. The eletron microscopy was kindly performed by Dr. P. L. Chang, Department of Anatomy, College of Medicine, National Taiwan University.

REFERENCES

1. Tobias, J. M., *J. Cell. comp. Physiol.*, 46: 183 (1955).
2. Chang, C. C., Chuang, S. T., Lee, C. Y. and Wei, J. W., *Br. J. Pharmac.* 44: 752 (1972).
3. Slotta, K. H. and Fraenkel-Conrat, H. L., *Ber. dtsch. Chem. Ges. 71*: 1076 (1938).

4. Rübsamen, K., Breithaupt, H. and Habermann, E., *Naunyn Schmiedebergs Arch. Pharmak. 270*: 274 (1971).

5. Hendon, R. A. and Fraenkel-Conrat, H., *Proc. Nat. Acad. Sci.*(Wash.) *68*: 1560 (1971).

6. Fohlman, J., Eaker, D., Karlsson, E. and Thesleff, S., *Europ. J. Biochem. 68*: 457 (1976).

7. Chang, C. C. and Lee, C. Y., *Arch. int. Pharmacodyn. 144*: 214 (1963).

8. Lee, C. Y., Chang, S. L., Kau, S. T. and Luh, S. H., *J. Chromatog. 72*: 71 (1972).

9. Kondo, K., Narita, K. and Lee, C. Y., *J. Biochem.*(Tokyo) *83*: 91 (1978).

10. Karlsson, E., Eaker, D. and Rydén, L., *Toxicon 10*: 405 (1972).

11. Harris, J. B., Johnson, M. A. and Karlsson, E., *Clin. exp. Pharmacol. Physiol. 2*: 383 (1975).

12. Dumarey, C., Sket, M. D., Joseph, D. and Boquet, P., *C. R. Acad. Sci.*(Paris) *208D*: 1633 (1975).

13. Chang, W. C., Chang, C. S., Hsu, H. P. and Lo, T. B., *J. Chinese Biochem. Soc. 4*: 62 (1975).

14. Lo, T. B. and Chang, W. C., *J. Formosan Med. Assoc. 70*: 644 (1971).

15. Wu, S. H., M. S. Thesis. Nat. Taiwan University, Taipei (1976).

16. Wu, S. H., Chang, W. C., and Lo, T. B., *J. Chinese Biochem. Soc. 6*: 9 (1977).

17. Bon, C. and Changeux, J. P., *FEBS Letters 59*: 212 (1975).

18. Strong, P. N., Goerke, J., Oberg, S. G. and Kelly, R. B. *Proc. Nat. Acad. Sci.*(Wash.) *73*: 178 (1976).

19. Ginsborg, B. L. and Warriner, J., *Br. J. Pharmac. 15*: 410 (1960).

20. Fatt, P. and Katz, B., *J. Physiol.*(London) *115*: 320 (1951).

21. Lee, C. Y., *Ann. Rev. Pharmacol. 12*: 265 (1972).

22. Lee, C. Y., Ho, C. L. and Eaker, D., *Toxicon 15*: 355 (1977).

23. Lee, C. Y., Chen, Y. M. and Karlsson, E., *Toxicon 14*: 493 (1976).

24. Chang, C. C., Chen, T. F. and Lee, C. Y., *J. Pharmacol. exp. Ther. 184*: 339 (1973).

25. Rosenberg, P., in "Snake Venoms" (C. Y. Lee, ed.), Handbook exp. Pharmacol. Vol. 52, Springer, Heidelberg, (in press).

26. Strydom, D. J., in "Snake Venoms" (C. Y. Lee, ed.), Handbook exp. Pharmacol. Vol. 52, Springer, Heidelberg (in press).

27. Halpert, J., Eaker, D. and Karlsson, E., *FEBS Letters 61*: 72 (1976).

Synthesis of Snake Venom Toxins

Kung-Tsung Wang

Institute of Biological Chemistry
Academia Sinica
Taipei

Cobrotoxin is the first toxin isolated from Taiwan cobra (<u>Naja naja atra</u>) in pure form and its structure was elucidated by Yang (1), Fig. 1. It consists of 62 amino acid residues and is linked by 4 disulfide bonds. This toxin is also the first toxin being tried its synthesis (2). Aoyagi in 1972 claimed they synthesized a peptide with cobrotoxin activity and their product has 20% activity of natural cobrotoxin. They resynthesized cobrotoxin again in 1976 but obtained a product with only 6% activity (3). No detailed characterization of product was reported.

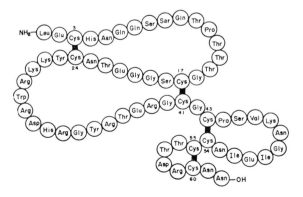

Fig. 1. The structure of Cobrotoxin

The work was supported by National Science Council, Republic of China.

447

We synthesized cobrotoxin with solid phase method of Merrifield (4) recently as shown in the reaction schedule (Table 1).

Table 1. Synthetic Schedule of Cobrotoxin

Step	Reagents and Operations	Mix times (min)
1.	CH_2Cl_2 wash, 30 ml (4 times)	2
2.	50% TFA in CH_2Cl_2, 30 ml (1 time[b])	3
3.	50% TFA in CH_2Cl_2, 30 ml (1 time)	15
4.	CH_2Cl_2 wash, 30 ml (3 times)	2
5.	33% dioxane in CH_2Cl_2, 30 ml (3 times)	2
6.	CH_2Cl_2 wash, 30 ml (3 times)	2
7.	10% Et_3N in CH_2Cl_2, 30 ml (1 time)	10
8.	CH_2Cl_2 wash, 30 ml (3 times)	2
9.	CH_2Cl_2 wash, 30 ml (3 times[c])	2
10.	Boc-amino acid in 20 ml CH_2Cl_2 (1 time[d])	10
11.	DCC in CH_2Cl_2, 10 ml (1 time)	240
12.	CH_2Cl_2 wash, 30 ml (3 times)	2
13.	Abs. EtOH wash, 30 ml (3 times)	2

[a] Starting from 4g of Boc-Asn-resin (0.21 mmol Asp/g).

[b] In Boc-Trp-OH incorporation, steps 2 and 3 were deleted and replaced by: 30 ml of 25% TFA in CH_2Cl_2 1 min; 30 ml of 25% TFA in CH_2Cl_2, 30 min; 50% TFA in CH_2Cl_2, 4 min. 0.2% of ethanedithiol was contained in these steps and thereafter in steps 2 and 3 in the schedule.

[c] DMF was used in steps 9-12 in Boc-Asn-ONp and Boc-Gln-ONp couplings and the reaction time at step 11 was 12 h.

[d] Boc-Arg(Tos)-OH and Boc-Tro-OH were dissolved in 10% DMF in CH_2Cl_2.

[e] Boc-Val-OH was double coupling, steps 9-12 were repeated.

After removal of peptide from resin by liquid
HF treatment, we oxidized the peptide in the same
condition as in cardiotoxin refolding studies (5).
We obtained 30mg of biologically active peptide
after purification. Fig. 2 shows the properties of
the product. LD$_{50}$ of the synthetic product is 0.2
μg/g compared to 0.065 μg/g for natural cobrotoxin.
Furthur purification to obtain pure cobrotoxin is
under progress.

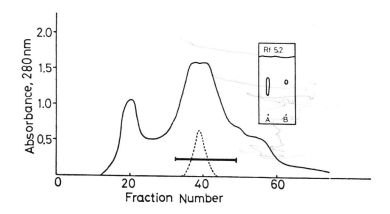

A.A.	Calc.	Native	Found
Asx	8	8.2	8.6
Thr	8	7.6	7.1
Ser	4	3.4	3.1
Glx	7	6.8	5.4
Pro	2	1.2	2.0
Gly	7	7.0	7.8
Cys	8	7.6	7.2
Val	1	0.7	1.3
Ile	2	1.6	2.2
Leu	1	0.8	1.0
Tyr	2	1.6[a]	0.7
Lys	3	3.2	2.2
His	2	1.6	1.1
Arg	6	5.9	6.6
Trp	1	0.7	0.2

Hydrolysis by 3N p-toluenesulfonic acid

[a] In the presence of 0.2% phenol.

Fig. 2. Properties of synthetic cobrotoxin.

Next toxin we synthesized is cardiotoxin from
Taiwan Cobra (<u>Naja</u> <u>naja</u> <u>atra</u>). There are four iso-
toxins of cardiotoxin in Taiwan cobra venom and this
is the major isotoxin (6, 7, 8, 9, 10). Its structure was elucidated as in Fig. 3.

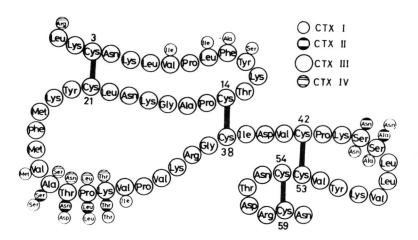

Fig. 3. The structure of cardiotoxin

First, we followed Merrifield's method (4) as
shown in Table 2. After purification, it gave a
product with LD$_{50}$ 3.3 μg/g. In the second synthesis by the same procedure, the final purification
was done by affinity chromatography on anticardio-
toxin serum column. This raise the LD$_{50}$ to 2.6 μg/g.
The properties were shown in Fig. 4. There is
slight difference in CD spectra but we do not know
the reason.

Cardiotoxin contains many lysine and arginine
residues. We thought that missing fragment with the
amino acids in cardiotoxin might give great changes
in properties of the missing sequence peptides.
This idea was conceived by Anfinsen (11) and luteni-
zing hormone-releasing hormone (12) and trypsin inhibitor (13) were successfully synthesized. It will
be easier to purify the synthetic product contaminated with missing sequence peptides. We planned a
fragment solid phase synthesis as in Fig. 5. We
first synthesized these fragments with protection
are crystalline compounds (14).

Table 2.   The Synthetic Schedule (Stepwise Solid
           Phase)

Step	Reagents and Operations	Mix times (min)
1.	CH_2Cl_2 wash, 20ml (3 times)	1
2.	50% TFA in CH_2Cl_2, 20ml	3
3.	50% TFA in CH_2Cl_2, 20ml	15
4.	CH_2Cl_2 wash, 20ml (3 times)	1
5.	33% Dioxane in CH_2Cl_2 wash, 20ml (3 times)1	
6.	CH_2Cl_2 wash, 20ml (3 times)	1
7.	5% Et_3N in CH_2Cl_2, 20ml	10
8.	CH_2Cl_2 wash, 20ml (3 times)	1
9.	CH_2Cl_2 wash, 20ml (3 times)	1
10.	Boc–AA (1.8 mmol) in CH_2Cl_2, 15ml[b]	10
11.	DCC (1.8 mmol) in CH_2Cl_2, 5ml	180[c]
12.	CH_2Cl_2 wash, 20ml (3 times)	1
13.	Anhydrous MeOH wash, 20ml (3 times)	1

[a]Starting from 2 g of Boc-Asn-0-Resin (0.28 mmol Asp/g resin)

[b]Boc-Arg(Tos)-OH was dissolved in 10% DMF in CH_2Cl_2, Boc-Asn-OH was dissolved with HOBt (2 mmol) in DMF and washed with the same solvent in steps 9-11. Valine was double coupling (steps 9-13 repeated).

[c]Coupling to proline was carried out for 240 min.

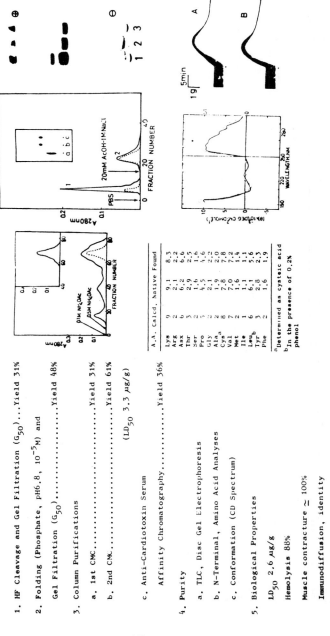

1. HF Cleavage and Gel Filtration (G_{50})...Yield 31%

2. Folding (Phosphate, pH6.8, 10^{-5}M) and

   Gel Filtration (G_{50}).................Yield 48%

3. Column Purifications

   a. 1st CMC........................Yield 31%

   b. 2nd CMC........................Yield 61%

               (LD_{50} 3.3 μg/g)

   c. Anti-Cardiotoxin Serum

      Affinity Chromatography.............Yield 36%

4. Purity

   a. TLC, Disc Gel Electrophoresis

   b. N-Terminal, Amino Acid Analyses

   c. Conformation (CD Spectrum)

5. Biological Properties

LD_{50} 2.6 μg/g

Hemolysis 88%

Muscle contracture ≃ 100%

Immunodiffusion, identity

A.A.	Calcd.	Native	Found
Lys	9	9.1	8.3
Arg	2	2.1	2.2
Asx	6	6.2	6.6
Thr	3	2.9	2.5
Ser	5	4.5	4.6
Pro	5	4.1	4.6
Gly	2	2.1	2.2
Ala	2	1.9	2.0
Cys[a]	8	7.8	7.8
Val	7	7.0	7.2
Met	2	1.6	1.4
Ile	1	1.1	1.1
Leu[b]	6	6.1	5.6
Tyr	3	2.8	2.3
Phe	2	1.6	1.9

[a] Determined as cysteic acid
[b] In the presence of 0.2% phenol

Fig. 4. The properties of synthetic cardiotoxin (stepwise solid phase)

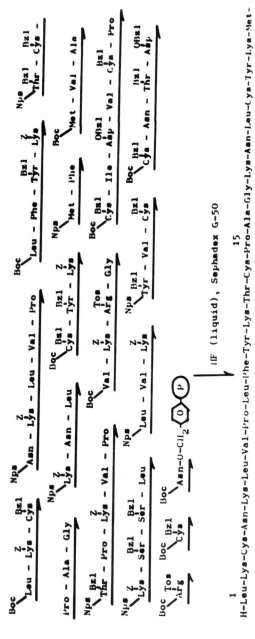

Fig. 5. Fragments of cardiotoxin for fragment solid phase synthesis.

453

The schedule of fragment solid phase synthesis is shown in Table 3.

Table 3.   Schedule of fragment solid phase

Step	Reagents and operations	Mix times(min)
1.	CH_2Cl_2 wash, 20ml (3 times)	1
2.	Ac_2O (3 mmol) and Et_3N (3 mmol) in CH_2Cl_2, 20ml	15
3.	CH_2Cl_2 wash, 20ml (4 times)	1
4.	50% TFA in CH_2Cl_2, 20ml[b]	3
5.	50% TFA in CH_2Cl_2, 20ml	15
6.	CH_2Cl_2 wash, 20ml (3 times)	1
7.	33% dioxane in CH_2Cl_2 wash, 20ml (3 times)	1
8.	CH_2Cl_2 wash, 20ml (3 times)	1
9.	5% Et_3N in CH_2Cl_2, 20ml	10
10.	CH_2Cl_2 wash, 20ml (3 times)	1
11.	DMF wash, 20ml (3 times)	1
12.	Boc-peptide (1.8 mmol) and HOBt (2 mmol) in DMF, 15ml[c]	5
13.	DCC (1.8 mmol) in CH_2Cl_2, 5ml	_D
14.	DMF wash, 20ml (3 times)	1
15.	Anhydrous MeOH wash, 20ml (3 times)	1

a. Starting from 2 g of Boc-Asn-O-Resin (0.28 mmol Asp/g resin)
b. If Nps- was for α-N protections, steps 4 and 5 were replaced by 0.1N HCl in $AcOH-CH_2Cl_2$ (1:10)
c. The three amino acid residues in C-terminal were condensed stepwise by DCC method and then followed by fragment condensation as the program described above.
d. When the reaction was carried out for 2 days, steps 14-15 were undertaken and sample of resin was removed to test the completeness of reaction as described in experiment. If the result was in agresment with desire, continued the next cycle from step 1; if not, steps 11-15 were repeated.

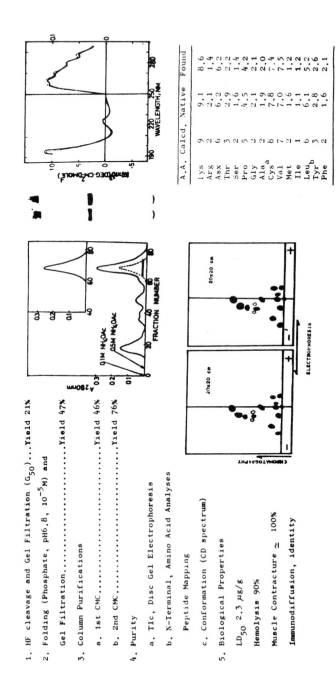

1. HF cleavage and Gel Filtration (G$_{50}$)...Yield 21%

2. Folding (Phosphate, pH6.8, 10^{-5}M) and

   Gel Filtration.........................Yield 47%

3. Column Purifications

   a. 1st CMC....................Yield 46%

   b. 2nd CMC....................Yield 76%

4. Purity

   a. Tlc, Disc Gel Electrophoresis

   b. N-Terminal, Amino Acid Analyses

      Peptide Mapping

   c. Conformation (CD spectrum)

5. Biological Properties

   LD$_{50}$ 2.3 μg/g

   Hemolysis 90%

   Muscle Contraction ≃ 100%

   Immunodiffusion, identity

A.A.	Calcd.	Native	Found
Lys	9	9.1	8.6
Arg	2	2.1	1.4
Asx	6	6.2	6.2
Thr	3	2.9	2.2
Ser	2	1.6	1.4
Pro	5	4.5	4.2
Gly	2	2.1	2.1
Ala	2	1.9	2.0
Cys	8	7.8	7.4
Val	7	7.0	7.5
Met	2	1.6	1.2
Ile	1	1.1	1.2
Leu$_b$	6	6.1	5.2
Tyr	3	2.8	2.6
Phe	2	1.6	2.1

Fig. 6. The properties of synthetic cardiotoxin by fragment solid phase.

After treatment of peptide resin with liquid
HF purification was done as in first stepwise synthesis. The final product after 2nd CMC column
chromatography gave fully active cardiotoxin.
Several properties support the purity of synthetic
product, Fig. 6. The LD_{50} is 2.3 μg/g and identical
with natural cardiotoxin (LD_{50} 2.4 μg/g). The comparison of some biological activities of synthetic
cardiotoxin to the natural one is shown in Table 4.
The haemolysis of synthetic cardiotoxin is slightly
low but may be this is the true activity.

Table 4. Biological properties of cardiotoxin

Tests	Natural	Synthetic stepwise	fragment
Immunodiffusion	——————— identical ———————		
LD_{50} (μg/g)	2.4	2.6	2.3
Haemolysis	100	88	90
Muscle contracture	——————— identical ———————		

```
1                                         21
L K C N K L V P L F Y K T C P A G K N L C
    |_____|

14                                        38
C P A G K N L C Y K M F M V A T P K V P V K R G C
|_____|

39                                        60
I D V C P K S S L L V K T V C C N T D R C N
      |_____| |_____|

39                        53
I D V C P K S S L L V K Y V C
      |_____|

54        60
C N T D R C N
|_____|
```

Fig. 7. The designed synthetic fragments of anti-
         body determinant.

Then we turn our attention to the synthesis of possible small fragments which might induce antibody formation but are devoid of toxicity. Since disulfide bonds in cardiotoxin are indispensible for its toxicity, we planned the synthesis of 4 disulfide loops as shown in Fig. 7 and have completed the synthesis of first fragment which occupy the sequence 54 to 60 Fig. 8. Unfortunately, this fragment showed no antigenic activity by inhibition test, but furthur synthesis of other fragments might be able to pin-point the antigenic determining site of cardiotoxin. At the moment, this should be a progress report but I want to point out the synthesis of possible antigenic determinant is important for medical application.

SOLID PHASE SYNTHESIS

OXIDATION: $K_3Fe(CN)_6$

54   60
C N T D R C N

Fig. 8. The synthesized fragment.

Here, comment should be made on the synthesis of α-Bungarotoxin by a Russian (15). They synthesized 11 gragments of this toxin. No furthur work was published and it seems they have a problem of refolding. α-Bungarotoxin has five disulfide bonds so that it probably needs different approach to achieve the total synthesis.

The third part of my report is the synthesis of cardiotoxin fragments by proteolytic enzymes catalyzing peptide bond formation. We have engaged in these studies since last two years and found this reaction is quite useful. Recently, we successfully synthesized opiatic peptide, Met-Enkephalin, by pepsin catalysis. As shown in Table 5, we used papain to synthesize the fragments in cardiotoxin. The yields of crude product with single spot on tlc are as high as 95%. After recrystallization, the yield are still good. The possibility of fragment coupling is shown by the synthesis of protected tetrapeptide. We also tried the synthesis of serine, threonine and arginine peptides without protecting the side group. The most interesting example is the synthesis of protected tripeptide Z–Thr–Asp–Arg–OMe. The yield is still low but side group protections are unnecessary for this synthesis.

Summary and Concluding Remarks. It is possible to synthesize the chemically pure snake venom toxins by fragment solid phase synthesis. Although the yield is low, the synthesis of fully active cardiotoxin was achieved. The synthesis of antigenic determinants is challenging area of snake venom toxin synthesis. It is also important for medicinal application. The enzyme–catalyzed peptide bond formation could accelerate the fragment synthesis of toxins and can be expanded to the synthesis of other peptides especially for medicinal purposes.

ACKNOWLEDGMENTS

I would express my sincere gratitude to Dr. C. H. Li of San Francisco, who encouraged me to start the peptide synthesis 9 years ago. I also thank Mr. C. H. Wong and Mr. S. T. Chen for their hard work in these investigation.

Table 5. The enzymatic synthesis of fragment peptides.

Reactants	Products	Yields
Z-Lys(Z)-OH+H-Leu-OTMB	Z-Lys(Z)-Leu-OTMB	70%
Z-Val-Cys(Bzl)-OH+H-Cys(Bzl)-OTMB	Z-Val-Cys(Bzl)-Cys(Bzl)-OTMB	44%
Z-Val-Cys(Bzl)-OH+H-Cys(Bzl)-Asn-OTMB	Z-Val-Cys(Bzl)-Cys(Bzl)-Asn-OTMB	65%
Z-Arg-OH+H-Cys(Bzl)-OTMB	Z-Arg-Cys(Bzl)-OTMB	75%
Z-Ser-OH+H-Leu-OTMB	Z-Ser-Leu-OTMB	75%
Z-Asp-OH+H-Arg-OMe	Z-Asp-Arg-OMe	70%
Z-Thr-OH+H-Asp-Arg-OMe	Z-Thr-Asp-Arg-OMe	50%

Reaction Condition

ENZYME   : **PAPAIN**

REACTANTS: Z-X-OH + H-Y-OTMB
           (0.1-1M) (0.1-1M)

SOLUTION : 0.2 M CITRATE, 5 mM β-MERCAPTOETHANOL
           pH 4-5

TEMP   : 37-40°C

X,Y: Amino Acid or Peptide

REFERENCES

1. Yang, C. C., Yang, J. J. and Chiu, R. H. C.,
    Biochim. Biophys. Acta, 214, 355-363 (1970)
2. Aoyagi, H., Yanezawa, H., Takahashi, N., Kato,
    T., Izumiya, N. and Yang, C. C., Biochim.
    Biophys. Acta, 263, 823-826 (1972)
3. Izumiya, N., Kato, T., Aoyagi, H., Takahashi,
    N., Yasutake, A. and Yang, C. C. in "Aminal
    Plant and Microbial Toxins" (A. Ohsaka, ed.),
    Vol.II p.89-92, plemm Publishing Corp., New
    York (1976)
4. Merrifield, R. B., J. Am. Chem. Soc., 85, 2149-
    2154 (1963)
5. Wong, C. H., Chen, Y. H., Hung, M. C., Wang,
    K. T., Ho, C. L. and Lo, T. B., Biochim.
    Biophys. Acta, in press (1978)
6. Lo, T. B., Chen, Y. H. and Lee, C. Y., J.
    Chinese Chem. Soc., 13, 25-37 (1966)
7. Hayashi, K. Takechi, M., Sasaki, T. and Lee,
    C. Y., Biochem. Biophys. Res. Comm., 64, 360-
    366 (1975)
8. Narita, K. and Lee, C. Y., Biochem. Biophys.
    Res. Comm., 41, 339-344 (1970)
9. Takechi, M., Sasaki, T., Kaneda, N. and Hayashi,
    K., FEBS Lett., 70, 217-222 (1976)
10. Narita, K., Cheng, K. L., Chang, W. C. and Lo,
    T. B., Int. J. Pept. Prot., in press (1978)
11. Anfinsen, C. B., Pure, Appl. Chem., 17, 461-470
    (1968)
12. Matsueda, R., Maruyama, H., Kitazawa, E. and
    Mukaiyama, T., Bull. Chem. Soc. Japan, 46,
    3240-3247 (1973)
13. Yajima, H., Kiso, Y., Okada, Y. and Watanabe,
    H., J. Chem. Soc. Chem. Comm., 106-107 (1974)
14. Wong, C. H. and Wang, K. T., unpublished.
15. Ivanov, V. T., Toxicon, 13, 100-101 (1975)

Index